Information Science and Statistics

Information Science and Statistics

For other titles published in this series, go to
www.springer.com/series/3816

Michael E. Schuckers

Computational Methods in Biometric Authentication

Statistical Methods for Performance Evaluation

Springer

Prof. Michael E. Schuckers
Dept. Mathematics, Computer
Science & Statistics
St. Lawrence University
Canton, NY 13617, USA
schuckers@stlawu.edu

Series Editors
Michael Jordan
Division of Computer Science
and Department of Statistics
University of California, Berkeley
Berkeley, CA 94720, USA

Bernhard Schölkopf
Max Planck Institute
for Biological Cybernetics
Spemmannstrasse 38
72076 Tübingen
Germany

Robert Nowak
Department of Electrical
and Computer Engineering
University of Wisconsin-Madison
3627 Engineering Hall
1415 Engineering Drive
Madison, WI 53706, USA

ISBN 978-1-4471-2566-2 ISBN 978-1-84996-202-5(eBook)
DOI 10.1007/978-1-84996-202-5
Springer London Dordrecht Heidelberg New York

British Library Cataloguing in Publication Data
A catalogue record for this book is available from the British Library

Cover design: deblik, Berlin

Printed on acid-free paper

Springer is part of Springer Science+Business Media (www.springer.com)

For Stephanie

Preface

The impetus for this book came from the 'perfect' storm of events. As I was preparing for my recent sabbatical, I was involved in several biometrics projects that confirmed for me the need for some basic statistical methods for testing and evaluation of biometric devices. During my sabbatical, in the spring of 2007, I was asked by our host, Fabio Roli of L'Università di Cagliari in Sardinia, to teach a graduate-level short course on my research. In preparing for that course, I was struck by how little quality research had been done for the practicing biometric researcher, tester or engineer in the way of basic statistical methods for measuring the performance of biometric authentication systems or for comparing the performance of two or more such systems. Most of the work that does exist is focused on methods for a single false match rate or a single false non-match rate. But those two metrics are only part of the story for a bioauthentication system and there is a need for methods that appropriately compare two or more methods. This book is a direct result of that short course. For inviting us to visit, for hosting us and for asking me to teach that short course, I am indebted to Fabio Roli and this research group. *Grazie mille di cuore. Ora siamo sardi.*

The goal of this book is to provide *basic* statistical methodology for practitioners and testers of bioauthentication devices. (I will use both the term biometrics and the bioauthentication to describe the devices that take physiological measurements and make decisions about access from those measurements.) The book also covers general classification performance since the data collection for most classification systems is similar to approaches used in biometric authentication. I do *not* aim to present a complete and thorough set of methodology for all possible applications and all possible tests. Instead, the aim is on a basic framework of methods that cover a variety of circumstances but that can be extended and generalized as needed.

The research that led to the publication of this book was generously funded by the Center for Identification Technology Research (CITeR) and the National Science Foundation (NSF). Funding for this work comes from NSF grants CNS-0325640 and CNS-0520990. CNS-0325640 is cooperatively funded by the National Science Foundation and the United States Department of Homeland Security (DHS). *Any opinions, findings, and conclusions or recommendations expressed in this book are those of the author(s) and do not necessarily reflect the views of the National Science Foundation.* I have been fortunate enough to obtain several grants from CITeR over the years. The support and the interaction that CITeR has provided have been pivotal to my research and my career. Special thanks to LaRue Williams at WVU for keeping CITeR running smoothly.

A book like this is never completely the effort of one individual. I have been blessed to have many giving mentors and teachers over the years who inspired as well as gave much of themselves and their time including: Joe Walker of Cumberland Valley High School, Robert Hultquist of Penn State, Brenda Gunderson at the University of Michigan, Sallee Anderson at Kellogg Company, Diane Sly Tielbur at Kellogg Company, Dean Isaacson of Iowa State University, Hal Stern, now at UC Irvine, Steve Vardeman of Iowa State University, Larry Hornak of West Virginia University, Patti Lock of St. Lawrence University and Robin Lock of St. Lawrence University.

I am very fortunate that the Department of Mathematics, Computer Science and Statistics at St. Lawrence University is an energetic and vibrant one. Patti Lock, the current chair, has been extremely flexible and encouraging throughout this process. My colleague, Robin Lock, is still willing to dispense his sage advice to his statistical junior. Jessica Chapman and Ivan Ramler have brought fresh energy to statistics at St. Lawrence. Collen Knickerbocker helped me down this path and I have not forgotten his wit and his wisdom. Jim DeFranza read the proposal for this book and offered helpful suggestions for improvements.

Many of my *undergraduate* students at St. Lawrence worked on this as part of a summer research grant fellowship or as part of their senior or honors projects. They are: Travis Atkinson, Hilary Hartson, Anne Hawley, Katie Livingstone, Dennis Lock, Nikki Lopez, Yordan Minev, Nona Mramba, Matt Norton, Amanda Pendergrass, and Emily Sheldon. (I am particularly indebted to Yordan for his assistance developing the R code for the ROC/EER chapter.) Marcus Tuttle assisted greatly in the creation of the tables and with the last minute minutia of book completion on some snowy days in December 2009. Ben von Reyn gave up part of his Thanksgiving and semester breaks to proofread and provide detailed feedback and comments on a draft of this book. Thanks to all.

Simon Rees and Wayne Wheeler at Springer have been tremendously supportive and patient with me as a first-time author. Thank you both for your time and efforts on behalf of this monograph.

I received many helpful comments and suggestions from colleagues and collaborators along the way. Rick Lazarick of CSC and Chris Miles at DHS have been encouraging throughout this process. Larry Hornak of West Virginia University and Bojan Cukic of West Virginia University have been tremendous mentors and wonderful friends. Andy Adler of Carleton University, Jessica Chapman of St. Lawrence University, Zach Dietz of Hamilton College, Eric Kukula of Purdue University, Gian Luca Marcialis dell'Università di Cagliari, Shimon Modi of Purdue University, Norman Poh of University of Surrey, George W. Quinn at the National Institute of Standards and Technology, and Stephanie Schuckers of Clarkson University each read a chapter and provided detailed feedback. Daqing Hou, my collaborator on the software that will accompany this book, has asked many questions that the target reader would, and so, has made this text better. Their contributions improved all facets of this book and I appreciate all of their useful comments. Thank you all. The errors that remain are solely mine.

Eric Kukula and Shimon Modi of Purdue University generously allowed me to use their failure to enrol and failure to acquire data, respectively. I must also thank Norman Poh for providing the bioauthentication research community with the BANCA match score database and the XM2VTS match score database. Thanks to National Institute for Standards and Technology and Patrick Grother, in particular, for making the Biometrics Score Set Release 1 available. All of these are valuable tools for benchmarking and testing.

To my family, Stephanie, Daniel, Christopher and Gabriella, thank you so much for your love and support. I can never repay your patience amid the writing of this book but I will try. My parents gave up chunks of their retirement to be with their grandchildren and to allow me extra time in my office. Over the years they have also been incredibly generous and enthusiastic. I am very fortunate to have had that support. Thank you, Mom and Dad. My sisters, Lisa and Carly, as always, provided cheer, humor and encouragement during this process.

Finally, to Stephanie, for being such an extraordinary person and wife, for giving your support to this endeavor and for giving me so many *halcyon* days; I adore you and I am so lucky.

Canton, NY, USA Michael E. Schuckers

Contents

Symbols

Symbol	Usage in this text
n	Number of individuals
$n_{\mathbb{P}}$	Number of individuals in the probe
$n_{\mathbb{G}}$	Number of individuals in the gallery
$I_{\{\}}$	Indicator function
Σ	Variance matrix
Φ	Correlation matrix
B	Specified width of a confidence interval
e	Bootstrapped error random variable
Δ	Randomized error random variable
M	Number of bootstrap repetitions
ς	Counter for bootstrap repetitions
Y_{iij}	False non-match score
D_{iij}	False non-match decision
m_i	Number of non-match decisions on the ith individual
π	False non-match rate (FNMR)
ρ	Correlation parameter for the FNMR
$Y_{ik\ell}$	False match score $i \neq k$
$D_{ik\ell}$	False match decision $i \neq k$
ν	False match rate (FMR)
m_{ik}	Number of match decisions on the pair of individuals (i, k)
η, ω, ξ	correlation parameters for the FMR
$Y_{ik\ell}$	Match score
r_θ	Radius of a receiver operating characteristic (ROC) curve at angle θ
Θ	Set of angles
Θ^*	Finite set of angles at which ROC is measured
T	Number of angles in Θ^*

τ	Match decision threshold
R	An ROC curve at all angles
$\hat{R}$	An estimated ROC curve
$\Re$	Confidence region for the ROC
ι_θ	ROC curve adjustment
w_m	Maximal z-score difference from observed ROC
δ_U, δ_L	Upper and lower bounds for ROC confidence region
χ	Equal error rate (EER)
E_i	Enrolment decision
ε	Failure to enrol rate (FTE)
$A_{i\kappa}$	Acquisition decision
γ	Failure to acquire rate (FTA)
ψ	Correlation parameter for the FTA
$\diamond$	Future observation
$N^\dagger$	Effective sample size

List of Figures

List of Tables

Part I
Introduction

Chapter 1
Introduction

Biometric authentication systems take physiological or behavioral measurements of individuals and attempt to use those to authenticate an individual's identity. The aim of this book is to provide rigorous statistical methods for the evaluation of biometric authentication systems. Biometric authentication, bioauthentication, or simply biometrics has received a great deal of attention and interest lately resulting in the development and improvement of these systems. As this has occurred, there has *not* been an equivalent improvement in the statistical methods for evaluating these systems. The goal of this book is to fill this need. The following are the primary highlights of this text:

- Statistical methodology for failure to enrol (FTE), failure to acquire (FTA, receiver operating characteristic (ROC) curves, false non-match rate (FNMR) and false match rate (FMR)
- Methods for comparison of two or more biometric performance metrics
- New bootstrap methodology for FMR estimation
- New bootstrap methodology for ROC curve estimation
- Over 120 examples using publicly available biometric data
- Addition of prediction intervals to the bioauthentication statistical toolset
- Sample size and power calculations for FTE, FTA, FNMR and FMR

1.1 Why This Book?

Performance is an important aspect for the selection and implementation of a biometric authentication system. Statistical summaries for the metrics of biometric system performance are essential for good decision making regarding these systems. While appropriate metrics regarding these systems and their accompanying subsystems have been known for some time, statistical methods for these metrics have lagged. The biometric performance metrics that are most commonly used are: failure to enrol (FTE), failure to acquire (FTA), false non-match rate (FNMR), false match rate (FMR), equal error rate (EER), and receiver operating characteristic (ROC)

M.E. Schuckers, *Computational Methods in Biometric Authentication*, Information Science and Statistics,
DOI 10.1007/978-1-84996-202-5_1,

curves. Statistical methods for these metrics have been proposed for estimation of a single false match rate (FMR) or a single false non-match rate (FNMR), but not for these other metrics and not for methods that compare systems. This book is meant to be a single resource of statistical methods for estimation and comparison of biometric authentication system performance. In particular, this book will focus on statistical methods for FTE, FTA, FNMR, FMR, EER, and ROC curve estimation. For each of these metrics, we derive statistical methods for a single system that includes confidence intervals, hypothesis tests, sample size calculations, power calculations and prediction intervals. We also extend these same methods to allow for the statistical comparison and evaluation of multiple systems for both independent and paired data. Examples are given for each methodology that we introduce. Where possible we have used publicly available data for these examples. The target audience for this book is someone interested in testing and evaluating the performance of biometric systems. This book is written at an upper-level undergraduate or master's level with a quantitative background. It is expected that the reader will have an understanding of the topics in a typical undergraduate statistics course.

While the main focus of this book is biometric authentication systems, the methods here have the potential for broader applications. In particular, the methods apply generally to the performance of classification systems. Most classification systems are evaluated using false accept rates (FAR's), false reject rates (FRR's) and ROC curves. Biometric systems differentiate between the false accept rate and the false match rate and between the false reject rate and the false non-match rate since biometric systems include image or signal acquisition subsystems which can fail. The statistical methods given in this book can be modified for all three of the traditional classification performance metrics: FAR's, FRR's and ROC curves. The nature of biometric authentication is rather distinct due to the comparison of two instances rather than the classification of a single instance. Consequently, biometric authentication matching decisions are inherently *non-iid* where *iid* is defined in Definition 2.33. The non-independent correlation structure for biometric matching is also a result of having multiple captures or signals from a given individual. As with any statistical application it is important to ensure that the method of analysis is appropriate for a particular set of data.

1.2 Process of a Biometric Authentication System

A biometric authentication system is comprised of several component subsystems. Wayman et al. [102] break down a generic system into five subsystems. These are: data collection component, transmission component, signal processing component, storage component and a decision component. The basic process using these subsystems begins with an attempt for an individual to enrol. Each individual is either able or unable to enrol in the system. One measure of a bioauthentication system is the rate at which individuals fail to enrol. When an individual presents themselves to the system, the system will either acquire a biometric signal of sufficient quality or it will not. Here we use the term signal to be very broadly defined to encompass signals, images, etc. Throughout this book we will use the terms signal and capture interchangeably. The percent of times that the system does not acquire such a signal is another performance metric. Once a system has obtained a signal, then it attempts to match that presented signal to one or more stored signals. Often this matching process is done, not on the raw signal, but on a transformed version of the signal called a template. Bioauthentication systems typically decide whether or not a match is made between two signals based upon a match score. This real-valued score is the outcome of comparing the two signals. Without loss of generality, we'll assume that a match is declared, if the match score exceeds some threshold. A well designed system will process these signals and come to a correct decision a vast majority of the time. However, errors can occur. If the system incorrectly matches the presented signal to a stored signal from another individual, then that is a false match. If the system incorrectly fails to match the presented signal to a stored signal from the same individual, then that is a false non-match. The rate at which these errors are made are known as the false match and false non-match rates, respectively. The receiver operating characteristic (ROC) curve is a plot of the changes in the error rates as we change the threshold for matching. This plot is an encapsulation of the performance of the biometric authentication system at all of the possible operating thresholds. The equal error rate (EER) is a single summary of system performance that gives the value of the FNMR and the FMR when they are equal, i.e. at the threshold which yields FNMR = FMR. One important metric that we will not discuss in-depth here is the mean transaction time (MTT) which is the length of time that it takes for a system to acquire, match and decide about a given signal. We give a brief overview of the MTT in Sect. 8.2.6.

1.3 Performance Metrics

There are several performance metrics that are commonly used to evaluate bioauthentication systems that will be the focus of this book. Each of these metrics is briefly described below. Table 1.2 has a short summary of these measures along with the notation that we will use for each. In addition, that table contains the fundamental unit of measurement, the notation for a single measurement, and the equation for an estimate. For example, our notation for an FTA is γ, the unit of measurement is the κth acquisition attempt by the ith individual and we denote that attempt by $A_{i\kappa}$. Note that for the ROC and the EER that our estimation methods are based upon polar coordinates and those estimates are too complicated for this table.

Failure to Enrol (FTE) Rate The failure to enrol measures the percent of individuals that are unable to enrol in a given system.

Failure to Acquire (FTA) Rate The failure to acquire is the proportion of acquisition decisions from presentations of a biometric or biometrics that are not able to be processed by the system.

Table 1.1 Table of matchers used in this book

Matcher	Modality	Database	Notation
IDIAP_voice_gmm_auto_scale _25_10	Voice	BANCA	IDIAP-25-10
IDIAP_voice_gmm_auto_ scale_25_10_pca	Voice	BANCA	IDIAP-25-10-PCA
IDIAP_voice_gmm_auto_ scale_25_25	Voice	BANCA	IDIAP-25-25
IDIAP_voice_gmm_auto_ scale_25_25_pca	Voice	BANCA	IDIAP-25-25-PCA
IDIAP_voice_gmm_auto_ scale_25_50	Voice	BANCA	IDIAP-25-50
IDIAP_voice_gmm_auto_ scale_25_50_pca	Voice	BANCA	IDIAP-25-50-PCA
IDIAP_voice_gmm_auto_ scale_25_75	Voice	BANCA	IDIAP-25-75
IDIAP_voice_gmm_auto_ scale_25_100	Voice	BANCA	IDIAP-25-100
IDIAP_voice_gmm_auto_ scale_25_100_pca	Voice	BANCA	IDIAP-25-100-PCA
IDIAP_voice_gmm_auto_ scale_25_200_pca	Voice	BANCA	IDIAP-25-200-PCA
IDIAP_voice_gmm_auto_ scale_25_300	Voice	BANCA	IDIAP-25-300
IDIAP_voice_gmm_auto_ scale_33_10	Voice	BANCA	IDIAP-33-10
IDIAP_voice_gmm_auto_ scale_33_10_pca	Voice	BANCA	IDIAP-33-10-PCA
IDIAP_voice_gmm_auto_ scale_33_100	Voice	BANCA	IDIAP-33-100
IDIAP_voice_gmm_auto_ scale_33_100_pca	Voice	BANCA	IDIAP-33-100-PCA
IDIAP_voice_gmm_auto_ scale_33_200	Voice	BANCA	IDIAP-33-200
IDIAP_voice_gmm_auto_ scale_33_300	Voice	BANCA	IDIAP-33-300
IDIAP_voice_gmm_auto_ scale_33_300_pca	Voice	BANCA	IDIAP-33-300-PCA
UC3M_voice_gmm_auto_ scale_10_100	Voice	BANCA	UC3M-10-100
UC3M_voice_gmm_auto_ scale_10_200	Voice	BANCA	UC3M-10-200
UC3M_voice_gmm_auto_ scale_10_300	Voice	BANCA	UC3M-10-300
UC3M_voice_gmm_auto_ scale_34_300	Voice	BANCA	UC3M-34-300
UC3M_voice_gmm_auto_scale _34_500	Voice	BANCA	UC3M-34-500
UCL_face_lda_man	Face	BANCA	UCL-LDA-MAN
SURREY_face_nc_man_ scale_50	Face	BANCA	SURREY-NC-50
SURREY_face_nc_man_ scale_100	Face	BANCA	SURREY-NC-100
SURREY_face_nc_man_ scale_200	Face	BANCA	SURREY-NC-200
SURREY_face_svm_auto	Face	BANCA	SURREY-AUTO
SURREY_face_svm_man	Face	BANCA	SURREY-SVM-MAN
SURREY_face_svm_man_ scale_0.50	Face	BANCA	SURREY-SVM-MAN-0.50
SURREY_face_svm_man_ scale_0.71	Face	BANCA	SURREY-SVM-MAN-0.71
SURREY_face_svm_man_ scale_1.00	Face	BANCA	SURREY-SVM-MAN-1.00
SURREY_face_svm_man_ scale_1.41	Face	BANCA	SURREY-SVM-MAN-1.41
SURREY_face_svm_man_ scale_2.00	Face	BANCA	SURREY-SVM-MAN-2.00
SURREY_face_svm_man_ scale_2.83	Face	BANCA	SURREY-SVM-MAN-2.83
SURREY_face_svm_man_ scale_4.00	Face	BANCA	SURREY-SVM-MAN-4.00
(FH, MLP)	Face	XM2VTS	(FH, MLP)
(DCTs, GMM)	Face	XM2VTS	(DCTs, GMM)
(DCTb, GMM)	Face	XM2VTS	(DCTb, GMM)
(DCTs, MLP)	Face	XM2VTS	(DCTs, MLP)
(DCTb, MLP)	Face	XM2VTS	(DCTb, MLP)
(LFCC, GMM)	Speaker	XM2VTS	(LFCC, GMM)
(PAC, GMM)	Speaker	XM2VTS	(PAC, GMM)
(SSC, GMM)	Speaker	XM2VTS	(SSC, GMM)
Face_C	Face	NIST	Face_C
Face_G	Face	NIST	Face_G

Table 1.2 Summary of biometric parameters, units of measurement and notation

Biometric parameter	Symbol	Unit of measurement	Measurement notation	Parameter estimate
FNMR	π	ℓth decision on individual i	$D_{ii\ell}$	$\hat{\pi} = \frac{\sum_{i=1}\sum_{\ell=1}^{m_i} D_{ii\ell}}{\sum_i m_i}$
FMR	ν	ℓth decision on the distinct pair of individuals i and k where $i \neq k$	$D_{ik\ell}$	$\hat{\nu} = \frac{\sum_{i=1}\sum_{k=1,k\neq i}\sum_{\ell=1}^{m_{ik}} D_{ik\ell}}{\sum_i \sum_k m_{ik}}$
ROC	r_θ, R	ℓth match scores from the pair of individuals i and k	$Y_{ik\ell}$	
EER	χ	ℓth match scores from the pair of individuals i and k	$Y_{ik\ell}$	
FTE	ε	ith individual	E_i	$\hat{\varepsilon} = \frac{\sum_{i=1}^{n_E} E_i}{n_E}$
FTA	γ	κth attempt by individual i	$A_{i\kappa}$	$\hat{\gamma} = \frac{\sum_{i=1}^{n_A}\sum_{\kappa=1}^{K_i} A_{i\kappa}}{\sum_i K_i}$

False Non-Match Rate (FNMR) The false non-match rate is the proportion of matching decisions from comparing two biometric captures from the same individual that result in a declaration of a non-match.

False Match Rate (FMR) The false match rate is the proportion of matching decisions from comparing two biometric captures from two different individuals that result in a declaration of a match.

Receiver Operating Characteristic (ROC) Curve The receiver operating characteristic curve is a graph that plots the false match rate against one minus the false non-match rate by changing the criterion for what constitutes a match or a non-match. This measure gives a summary of the matching performance of a system.

Equal Error Rate (EER) The equal error rate is the rate at which the false non-match and the false match rates are the same. This value is a single number that is sometimes used to compare matching performance between systems.

Note that for the FTA, FNMR and FMR, we are considering rates of decisions. An individual tester can set the number of attempts or transactions that lead to a given decision but inference about system performance needs to be based upon decisions. We spend one chapter of this book developing statistical methods for testing and estimation of each of the performance metrics given above except for ROC and EER which share a chapter. In each chapter, we provide a framework for estimation as well as a variety of statistical methods for comparing estimates.

1.4 Statistical Tools

Below we present a short summary of some of the statistical tools that we will use throughout this book. More rigorous definitions and additional information about these topics can be found in Chap. 2.

Confidence Intervals The goal of a confidence interval is to create a range of values for a parameter or quantity that we want to estimate. Since any method of making inference about a process from a sample will include uncertainty, a confidence interval is designed to reflect this uncertainty. Confidence intervals and their resulting width depend upon the amount of information in the sample as well as how much confidence the process parameter will be inside our interval. A confidence interval is typically used for a scalar parameter while a confidence region can be used for a multi-dimensional parameter. A confidence interval is the *preferred* methodology for estimation of a parameter.

Hypothesis Tests Hypothesis tests are a mechanism for testing whether a particular value for a process parameter is reasonable or whether the observations from the sample provide enough evidence to suggest that value is not reasonable. For example, we can use a hypothesis test for determining whether or not a matching process yields a false non-match rate that is significantly less than 2%.

Prediction Intervals A prediction interval is a confidence interval for the value of a performance metric based upon future, as yet unobserved, observations. As with a confidence interval, the goal is to create a range of values that are credible for that particular metric given observations that have been collected. Prediction intervals are a well-studied and well-known tool in the statistics literature. See, e.g., Vardeman [95]. We introduce these to a biometric authentication audience in this book.

Sample Size and Power Calculations One very common interest for those involved in testing of biometric authentication systems is the size of the sample. How many observations to collect is a function of the goals of the test. In this book, we provide sample size calculations for determining how many individuals are necessary to create a confidence interval for a particular process metric. Similarly, we provide power calculations for determining the number of individuals needed for evaluating a particular hypothesis test.

Independent and Paired Data Collections Throughout this book, when we are comparing two or more performance metrics, we distinguish between independent and paired data collections. Data collections that are independent are done separately such that the individuals involved are distinct. Data collections that are paired imply that each measurement in one group has a 'connected' or 'paired' measurement in the other groups. Data collections done at two different airports would be considered independent, while a test where every individual uses each of four devices at every visit would be a paired collection. The appropriate methodologies for each do not overlap. When comparing two or more performance metrics in this book, we will assume that the data collections are either independent or paired. Data collections with a more complicated structure are *not* covered in this book.

Large Sample, Bootstrap and Randomization Methods Statistical tools are often categorized into two groups: parametric and non-parametric. Parametric methods tend to be those that utilize a particular distribution for making inference about a parameter, while non-parametric methods, such as bootstrap or randomization methods, tend to make use of fewer distributional assumptions. Whenever possible in this book, we provide *both* parametric and non-parametric approaches for a given inferential scenario. Additionally, we provide guidance for when the distributional assumptions that we are making are appropriate.

Gallery and Probe In bioauthentication testing, we are often comparing two compilations of biometric captures. Generally, we think of the compilation of signals submitted to a system or a subsystem as the probe and the compilation of signals in the system against which the probe is compared as the gallery. We will denote the probe by $\mathbb{P}$ and the gallery by $\mathbb{G}$.

Signal, Image, Capture, Samples There are several terms that are used for the information that is obtained by a bioauthentication sensor. These include the four listed at the beginning of this paragraph. In this book, we will use signal, image and capture interchangeably. We will also use these terms broadly to represent the totality of information that is collected by a bioauthentication system. Thus, information from multiple sensors might be included as part of a signal or capture. Our use of the word *sample* will be different. We will use the word sample to represent any collection of observations. This will include a collection of signals but also includes a collection of enrolment decisions or a collection of acquisition attempts.

As was mentioned above, Chap. 2 is a discussion of the statistical results that underlie the methodology that we describe in this book. Readers who want to focus on the methods and not the underlying statistical theory can skip this chapter. Those interested in the fundamental statistical theory are encouraged to peruse Chap. 2. Those who are familiar with this background material should skip to the other numbered chapters.

1.5 Data Sources

Data are at the heart of any statistical method and statistical analysis. For all of the examples used in this text, we have sought to use publicly available datasets. Readers are encouraged to obtain these data and use them to replicate the results found throughout this book. Below we list, along with a brief description, the data collections that we have used for examples throughout this text. Links to the datasets described below are given at: http://myslu.stlawu.edu/~msch/biometrics/book/data/. Table 1.1 contains a summary of the biometric matchers that we have used from some of the sources listed below.

1.5.1 PURDUE-FTA and PURDUE-FTE Databases

The PURDUE-FTA and PURDUE-FTE databases are from collections at Purdue University. The first database, PURDUE-FTA, is a collection of failure to acquire observations that was part of the doctoral dissertation of Shimon Modi [67]. Overall, there are acquisition decisions on 7 different fingerprint devices from 186 different individuals who attempted to use these devices. Individuals in that data collection used two fingers from each hand as part of that collection. More details on the collection can be found in Modi et al. [68]. We will use data from that collection extensively in Chap. 7 on failure to acquire rates.

The PURDUE-FTE database is a collection of enrollment decisions on four sensors. Designed and collected as part of the Ph.D. dissertation research of Eric Kukula, 85 individuals contributed enrolment decisions to this collection. We will make extensive use of these data in our chapter on analyzing failure to enrol rates, Chap. 6. Additional details on this data collection can be found in Kukula [53].

1.5.2 FTE-Synth Database

The FTE-Synth database is a collection of synthetic enrolment attempts that were created for this book. In order to test some of the methods in Chap. 6, we needed datasets that were very large. Following the basic structure of the PURDUE-FTE, we created artificial collections on a handful of devices. These are used in Chap. 6 on failure to enrol rates.

1.5.3 XM2VTS Database

The XM2VTS database is a collection of match scores from data collected following the design discussed in Lüttin [58]. Since this is a database of match scores, we will utilize it in the chapters on false non-match rates, the false match rates and the receiver operating characteristic curves. Included in this database are eight different matchers. Five are face matchers and three are speaker matchers. Additional details on this database are available at: http://info.ee.surrey.ac.uk/Personal/Norman.Poh/web/fusion. The database was acquired within the M2VTS project (Multi Modal Verification for Teleservices and Security applications), a part of the EU ACTS programme, which deals with access control by the use of multimodal identification of human faces.

1.5.4 BANCA Database

The BANCA database, cf. Bailly-Baillière [5], is a database of match scores for 76 different matchers. There are both face and voice matchers in this database. We make use of portions of this database for our examples in Chaps. 3, 4 and 5. Additional information on this database can be found at: http://personal.ee.surrey.ac.uk/Personal/Norman.Poh/web/banca_multi/main.php?bodyfile=entry_page.html.

1.5.5 NIST Database

Another publicly available database of match scores is the NIST Biometrics Score Set Release 1 [94]. We will refer to this collection of match scores as the NIST database throughout this book and we will use only match scores from Set 1 of this release. This database has match scores from two fingerprint algorithms and two face algorithms. For each of those four algorithms, there are 517 individuals. Examples from these data are utilized in the chapters on false match rates, false match rates and receiver operating characteristic curves. More details are available at http://www.itl.nist.gov/iad/894.03/biometricscores/.

1.6 Organization

The rest of the book is organized in the following manner. We begin with a refresher of statistical methods and some of the theory that underlies these methods in Chap. 2. That chapter along with this one comprises Part I of this book. Part II is about statistical methodology for evaluating matching and classification performance. The methods in the chapters of Part II are specifically aimed at biometric matching subsystems but they also apply more broadly to classification systems. Classification methods usually use false reject and false accept rates to measure their performance. The inferential tools that are used in false non-match rates (Chap. 3) and false match rates (Chap. 4) are similar those for false reject and false accept rates. Additionally, both matching and classification systems use receiver operating characteristic curves as measures of overall performance. Thus, we combine our chapters on statistical methods for false non-match rates, false match rates and receiver operating characteristics curves into Part II. Part III is about statistical methods for performance metrics that are biometric specific. Thus, we put our chapters on failure to enrol and failure to acquire in this part of the book. We conclude this book with a discussion chapter, Chap. 8, that includes some additional topics not covered in the book and a chapter of statistical tables, Chap. 9.

Chapter 2
Statistical Background

The goal of this chapter is to lay the foundation for the rest of this book by presenting the statistical theory and methods that are utilized in the rest of this book. It is necessarily a mathematical chapter. The subsequent chapters are based largely on the general formulations and theory found here. It is possible to skip this chapter and simply apply the methodology found in the chapters that follow; however, the fundamentals of the underlying approaches are found here.

We begin this chapter with definitions of some statistical terms, primarily involving random variables. That is followed by introductory probability theory. We strive here for a level of comprehension that is at about the level of a senior undergraduate or master's student in a quantitative discipline. The assumption is that readers of this chapter are familiar with integration, basic notions of probability and limits, for example. Citations are given to more advanced works that underpin the mathematical and statistical structure here. In this chapter we present no examples, since our aim is not to explicate but rather to remind the reader of relevant results. The organization of this chapter begins with some background on probability and probability models. Having established these, we turn to statistical methods. The methods that we focus on here and throughout this text are hypothesis testing and confidence intervals. To that end, we next discuss resampling methods including bootstrap methods that will play a prominent role in later chapters. Finally, we discuss sample size and power calculations which are methods for determining the number of observations that need to be taken to achieve a particular set of statistical aims.

2.1 Some Preliminaries

Before moving to the mathematical materials, it is important to keep in mind the big picture of our goals. We would like to use statistical inference to study how well a given biometric authentication system performs. To that end, data is sampled from an ongoing process and we analyze that data in order to assess the system performance. As with any analysis, the quality of the resulting output is dependent upon the quality of the input. Statistical inference is not an exception to this dependence.

M.E. Schuckers, *Computational Methods in Biometric Authentication*,
Information Science and Statistics,
DOI 10.1007/978-1-84996-202-5_2,

In particular, many statistical methods *assume* a random sample from a process to permit unbiased inference. The following quote is one that best captures the relevance of such an assumption. In a section entitled "The Central Role of Practical Assumptions Concerning 'Representative Data' " Hahn and Meeker [43] on pp. 5–6 of their book *Statistical Intervals: A Guide for Practitioners* include the following discussion:

> Departures from these implicit assumptions are common in practice and can invalidate the entire analyses. Ignoring such assumptions can lead to a false sense of security, which, in may applications, is the weakest link in the inference chain. . . .
> In the best of situations, one can rely on physical understanding, or information from outside the study, to justify the practical assumptions. Such assessment is, however, principally the responsibility of the engineer—or "subject-matter expert." Often, the assessment is far from clear-cut and an empirical evaluation may be impossible. In any case, one should keep in mind that the intervals described in this book reflect on the statistical uncertainty, and, thus, provide a *lower bound* on the true uncertainty. The generally nonquantifiable deviations of the practical assumptions from reality provide an added *unknown* element of uncertainty. If there were formal methods to reflect this further uncertainty (occasionally there are, but often there are not), the resulting interval, expressing the *total* uncertainty, would clearly be longer than the statistical interval alone. This observation does, however, lead to a rationale for calculating a statistical interval for situations where the basic assumptions are questionable. In such case, if it turns out that the calculated statistical interval is long, we then know that our estimates have much uncertainty—even *if* the assumptions were all correct. On the other hand, a narrow statistical interval would imply a small degree of uncertainty *only if* the required assumptions hold.

The emphasis here is from the original. Thus, we recommend the methods in this book based upon the utility of the statistical method, but we caution users that the assumptions of a particular approach should not be overlooked or ignored.

Another notion that is less common in statistical discussions needs to be addressed at this juncture. Throughout the rest of this text we will discuss inference in terms of a process rather than in terms of a population. The relevant difference was first articulated by W. Edwards Deming and is an important one for biometric systems, Deming [22]. Deming, in particular, differentiated between an enumerative study and an analytic study. An enumerative study is one that is concerned with understanding a fixed population while an analytic one is concerned with understanding a process and, hence, prediction. As Deming wrote on p. 249 of [22] "[i]n analytic problems the concern is not this one bowl alone but the sequence of bowls that will be produced in the future. . . " where samples have been taken from a single bowl. Since biometric systems are developed and assessed with an aim toward implementation and with interest in future outcomes of these systems, we view these studies as analytic rather than enumerative. As a consequence, the language that we use will be of an ongoing *process* rather than a fixed population.

2.2 Random Variables and Probability Theory

In this section, we define random variables and some of the properties of these variables. Because we aim to keep this discussion at a relatively low mathematical level, we take a less formal approach to some of the definitions that follow. More formal (and complete) structure for these constructs can be found in, for example, Casella and Berger [14]. For those who are interested, a mathematically higher level introduction to these topics can be found in Billingsley [7] or Schervish [82]. More thorough background at approximately the same level as this book can be found in Wackerly et al. [99] or in DeGroot and Schervish [21].

Definition 2.1 An *experiment* is an event whose outcome is not completely known with certainty in advance.

Definition 2.2 A *sample space* is the collection of all possible outcomes for an experiment.

A sample space can be thought of as all of the possible results from an event whose outcome is uncertain. A classic example is the toss of a coin with the sample space being the set $\{Heads, Tails\}$. A more interesting example might be the high temperature in London, UK tomorrow. If we restrict ourselves to whole Celsius units, then the sample space will be some subset of the set of all integers, $\mathbb{Z}$. If, instead, we consider fractional Celsius values, e.g. 18.5 or 13.265 or 29.48733..., then the sample space becomes some subset of the real number line, $\mathbb{R}$.

Definition 2.3 A *parameter* is a real-valued numerical summary of a process or population.

Definition 2.4 A *random variable* (RV) is a real-valued function defined on a sample space.

We can think of a random variable as the outcome of some future experiment or of some as yet unobserved event. For mathematical integrity we say that the values that the random variable can take must be numerical.

Definition 2.5 A *discrete random variable* is a random variable that can only take a countable number of values.

Definition 2.6 A *continuous random variable* is a random variable that can take values on a subset of the real line $\mathbb{R} = (-\infty, \infty)$.

For the examples given above, the tossing of a coin is a discrete random variable—we can map heads and tails to the values 0 and 1—as is the measurement of the high temperature in London using whole Celsius units. If we use decimal units to make our measurements, then the random variable is (usually) treated as a continuous one. Most of the work in this book will focus on discrete random variables but some sections involve continuous random variables.

Definition 2.7 A *probability mass distribution* for a discrete random variable is a table, graph or formula that gives the probability, $P(v)$, of a particular outcome.

We have the following rules for probability mass distributions:

1. $0 \leq P(v) \leq 1$ for all values v, and
2. $\sum_v P(v) = 1$.

Definition 2.8 A *cumulative density function*, $F_X(x)$, for a random variable X is equal to the probability that the value of the random variable, X, is less than or equal to the specific value x.

$$F_X(x) = P(X \leq x). \tag{2.1}$$

Definition 2.9 A *probability density function*,

$$f_X(x) = \frac{dF_X(x)}{dx}, \tag{2.2}$$

is the derivative of the cumulative density function if the cumulative density function is differentiable.

Result 2.1 The probability that a random variable X takes values at or below x is written as

$$P(X \leq x) = \int_{-\infty}^{x} f_X(t)dt. \tag{2.3}$$

It is important to remember that a random variable (or its cumulative distribution function (cdf) or probability density function (pdf)) or its probability mass function is a mathematical/ statistical model of reality. The choice of which random variable to use is one that highly depends on the system or the process we are trying to model.

Definition 2.10 We will denote an *indicator function* of a statement A as the function that is defined to be one if the statement A is true and zero otherwise. We will denote this function by I_A and formally define it as

$$I_A = \begin{cases} 1 & \text{if } A \text{ is true} \\ 0 & \text{otherwise.} \end{cases} \tag{2.4}$$

Definition 2.11 The *expected value* for a random variable, V, written as $E[V]$ is the mean of that random variable. For a discrete random variable, V_d, the expected value is calculated as

$$E[V_d] = \sum_v vP(V = v). \tag{2.5}$$

For a continuous random variable, V_c, the expected value is calculated as

$$E[V_c] = \int_{-\infty}^{\infty} vf(v)dv. \tag{2.6}$$

For a given random variable, discrete or continuous, the expected value of that random variable is often known as the expectation of that random variable.

The expected value is a measure of central tendency meaning that is meant to be a measure of the 'center' or average of a distribution. The expected value is known as the first moment. We also note that for some collection of outcomes A, $E(I_A) = P(A)$.

Result 2.2 The expected value of a constant, a, is just the constant, i.e. $E[a] = a$.

Definition 2.12 The *expected value for a function $g()$ of a random variable, $E[g(V)]$*, is defined as

$$E[g(V)] = \sum_v g(v)P(V = v) \tag{2.7}$$

for a discrete random variable and as

$$E[g(V)] = \int_{-\infty}^{\infty} g(v)f(v)dv \tag{2.8}$$

for a continuous random variable.

The expected value is also known as the mean.

Definition 2.13 The *variance* of a random variable, V, is the expected squared distance from the mean, $E[(V - E(V))^2]$, and is written as $Var[V] = E[(V - E(V))^2]$. For a discrete random variable, V_d, the variance is calculated as

$$Var[V_d] = \sum_v (v - E[V_d])^2 P(V = v). \tag{2.9}$$

For a continuous random variable, V_c, the variance is calculated as

$$Var[V_c] = \int_{-\infty}^{\infty} (v - E[V_c])^2 f(v)dv. \tag{2.10}$$

We will use the notation $Var[\ ]$ as well as $V[\ ]$ for the variance throughout this book.

Definition 2.14 The *standard deviation* for a random variable, V, is the square root of the variance, $\sqrt{Var[V]}$.

We will often use the Greek letter (μ) "mu" to represent the expected value of a random variable. The variance and standard deviation are measures of the variability, or spread, of a distribution and are often denoted by the Greek characters σ^2 and σ, respectively. In particular the variance is the average squared difference of a random variable from its mean. The variance is also known as the 2nd central moment.

We are often interested in combinations of two or more random variables or the relationship between these random variables. As such, we need to define a structure that accommodates all of the possible relationships that may exist between random variables. Below we focus on the case of the relationships between two random variables. Generalization to more than two random variables can be found in Casella and Berger [14], for example.

Definition 2.15 The *joint probability density function*, f, for the continuous random variables $V_1, \ldots, V_n$ is $f(V_1, \ldots, V_n)$, if the function $f(V_1, \ldots, V_n)$ satisfies the following properties:

1. $f(v_1, \ldots, v_n) \geq 0$ for all values of $v_1, \ldots, v_n$,
2. $\int_{-\infty}^{\infty} \cdots \int_{-\infty}^{\infty} f(v_1, \ldots, v_n) dv_1 \ldots dv_n = 1$.

Definition 2.16 The *joint probability distribution* for the discrete random variables $V_1, \ldots, V_n$ is $P(V_1 = v_1, \ldots, V_n = v_n)$ and must satisfy the following properties following properties:

1. $P(V_1 = v_1, \ldots, V_n = v_n) \geq 0$ for all $v_1, \ldots, v_n$,
2. $\sum_{v_1} \cdots \sum_{v_n} P(V_1 = v_1, \ldots, V_n = v_n) = 1$.

It is possible to have a joint distribution of a continuous random variable and a discrete random variable but this case is not relevant for the methods described in this book.

Definition 2.17 Let $f(V_1, V_2)$ be the joint probability density function for V_1 and V_2. Then the *marginal density function* for V_1 is given by

$$f_1(v_1) = \int_{-\infty}^{\infty} f(v_1, v_2) dv_2. \tag{2.11}$$

Definition 2.18 Two continuous random variables, V_1 and V_2, with joint probability density function f are *independent* if and only if

$$f(v_1, v_2) = f_1(v_1) f_2(v_2) \tag{2.12}$$

for all pairs of values v_1 and v_2. Two discrete random variables, V_3 and V_4, are *independent* if and only if

$$P(v_3, v_4) = P(v_3) P(v_4) \tag{2.13}$$

for all values of v_3 and v_4.

Definition 2.19 Let V_1 and V_2 be two continuous random variables with joint probability density function $f(v_1, v_2)$. Further, let V_3 and V_4 be two discrete random variables with joint probability distribution $P(V_3, V_4)$. The *expected value of a function of more than one random variable* $E[g(V_1, V_2)]$ or $E[g(V_3, V_4)]$ is defined to be

$$E[g(V_1, V_2)] = \int_{-\infty}^{\infty} \int_{-\infty}^{\infty} g(v_1, v_2) f(v_1, v_2) dv_1 dv_2 \tag{2.14}$$

for the continuous random variables V_1 and V_2 and

$$E[g(V_3, V_4)] = \sum_{v_3} \sum_{v_4} g(v_3, v_4) P(v_3, v_4) \tag{2.15}$$

for the discrete random variables V_3 and V_4.

Definition 2.20 The *covariance* of two random variables, $Cov(V_1, V_2)$, is defined to be

$$Cov(V_1, V_2) = E(V_1 - E[V_1])(V_2 - E[V_2]). \tag{2.16}$$

We also note that

$$Cov(V_1, V_2) = Cov(V_2, V_1). \tag{2.17}$$

Result 2.3 We note here that $Cov(V, V) = Var[V]$. This is a direct result of the definition of these two quantities.

Definition 2.21 The *correlation* between two random variables, V_1 and V_2, is defined to be

$$Corr(V_1, V_2) = \frac{Cov(V_1, V_2)}{\sqrt{Var[V_1]}\sqrt{Var[V_2]}}. \tag{2.18}$$

As with the covariance, the correlation is symmetric, i.e.

$$Corr(V_1, V_2) = Corr(V_2, V_1). \tag{2.19}$$

Definition 2.22 We will say that two random variables V_1 and V_2 are *uncorrelated* if $Corr(V_1, V_2) = 0$.

Comment 2.1 We note specifically here that independence is a stronger condition than being uncorrelated and, hence, independence implies uncorrelated. That is, if we know that V_1 and V_2 are independent then we know they are uncorrelated. However, it is possible for two random variables to be uncorrelated and not independent.

The covariance and the correlation are measures of how two random variables vary together or how they *co-vary*. The correlation is a unitless measure that attempts to calibrate the way the two variables co-vary in a way that controls for the individual variability in each.

Here we have defined joint densities and joint expectations for two variables. There are equivalent multivariable extensions of these notions. Similarly there are multivariate notions of independence.

Definition 2.23 We will call a sequence of random variables $V_1, \ldots, V_n, \ldots$ *stationary* if the mean and the covariances of the V_i's do not depend on i. That is, we will call a process stationary if $E[V_i]$ and $Cov(V_i, V_j)$ do not depend on i and j.

Definition 2.23 is usually denoted as *weak stationarity*. It is the sense that we will use here. More detail on stationarity for processes can be found in Grimmett and Stirzaker [40] and Brockwell and Davis [10].

We offer the following results for linear combinations of random variables. Let $a_1, a_2, \ldots$ be fixed, known constants and let $E[V_i] = \mu_i$ and $Var[V_i] = \sigma_i^2$ for $i = 1, 2, 3, \ldots$.

Result 2.4

$$E[a_i V_i] = a_i \mu_i. \tag{2.20}$$

Result 2.5

$$Var[a_i V_i] = a_i^2 \sigma_i^2. \tag{2.21}$$

Result 2.6

$$E\left[\sum_i a_i V_i\right] = \sum_i a_i \mu_i. \tag{2.22}$$

Result 2.7

$$Var\left[\sum_i a_i V_i\right] = \sum_i a_i^2 \sigma_i^2 + \sum_i \sum_{j \neq i} a_i a_j \sigma_i \sigma_j Cov(V_i, V_j). \tag{2.23}$$

Comment 2.2 We note that for Result 2.7 the second term on the right hand side is eliminated if all of the variables are pairwise uncorrelated. If some correlation exists, that is, if there is at least one non-zero correlation, then that term cannot be eliminated.

Result 2.8 If $V_1, \ldots, V_n$ have the same mean, $E[V_i] = \mu$, and the same variance, $Var[V_i] = \sigma^2$, then for a given set of constants $a_1, \ldots, a_n$ Result 2.7 can be written as

$$Var\left[\sum_i a_i V_i\right] = \sum_i a_i^2 \sigma_i^2 + \sum_i \sum_{j \neq i} \sigma_i \sigma_j a_i a_j Corr(V_i, V_j) \tag{2.24}$$

which becomes

$$Var\left[\sum_i a_i V_i\right] = \sigma^2 \sum_i a_i^2 + \sigma^2 \sum_i \sum_{j \neq i} a_i a_j Corr(V_i, V_j). \tag{2.25}$$

Comment 2.3 Result 2.8 is *crucial* to the rest of this book because we will be dealing with a large number of random variables that we can treat as having a common mean and common variance but which are correlated. We will use the assumption of a stationary process to justify our assumption about a common mean and a common variance. Thus, we will see Result 2.8 many more times.

Result 2.9 We can use the above results to derive the variance for a mean of a process that is stationary. If $E[V_i] = \mu$ and $Var[V_i] = \sigma^2$, for all i, then with $a_i = \frac{1}{n}$ Result 2.7 becomes

$$Var\left[\frac{1}{n}\sum_i V_i\right] = \frac{\sigma^2}{n^2}\left[n + \sum_i \sum_{j \neq i} Corr(V_i, V_j)\right]. \tag{2.26}$$

Note that based upon Result 2.9 for a stationary process, we need only specify the mean, variance, and correlation of the random variables that encompass the process in order to be able to determine the variance of the mean. We further note that a proportion is simply a mean of random variables that take the value 0 or 1. So this result also applies to proportions.

Result 2.10 For two sets of random variables $V_1, \ldots, V_n$ and $R_1, \ldots, R_m$, we have that

$$Cov\left(\sum_{i=1}^{n} V_i, \sum_{j=1}^{m} R_j\right) = \sum_{i=1}^{n}\sum_{j=1}^{m} Cov(V_i, R_j). \tag{2.27}$$

It is often the case that it is easier to write a linear combination of random variables by using vector notation. Consequently, let $\mathbf{V} = (V_1, V_2, \ldots, V_n)^T$ be a random (column) vector (where ^{T} represents the transpose of the vector). For example, the linear combination $\sum a_i V_i$ can be written as $\mathbf{a}^T\mathbf{V}$ where $\mathbf{a} = (a_1, a_2, \ldots, a_n)^T$. The definitions below further develop these ideas.

Definition 2.24 The *expected value of a random vector* is the expected value of the elements in that vector. That is,

$$E[\mathbf{V}] = (E[V_1], E[V_2], \ldots, E[V_n])^T = (\mu_1, \mu_2, \ldots, \mu_n)^T. \tag{2.28}$$

Definition 2.25 The *covariance of a random vector* is a matrix whose individual terms, σ_{ij} are the individual covariances, $\sigma_{ij} = Cov(V_i, V_j)$. We will use the notation $\boldsymbol{\Sigma}$ to represent covariance matrices. $\boldsymbol{\Sigma}$ is a symmetric matrix defined as

$$\boldsymbol{\Sigma} = Cov(\mathbf{V}) = \begin{pmatrix} \sigma_{11} & \sigma_{12} & \cdots & \sigma_{1n} \\ \sigma_{21} & \sigma_{22} & \cdots & \sigma_{2n} \\ \vdots & \vdots & \vdots & \vdots \\ \sigma_{n1} & \sigma_{n2} & \cdots & \sigma_{nn} \end{pmatrix}. \tag{2.29}$$

If we let $\mathbf{a}$ be a (column) vector of constants that is the same length as $\mathbf{V}$, then the expected value and variance of $\mathbf{a}^T\mathbf{V}$ is:

Result 2.11

$$E[\mathbf{a}^T\mathbf{V}] = \mathbf{a}^T E[\mathbf{V}] = \mathbf{a}^T \boldsymbol{\mu} \tag{2.30}$$

where $\boldsymbol{\mu} = (\mu_1, \mu_2, \ldots, \mu_n)^T$,

and

Result 2.12

$$Var[\mathbf{a}^T\mathbf{V}] = \mathbf{a}^T \boldsymbol{\Sigma}\mathbf{a}. \tag{2.31}$$

2.2.1 Specific Random Variables

Definition 2.26 We will call a random variable, V, *a Bernoulli random variable* if it has the following characteristics:

1. V only takes two values, say zero (failure) and one (success),
2. $P(V = 1) = p$ and $P(V = 0) = 1 - p$.

We will denote a random variable, V, as being a Bernoulli random variable with probability of success p as $V \sim Bern(p)$.

Result 2.13 If V is a Bernoulli random variable ($V \sim Bern(p)$), then $E[V] = p$ and $Var[V] = p(1 - p)$.

Definition 2.27 We will refer to a random variable as a *binomial random variable* if it is the sum of n independent Bernoulli random variables, each with the same probability of a success, p. Let $V_i \sim Bern(p)$ for $i = 1, \ldots, n$ and assume that the V_i are independent. We will denote a random variable $V = \sum_{i=1}^{n} V_i$ as being a binomial random variable with parameters n and p as $V \sim Bin(n, p)$.

Result 2.14 If V is a Binomial random variable ($V \sim Bin(n, p)$), then $E[V] = np$ and $Var[V] = np(1 - p)$.

Comment 2.4 The variance of the sum of n uncorrelated Bernoulli random variables is the same as that for a binomial random variable.

Definition 2.28 We will call V *a Gaussian or normal random variable* with mean μ and variance σ^2 if it has the following density function:

$$f_V(v) = \frac{1}{\sqrt{2\pi\sigma^2}} e^{-\frac{1}{2\sigma^2}(v-\mu)^2}. \tag{2.32}$$

We will denote a Gaussian random variable, V, with mean μ and variance σ^2 as $V \sim N(\mu, \sigma^2)$. The terms Gaussian and normal are interchangeable for this distribution but we will use Gaussian in most cases in this book.

Result 2.15 Let $V \sim N(\mu, \sigma^2)$. then $Z = \frac{V-\mu}{\sigma} \sim N(0, 1)$. The distribution $N(0, 1)$ will be referred to as a standard Gaussian distribution.

Tables 9.1 and 9.2 give percentiles for a standard Gaussian distribution and will be used extensively throughout this text. A Gaussian random variable is the one that is typically associated with a bell-shaped curve.

2.2.2 Estimation and Large Sample Theory

In this section, we review some of the results from large sample probability theory. Large sample probability theory investigates functions of sequences of random variables, $f(V_1, V_2, \ldots)$.

Definition 2.29 An *estimator*, $\hat{\theta}$, of a parameter θ is a function of random variables, $\hat{\theta}(V_1, \ldots, V_n)$. We will use a ^ to denote an estimator of a parameter.

Definition 2.30 One estimator of importance is the *sample mean*

$$\bar{V} = \frac{1}{n} \sum_{i=1}^{n} V_i. \tag{2.33}$$

Though, perhaps obvious, the sample mean is an estimator of the mean of a process.

Definition 2.31 We will call an estimator, $\hat{\theta}$, of θ *unbiased* for θ if $E[\hat{\theta}] = \theta$.

Definition 2.32 For an estimator $\hat{\theta} = \hat{\theta}_n(V_1, V_2, \ldots, V_n)$ where $V_1, V_2, \ldots, V_n$ are random variables, the probability distribution of θ is called the *sampling distribution* of θ.

Definition 2.33 If random variables $V_1, V_2, \ldots, V_n$ each have the same probability distribution and they are all independent of each other, then we will say that these random variables are *independently and identically distributed* (*iid*).

We will use the notation *iid* to represent a collection of random variables that are independently and identically distributed.

Theorem 2.1 *Let $V_1, \ldots, V_n$ be a sequence of* iid *random variables with expected value μ and variance σ^2. Then,*

$$\lim_{n\to\infty} \frac{1}{n} \sum_{i=1}^{n} V_i = \lim_{n\to\infty} \bar{V}_n = \mu. \tag{2.34}$$

Theorem 2.2 *Let $V_1, V_2, \ldots, V_n$ be* iid *Gaussian random variables with expected value (mean) μ and variance $\sigma^2 < \infty$. That is, $V_i \sim N(\mu, \sigma^2)$. Then*

$$\lim_{n\to\infty} P\left(\frac{\bar{V}_n - \mu}{\sigma} \leq x\right) \to P(Z \leq x) \tag{2.35}$$

where $Z \sim N(0, 1)$.

The following result relaxes the assumption of Gaussianity necessary to achieve a mean which follows a Gaussian distribution.

Theorem 2.3 *Let $V_1, V_2, \ldots, V_n$ be* iid *random variables with expected value (mean) μ and variance $\sigma^2 < \infty$. Then*

$$\lim_{n\to\infty} f_{\frac{\bar{V}_n - \mu}{\sigma/\sqrt{n}}} \to f_Z \tag{2.36}$$

where Z is a standard Gaussian random variable.

Comment 2.5 Theorem 2.3 is known as the Central Limit Theorem (CLT). In more general language it states that if we average a large number of random variables, then the distribution of the sample mean will be a Gaussian distribution. This version of the central limit theorem is based upon *iid* data. There are numerous other central limit theorems that relax these conditions. We will use those as necessary. The interested reader is directed to Jacod and Shiryaev [50], for example. Below we offer an extension to Result 2.3. Recently, Dietz and Schuckers [23] proposed some biometric specific extensions to the CLT given above.

Theorem 2.4 *Let $V_1, V_2, \ldots,$ be a sequence of independent random variables that are each the sum of m_i correlated binary random variables such that $E[V_i] = m_i\pi$ and $\sigma^2 = Var[V_i] = m_i\pi(1-\pi)(1+\rho(m_i - 1))$ then by Moore [69] we have the following result. Let*

$$\hat{\pi} = \frac{\sum_{i=1}^{n} V_i}{\sum_{i=1}^{n} m_i}, \tag{2.37}$$

then

$$\lim_{n\to\infty} f_{\frac{\hat{\pi}_n - \pi}{\sigma_{\hat{\pi}}}} \to f_Z \tag{2.38}$$

where $Z \sim N(0, 1)$,

$$\sigma_{\hat{\pi}} = \sqrt{\frac{\pi(1-\pi)(n\bar{m} + \rho\sum_{i=1}^{n}(m_i - 1))}{n^2\bar{m}^2}} \tag{2.39}$$

and

$$\bar{m} = \frac{\sum_{i=1}^{n} m_i}{n}. \tag{2.40}$$

The mean and variance that we use in Theorem 2.4 are the result of a particular correlation structure that we will see in Chaps. 3 and 7. This result is another form of a central limit theorem.

Theorem 2.5 *Let* $V_1, V_2, \ldots$ *be a sequence of independent random variables with* $E[V_i] = \mu_i$, $Var[V_i] = \sigma_i^2$ *and* $r_i^3 = E[\mid V_i - \mu_i \mid^3] < \infty$. *Also let* $S_n = \sum_{i=1}^{n} V_i$. *If*

$$\lim_{n\to\infty} \frac{(\sum_{i=1}^{n} r_i^3)^{1/3}}{(\sum_{i=1}^{n} \sigma_i^2)^{1/2}} = 0, \tag{2.41}$$

then

$$f_{S_n^*} \to N(0, 1) \tag{2.42}$$

where

$$S_n^* = \frac{S_n - \sum_{i=1}^{n} \mu_i}{(\sum_{i=1}^{n} \sigma_i^2)^{1/2}}. \tag{2.43}$$

Equation (2.41) is known as the *Lyapunov Condition* and provides a central limit theorem for *non-iid* random variables.

Definition 2.34 The *effective sample size* is the equivalent number of observations that would result in the same variance of the estimator if the data were *iid*. We will generally denote an effective sample size with a superscript dagger, i.e. †. The effective sample size is calculated by taking the actual sample size say, n, and multiplying it by the ratio of the variance if the data were *iid* to the observed variance. The equation is

$$n^\dagger = n\left[\frac{V_{iid}[\hat{\theta}]}{V[\hat{\theta}]}\right] \tag{2.44}$$

where V_{iid} is the variance assuming the data is collected via a simple random sample or assuming the data is *iid*.

The effective sample size is used to measure the amount of independent information in a sample. It is something that we will consistently use later as we discuss methods for evaluating the performance of a classification or a biometric system.

2.3 Statistical Inference

In this section, we discuss statistical inference. Statistical inference draws conclusions or makes statements about a process parameter based upon a sample. We treat a sample as composed of n random variables. For statistical inference throughout this book, we will focus on confidence interval estimation and hypothesis testing. We assume that the reader has been exposed previously to confidence intervals and hypothesis tests. A brief refresher of these topics is given. An introduction is then given to some less well known statistical methods. These include randomization tests, bootstrap , jackknife methods, sample size calculations, power calculations and prediction intervals. Bootstrap, jackknife and randomization methods all fall under the category of non-parametric or distribution-free approaches. Prediction intervals are inferential intervals for some function of future observations from a process. This is in contrast to confidence intervals which are generally for a single parameter of that process.

Non-parametric here will mean that no attempt is made to model the distribution or shape of the observations. Our methods will attempt to maintain the covariances and correlations between observations. The bootstrap and jackknife approaches are methods for estimating the sampling distribution by resampling the collected data. Randomization tests follow a similar idea to resampling methods in that they are non-parametric methods. The basic approach of a randomization test is to combine the data from two or more groups and then randomly re-assign those observations back to those groups and recalculate a given test statistic. This process is then repeated multiple times creating a distribution for the given test statistic assuming a null hypothesis of equality of some parameter among the groups. These methods are an alternative to the large sample methodology which *does* depend upon a specific distribution.

All of the methods in this section are dependent upon having a quality sample from a population or from a process. The following comment addresses some concerns about the general use of statistical methods.

Comment 2.6 Most statistical methods start with an assumption that the sample upon which inference is based is a random sample. We note here that there are statistically appropriate methods for collection observations beyond a simple random sample. (See Definition 2.35.) Frequently, in biometric systems testing the sample that is used is a convenience sample, i.e. one that is readily available. Some in the biometrics community have stated that this disqualifies the use of statistical tools. Obviously, a book on statistical methods will argue against such a broad statement. Here we note that this does not completely preclude the use of statistical methods. We quote now from p. 17 of Hahn and Meeker's influential text *Statistical Intervals: A Guide for Practitioners* [43]:

> Because one is not sampling randomly, statistical intervals strictly speaking are not applicable for convenience sampling. In practice, however, one uses experience and understanding of the subject matter to decide on the applicability of applying statistical inferences to the results of convenience sampling. Frequently, one might conclude that the convenience sample will provide data that, for all practical purposes, are as "random" as those that one would obtain by a simple random sample. . . . Our point is that, treating a convenience sample as if it were a random sample *may sometimes* be reasonable from a practical point of view. However, the fact that this assumption is being made needs to be recognized, and the validity of using statistical intervals, as if a random sample had been selected needs to be critically assessed based upon the specific circumstances.

The emphasis in the above paragraph is from the original text.

The larger point that we wish to make here is that there are limitations to the inference that can be made from statistical methods. These limitations should not be ignored. These limitations need to be thoughtfully considered as part of any conclusions that one wants to draw from a data analysis.

Definition 2.35 A *simple random sample* or SRS is a sample of size n taken such that each possible combination of n units is equally likely.

Definition 2.36 For a sample of n random variables, $V_1, \ldots, V_n$, we will call s the *sample standard deviation* where

$$s = \sqrt{\frac{\sum_{i=1}^{n}(V_i - \bar{V})^2}{n-1}} \tag{2.45}$$

and

$$\bar{V} = n^{-1}\sum_{i=1}^{n} V_i. \tag{2.46}$$

Result 2.16 If $V_1, V_2, \ldots, V_n$ are *iid* random variables with $Var(V_i) = \sigma^2$ for all i, then s^2 is unbiased for σ^2 where s is given by (2.45). We refer to s^2 as the sample variance.

Definition 2.37 If we have n samples from a distribution, $V_1, \ldots, V_n$ and we estimate the population standard deviation, σ, using the sample standard deviation, s, then

$$\frac{\bar{V} - \mu}{s/\sqrt{n}} \sim t(n-1) \tag{2.47}$$

where $t(n-1)$ represents a *t-distribution* with $n-1$ degrees of freedom (df).

A table of percentile of the *t-distribution* can be found in Table 9.3 for a variety of degrees of freedom. We use *degrees of freedom* here in the statistical sense, not the biometrics sense that is used by Daugman [19].

Definition 2.38 The *Student's t-distribution or t-distribution probability density function* with ν degrees of freedom is

$$f_T(t) = \frac{\Gamma((\nu+1)/2)}{\sqrt{\nu\pi}\,\Gamma(\nu/2)}(1 + t^2/\nu)^{-(\nu+1)/2} \tag{2.48}$$

for a random variable T.

Definition 2.39 The *Chi-squared distribution* with ν degrees of freedom is

$$f_X(x) = \frac{x^{(\nu/2)-1}e^{-x/2}}{2^{\nu/2}\Gamma(\nu/2)} \tag{2.49}$$

for a random variable X when $x > 0$ and 0 otherwise.

Definition 2.40 The *standard error* for an estimator $\hat{\theta}$ is the estimated standard deviation of the sampling distribution of that estimator.

Result 2.17 If we take a random sample of size n from a process or population with mean μ and finite variance σ^2 and n is large, then the sampling distribution for the normalized sample mean

$$\frac{\bar{V}_n - \mu}{s/\sqrt{n}} \tag{2.50}$$

is a *t-distribution.*

Result 2.18 If we take a random sample of size n from a process or population with proportion p and large n, then the sampling distribution for the sample proportion is a Gaussian distribution.

Result 2.19 The standard error for the mean is $\frac{s}{\sqrt{n}}$ for a simple random sample and is denoted by $s_{\bar{X}}$.

Result 2.20 The standard error for a proportion, $\hat{p} = \frac{X}{n}$ taken from a simple random sample is $\sqrt{\frac{\hat{p}(1-\hat{p})}{n}}$ and is denoted by $s_{\hat{p}}$.

Results 2.17 and 2.18 will also hold if we do not have correlated samples. The sample size needed for convergence to the reference distributions, t and Gaussian, respectively, increases as the amount of dependence increases. If we have independent samples, then the general guidance that is given is that the sample size for the mean needs to be at least $n \geq 30$ and the sample size for a proportion needs to be large enough so that $np \geq 10$ and $n(1 - p) \geq 10$, see, for example, Deveaux et al. [96].

2.3.1 Confidence Intervals

Statistical methods are predicated on the idea that information that we have collected from a population or process is incomplete. Consequently, the estimated values for a population or process parameter are unlikely to be equal to the parameters themselves. To reflect this uncertainty, intervals or ranges of values are created. In the case of multivariate parameters, confidence regions or bands are created. Below we discuss in some detail confidence intervals for parameter estimation. In Sect. 2.3.2, we introduce the idea of prediction intervals for future observations. A complete list of statistical intervals along with a thorough description of the issues involved in each can be found in Hahn and Meeker [43].

Confidence intervals are statistical tools for making inference about a parameter. We defined a parameter above in Definition 2.3. A point estimate is a single point that is an estimate of a parameter. Because there is variability in sampling, our estimate rarely has the same value as the *estimand*, the quantity we are trying to estimate. Therefore, we recognize that a range of values is necessary to summarize our knowledge of the parameter based upon the information in our sample. Note that the width of a particular interval is indicative of how much information we have about the parameter. That is, if we believe that a rate or proportion falls in the interval $(0.20, 0.40)$ we have far less information than if we believe it is in the interval $(0.29, 0.32)$. Thus, the width of an interval is a measure of how much information we have about the parameter. Wider intervals represent less information; narrower intervals represent more information about the process under consideration.

Definition 2.41 A *confidence level*, $1 - \alpha$, is defined to be the proportion of times that a confidence interval 'captures' the parameter in repeated sampling from a process or a population. Here by capture we mean that the parameter value in contained in the confidence interval.

Typical values for confidence levels are 90%, 95% and 99% which correspond to $\alpha = 0.10, 0.05$ and 0.01, respectively. As mentioned above, there are multidimensional equivalents to a confidence interval. These are commonly referred to as confidence regions.

Definition 2.42 The $100 \times k$th *percentile* is the point, υ_k, in the distribution of a random variable V where $P(V \leq \upsilon_k) \leq k$ and $P(V \geq \upsilon_k) \geq 1 - k$.

Definition 2.43 A $100 \times (1 - \alpha)\%$ *confidence interval* for a parameter θ is defined to be an interval formed by two real-valued functions of the data $U(V_1, \ldots, V_n)$ and $L(V_1, \ldots, V_n)$ such that

$$P(L(V_1, \ldots, V_n) \leq \theta \leq U(V_1, \ldots, V_n)) \geq 1 - \alpha. \tag{2.51}$$

U and L are meant to denote the upper and lower endpoints of the interval and are themselves random variables. Note that it is possible for U or L to be defined as constants.

The typical methodology for making a confidence interval is to take the $\alpha/2$th percentile and the $1 - \alpha/2$th percentile of the sampling distribution for an estimator and use those to make a confidence interval. This is the typical *two-sided* confidence interval. It is the most commonly used because for estimation of means and proportions and it yields the narrowest interval for a given confidence level. Other choices are possible (and sometimes advisable). One other common choice is to create an interval that is one-sided, i.e. that has either a set upper or a set lower bound but not both. This is done by setting $U(V_1, \ldots, V_n) \stackrel{set}{=} \infty$ or $L(V_1, \ldots, V_n) \stackrel{set}{=} -\infty$. This is appropriate if we are only interested in estimation of one tail for the parameter. For example, we might only be interested in the worst case false accept rate that is credible based on a sample from the process. In that case we are uninterested in the lower bound and we would set $L(V_1, \ldots, V_n) \stackrel{set}{=} -\infty$. (For a proportion we could equivalently, and without loss of information, set $L(V_1, \ldots, V_n) \stackrel{set}{=} 0$.)

Below we present methodology for two confidence intervals assuming random samples from an *iid* process. It is not always the case in classification performance that one has independent (or uncorrelated) samples. We'll spend a good deal of the rest of this text on cases that are not *iid* but here as a heuristic foundation we present *iid* cases. Thus, we provide these tools as a basis for discussion and as a summary of the background that is assumed for the rest of this text. We will provide generalizations for these methods, as appropriate, in later chapters.

Result 2.21 Let $R_1, \ldots, R_n$ be *iid* samples from a population with fixed mean, say μ_R. If n is large (generally $n \geq 30$) then we can use the following to make a $(1-\alpha)100\%$ *confidence interval for* μ_R:

$$\bar{R} \pm t_{\alpha/2;n-1} \frac{s_R}{\sqrt{n}} \tag{2.52}$$

where

$$\bar{R} = \frac{1}{n} \sum_{i=1}^{n} R_i \tag{2.53}$$

and

$$s_R = \sqrt{\frac{\sum_{i=1}^{n} (R_i - \bar{R})^2}{n-1}} \tag{2.54}$$

and $t_{\alpha/2;n-1}$ is the $(1-\alpha/2) \times 100$th percentile of a t-distribution with $n-1$ degrees of freedom. Table 9.3 gives percentiles for some t-distributions.

Some comments on the use of the interval above are important. We have two possible sources of a Gaussian distribution for the sampling distribution of the sample mean. The first happens if we are sampling from a Gaussian distribution and then we get a Gaussian sampling distribution for any sample size. The second happens if we are sampling from a non-Gaussian distribution and in that case we need our sample to be sufficiently large. As a consequence, it is often possible to use the methods described in Result 2.21 when the sample size is less than 30. This will be true if the observed values in the sample are roughly bell-shaped and symmetric.

Result 2.22 Let R be a Binomial random variable with probability of success p and number of trials n. $R \sim Bin(n, p)$. If n is large (generally $n\hat{p} \geq 10$ *and* $n(1-\hat{p}) \geq 10$) then we can use the following to make a $(1-\alpha)100\%$ *confidence interval for* p*, the success rate or probability for* R:

$$\hat{p} \pm z_{\alpha/2} \sqrt{\frac{\hat{p}(1-\hat{p})}{n}} \tag{2.55}$$

where

$$\hat{p} = \frac{r}{n} \tag{2.56}$$

which is the proportion observed in the sample and $z_{\alpha/2}$ is the $(1 - \alpha/2) \times 100$th percentile from a Gaussian distribution. r is the observed value of the RV R.

Note that a proportion is just an average of zero's and one's. Thus Theorem 2.3 holds for proportions. Confidence intervals can be created for a wide range of process quantities, including medians, variances and percentiles. We will not cover these confidence intervals in this book.

Comment 2.7 In order to use the confidence interval given in Result 2.22, we must have a large sample size, that is, n needs to be sufficiently large. Here we say that n is large if $n\hat{p} \geq 10$ *and* $n(1 - \hat{p}) \geq 10$. We follow De Veaux et al. [96] among other introductory statistics texts in using the value 10 as a cutoff. We do note that this is an area that has seen some recent attention in the statistics literature. The interested reader should consult, for example, Agresti and Coull [4], Brown et al. [11], or Brown et al. [12] for more details.

2.3.2 Prediction Intervals

Above we have defined the basics of confidence intervals. Again, see Hahn and Meeker [43] for a thorough treatment of these ideas. Here we point out that a confidence interval is meant to take information collected on a particular process and give us information (in the form of a range of values) about a parameter or parameters measured on that process. It is not meant to predict future performance of that process, although it can be a useful tool for this prediction. Prediction intervals, on the other hand, are meant to do exactly that—provide an inferential interval for a function of future observations.

Definition 2.44 A $100 \times (1 - \alpha)\%$ *prediction interval* for a real-valued function of future observations, $g(V_{n+1}, \ldots, V_{n+m})$, is defined to be an interval formed by two real-valued functions of n observed observations $U_p(V_1, \ldots, V_n)$ and $L_p(V_1, \ldots, V_n)$ such that

$$P(L_p(V_1, \ldots, V_n) \leq g(V_{n+1}, \ldots, V_{n+m}) \leq U_p(V_1, \ldots, V_n) \mid V_1, \ldots, V_n) \geq 1 - \alpha \tag{2.57}$$

where $V_1, \ldots, V_n$ are n observed values and $V_{n+1}, \ldots, V_{n+m}$ are m random variables from the same process representing future unknown values. U_p and L_p are meant to denote the upper and lower endpoints of the interval and are themselves random variables until $V_1, \ldots, V_n$ are observed. As above, it is possible for U or L to be defined as constants.

Prediction intervals are confidence intervals but they are confidence intervals for future values of a process. Hence they are called *prediction* intervals. Below we present prediction intervals for a mean and for a proportion both from *iid* processes.

Result 2.23 Let $R_1, \dots, R_n$ be an *iid* sample from a population with fixed mean, say μ_R, and let $\bar{R}_n^\diamond$ be the mean of the $n^\diamond$ unobserved future values, $R_{n+1}, \dots, R_{n+n^\diamond}$. If n is large (generally ≥ 30) then we can use the following to make a $(1-\alpha)100\%$ *prediction interval for* $\bar{R}_n^\diamond$:

$$\bar{R} \pm t_{\alpha/2;n-1} s_R \sqrt{\frac{1}{n^\diamond} + \frac{1}{n}} \tag{2.58}$$

where

$$\bar{R} = \frac{1}{n}\sum_{i=1}^{n} R_i \tag{2.59}$$

is the sample mean and

$$s_R = \sqrt{\frac{\sum_{i=1}^{n}(R_i - \bar{R})^2}{n-1}} \tag{2.60}$$

is the sample standard deviation following Definition 2.36 calculated on $R_1, \dots, R_n$.

Result 2.24 Let R be a Binomial random variable with probability of success p and n observed trials and let R_2 be a Binomial random variable with probability of success p from $n^\diamond$ unobserved trials. We assume that R_2 takes observations independently of R, but with the same probability of success p.

If n and $n^\diamond$ are large (generally $n\hat{p} \geq 10$ and $n(1-\hat{p}) \geq 10$) then we can use the following to make a $(1-\alpha)100\%$ prediction interval for the future proportion of successes $\frac{R_2}{n^\diamond}$:

$$\hat{p} \pm z_{\alpha/2}\sqrt{\hat{p}(1-\hat{p})\left(\frac{1}{n} + \frac{1}{n^\diamond}\right)} \tag{2.61}$$

where $\hat{p} = \frac{R_1}{n}$ which is the proportion of observed successes in the original sample.

Prediction intervals are wider than the equivalent confidence intervals because there are two sources of variability: one for estimation of the relevant parameter—which is the appropriate variability for a confidence interval—and one for the sampling variability of $g(V_{n+1}, \dots, V_{n+n^\diamond})$ a function of the future observations.

2.3.3 *Hypothesis Testing*

Hypothesis tests are a statistical counterpart to confidence intervals. Hypothesis testing is a methodology for determining whether a particular parameter value is supported by the data. This methodology is in opposition to confidence intervals which start with no particular value in mind. Hypothesis tests are particularly useful in biometrics for testing whether an error rate is below a particular value. For example, hypothesis testing would be used to determine if a false accept rate is less than 1.5% or the failure to enrol rate is less than 4%. The statistical theory also allows for the possibility that the hypothesis we want to test is simply that the parameter is not equal to a particular value. We outline the fundamental pieces of a hypothesis test below:

1. The *null hypothesis* H_0: is written first and represents the value of the parameter that we would like to test against.
2. The *alternative hypothesis* H_1 is written next and it represents the hypothesis regarding the parameter we would like to suggest is true.
3. The *test statistic* is what follows next and it is generally a method of measuring how large the discrepancy is between the parameter value specified by the null hypothesis and the observed estimator of the parameter calculated from the data.
4. The *p-value* is the probability that an observed test statistic will be more extreme (relative to the alternative hypothesis) than the test statistic that was observed, if H_0 were true.
5. The *decision rule* is what determines whether the null hypothesis is rejected or not rejected. *The decision rule is to reject the null hypothesis if the p-value is small.* If we reject the null hypothesis, we conclude that the parameter estimate based upon data collected represented a significant difference from the hypothesized parameter value given in the null hypothesis. The alternative conclusion that we can make is to 'fail to reject' the null hypothesis.

The logic of hypothesis testing is related to the idea of falsifiability, Popper [77]. The null hypothesis is usually a value for a parameter that is specified *a priori* data collection. The null hypothesis is the statement that it is possible to falsify. It is rarely possible to show that a particular hypothesis is true—that would require a census of our process outcomes—but rather we can find enough empirical evidence to conclude that it is not. Thus the language of our decision rule is to *reject* the null hypothesis or to *fail to reject* the null hypothesis. How much evidence is necessary is set by how small the p-value needs to be in order to reject. The common choices for the significance level are 0.10, 0.05 and 0.01. These values—significance levels usually denoted by the Greek letter α, see Definition 2.46 below—are instituted structurally by some disciplines and are backed by historical precedence. The exact value to be used in practice need not be any of these; however, if a value for α is chosen, it needs to be established before data is analyzed. A selected α should be clearly stated as part of any reports on the performance of a classification or matching system. Further, it is possible and increasingly the case in some fields, such as medicine, to report only the *p-value* without specifying α and letting readers determine the significance of a given result particularly for those results where different choices of α could lead to different conclusions.

In any decision making process that involves uncertainty or incomplete information, errors can result. One source of these errors is the incomplete nature of any data collection to obtain information about a process. Statisticians distinguish between two types of errors: Type I and Type II errors. We define these below as well as the notation for the probabilities of each.

Definition 2.45 A *type I error* occurs when an experimenter rejects the null hypothesis and, in fact, the null hypothesis is true.

Definition 2.46 The *significance level*, denoted by the Greek letter α, is the probability that we reject the null hypothesis when indeed the null hypothesis is true. Formally,

$$\alpha = P(\text{Reject } H_0 \mid H_0 \text{ is true}). \tag{2.62}$$

The significance level is the probability of making a type I error.

Definition 2.47 A *type II error* occurs when an experimenter fails to reject the null hypothesis if the null hypothesis is false. Thus,

$$P(\text{Fail to Reject } H_0 \mid H_0 \text{ is not true}). \tag{2.63}$$

Definition 2.48 Statistical *power* is the probability of rejecting the null hypothesis when the null hypothesis is false. This is one minus the probability of a type II error, and is usually denoted by $Power = 1 - P(\text{type II error}) = 1 - \beta$ where $\beta = P(\text{type II error})$.

The evaluation of classification and matching performance is often saddled by a difficulty of notation and definitions. This confusion occurs because we are trying to estimate and make inference about classification errors. Since these classification rates are themselves similar to the type I and type II error rates, confusion can arise. This misunderstanding becomes acute if we are testing the classification error rates. In this book, we aim to be as explicit as possible in identifying both the quantity of interest and the relevant statistical error rate.

Below we present two example hypothesis tests, one for a process mean and one for a process proportion. Both of these are large sample tests meaning that they use a central limit theorem to allow an approximation to a specific sampling distribution. Most of the measures that we will describe in this text do not assume an *iid* data collection but are general enough to include such a structure if it is appropriate. Additionally, we will present hypothesis tests that do not depend upon a particular limiting distribution from a large sample.

Result 2.25 Here we present a large sample hypothesis test for a process mean. Let $R_1, \ldots, R_n$ be an *iid* sample from a process with finite variance. If $n \geq 30$ or the observations $R_1, \ldots, R_n$ are sampled from a Gaussian distribution, then we can use the following hypothesis test to test whether the process mean is less than a hypothesized value μ_0.

$$H_0 : \mu = \mu_0$$
$$H_1 : \mu < \mu_0.$$

Test Statistic:

$$t = \frac{\bar{R} - \mu_0}{s_R/\sqrt{n}}. \tag{2.64}$$

p-value: $p = P(T < t)$ where T is a t-distribution random variable with df $n-1$.

We will reject the null hypothesis, H_0, if *p-value* is small. If a significance level, α, has been prespecified then we will reject the null hypothesis if $p < \alpha$. Table 9.3 has percentiles for certain t-distributions.

To test whether the population mean is greater than μ_0 we would change our alternative hypothesis to $H_1 : \mu > \mu_0$ and change our *p-value* calculation to be $p = P(T > t)$.

Result 2.26 This is a large sample hypothesis test for a process proportion. Let B_n be a binomial random variable such that $B_n \sim Bin(n, p)$. If $B_n \geq 10$ and $n - B_n \geq 10$, then we can use the following hypothesis test to determine whether the probability of interest for this process is less than p_0.

$$H_0 : p = p_0$$
$$H_1 : p < p_0.$$

Test Statistic:

$$z = \frac{\hat{p} - p_0}{\sqrt{\frac{p_0(1-p_0)}{n}}}. \tag{2.65}$$

p-value: $p = P(Z < z)$ where Z is a standard Gaussian random variable and we can use Table 9.1 to determine the *p-value*.

We will reject the null hypothesis, H_0, if $p < \alpha$ for a particular significance level. If a significance level has not been determined *a priori*, then we will reject the null hypothesis if the *p-value* is small.

For the hypothesis tests above, we used a one-sided alternative hypothesis of $H_1 : \mu < \mu_0$ and $H_1 : p < p_0$. The reason for focusing on these particular tests is that testing to determine if an error rate is below a particular boundary is a common goal for biometric authentication. We will focus on one-sided tests with an alternative that is 'less than' in the rest of this book; however, other alternative hypotheses are possible. The other one-sided alternative hypotheses would be $H_1 : \mu > \mu_0$ and $H_1 : p > p_0$. Respectively, the *p-values* for those tests are $p = P(T > t)$ and $p = P(Z > z)$. If we are unsure about the direction of the alternative hypothesis (either higher or lower), we can use a two-sided alternative hypothesis. For the two examples given above, we would use $H_1 : \mu \neq \mu_0$ for testing a process mean and $H_1 : p \neq p_0$ for testing a process proportion. The *p-values* that are associated with these two sided tests are $2P(T > |t|)$ and $2P(Z > |z|)$, respectively.

2.3.4 Sample Size and Power Calculations

Researchers often want to know how many observations they will have to collect for a particular sample. These type of determinations generally fall under a group of statistical methods known as sample size calculations. Some attempts have been made in biometrics to provide sample size calculations for false match and false non-match rates, e.g. Schuckers [87]. It is possible to take the confidence intervals and hypothesis testing methods that we have outlined above and solve them for the number of individuals that we want to test. We do this by establishing a criterion and then determining the number of samples needed for that criterion. In the case of a confidence interval, we are often interested in specifying the total width of a confidence interval for a given confidence level, while for hypothesis testing we often want to specify the power, or the probability of not making a type II error. In general, determining the amount of data to be collected for a confidence interval is considered a sample size calculation, while a similar determination for the power of a hypothesis test is considered a power calculation.

It is important to note that if the process changes, then any inferences about the process based upon pre-change observations would not be appropriate. Further, the calculations below as well as similar ones in the rest of the text will depend upon some estimation of the process parameters, especially those involving the variability of the process. Without those, it is not possible to derive methods for the amount of data needed. Care and thought need to be used in consideration of which estimates to use for these parameters. Schuckers [87] provides some insight in the context of biometrics for the ways to undertake this process. We provide a brief discussion of these strategies below.

2.3.4.1 Sample Size Calculations

We start with a determination of the number of observations necessary for creating a $(1-\alpha)100\%$ confidence interval for a mean to have a margin of error of B. The margin of error for a traditional confidence interval is the portion after the $\pm$. We note that the notation 6 ± 3 here means the interval $(6-3, 6+3) = (3, 9)$.

Result 2.27 It is possible to 'invert' a CI for mean to determine the number of observations needed to get a CI with a margin of error of B.

$$B = z_{\alpha/2}\frac{\sigma}{\sqrt{n}} \tag{2.66}$$

$$B^2 = z^2_{\alpha/2}\frac{\sigma^2}{n}. \tag{2.67}$$

Solving for n,

$$n = \left\lceil z^2_{\alpha/2}\frac{\sigma^2}{B^2} \right\rceil, \tag{2.68}$$

where $\lceil\ \rceil$ is the ceiling function or the 'next largest integer' function.

Comment 2.8 Note that here we have used the $1-\alpha/2$th percentile for a Gaussian or normal distribution here, $z_{\alpha/2}$. The formula for a confidence interval for the mean uses a t-distribution. Here we use the $z_{\alpha/2}$ as an approximation to the percentile of the t-distribution since the latter depends upon the degrees of freedom of the estimated variance, here $n-1$ which is one less than the observed sample size. It is possible to recursively determine the appropriate sample size by iteratively using the following

$$n_{m+1} = \left\lceil (t_{\alpha/s;n_m-1})^2\frac{\sigma^2}{B^2} \right\rceil \tag{2.69}$$

until n_m converges. For most applications the sample size using the Gaussian approximation is sufficient. Here we need to specify *a priori* an estimate for the variance σ^2, as well as the confidence level, $1-\alpha$, in order to be able to calculate n, the number of observations to collect.

Result 2.28 Here we derive the sample size for a confidence interval for a proportion from a process. As with the mean case above, we can solve the equation

$$B = z_{\alpha/2}\sqrt{\frac{p(1-p)}{n}} \tag{2.70}$$

$$B^2 = z^2_{\alpha/2}\frac{p(1-p)}{n} \tag{2.71}$$

for n to get a sample size for estimation of a process proportion to within a given width. We then get,

$$n = \left\lceil z^2_{\alpha/2}\frac{p(1-p)}{B^2} \right\rceil. \tag{2.72}$$

Comment 2.9 For both of the sample size calculations given here—Result 2.27 and Result 2.28—it is necessary to specify *a priori* some quantities. In the case of the mean it is necessary to specify the confidence level $(1-\alpha)$, the margin of error (B) and the standard deviation, (σ), or equivalently the variance, (σ^2). How to specify these quantities is often a nominal hurdle for some practitioners. In particular, the requirement to specify σ can be vexing since this is a process parameter and gaining knowledge about the process of interest is often the goal of a data collection. Thus, an investigator may not have a good sense of how to specify σ. Statisticians have developed a series of suggestions and guidelines for choosing these values.

For process parameters, there are three basic methods available to obtain useable values: a pilot study, a previous study on a similar process, or an educated guess. A pilot study is a small-scale data collection to ascertain information about both the process of interest and the data collection tools. To that end, it can be used to gain valuable information. In particular, the standard deviation from observations collected on a pilot study from the process of interest is a valid point estimate for σ. Similarly, a standard deviation reported by another study from a similar process would likewise be appropriate. Lastly, an approximation based upon available information is possible to use. One useful tool in this endeavor is to imagine the range of values (*maximum minus minimum*) that the process of interest can take and divide that range by 6 to get an estimate of σ. This procedure comes from approximating the total range by 6σ which is based upon the fact that over 99% of a Gaussian distribution will fall within three standard deviations of the mean. It is important to note that the sample size, n, varies with the square of the standard deviation (or with the variance), σ^2. Thus, if a practitioner wants to be conservative (in a statistical sense), then it is reasonable to inflate an observed or estimated standard deviation, σ, to compensate for the fact that most values for σ that are used in sample size calculations are estimates.

Comment 2.10 For determining the sample size for a proportion—Result 2.28—it is likewise necessary to specify a process parameter, the process proportion (p), in order to obtain a sample size calculation. This is tricky since the data collection is aimed at learning about p, but it is necessary to specify the estimand in order to determine the sample size. As with the sample size calculation for the mean, we can use information from a pilot study or from a previous study on a similar process to glean a plausible value for p. Additionally, in the case of inference about a process proportion, we can also use the value $p = 0.5$ since that value is the most conservative possible choice meaning that it will produce the largest possible n.

Comment 2.11 In addition to estimation of process parameter(s) for sample size calculations, it is also important to specify the confidence level $(1-\alpha)$ and the margin of error (B). The selection of these is often guided by external constraints such as those required by regulating agencies. We point out here that by decreasing the margin of error, we increase the sample size. Intuitively, more observations (and hence more information) allows for the construction of intervals with greater precision. Similarly, we note that for a given margin of error B, having a higher confidence level means a larger sample size. Thus, decreasing α (and, hence, increasing the confidence level) while simultaneously maintaining B, means that relatively more observations will be required.

2.3.4.2 Power Calculations

Power calculations are to hypothesis testing as sample size calculations are to confidence intervals. We use power calculations to determine the number of observations that are needed for a hypothesis test. For the sample size calculations above, our criterion was the width of the confidence interval. For power calculations we determine the sample size needed to achieve a given power, $1-\beta = 1-P$ (type II error), for a specific significance level of a test, α. As with sample size calculations, it is necessary to provide estimates about the process of interest in order to be able to utilize these calculations.

Power calculations start with a hypothesis test that we wish to carry out at the end of the data collection. With the desired test in mind, we specify the desired or required significance level of the test, α, and the desired power $= 1-\beta$ for the test. We must also specify a value for our estimand in the alternative hypothesis that will give us the desired power. This is done so that we can calculate the power explicitly. For example, if we might want to test the hypotheses: $H_0 : \theta = 0.05$ against $H_a : \theta < 0.05$, we would have to specify a value for $\theta < 0.05$, say $\theta = 0.03$. This value would then allow the calculation of power and the determination of the number of samples needed to achieve that power.

Recall that for a single proportion from *iid* observations that are Bernoulli, i.e. zero's or one's, we have the following hypothesis test:

$$H_0 : p = p_0$$
$$H_1 : p < p_0.$$

Suppose that we want to test at the $\alpha = 0.05$ significance level whether a particular process has a failure rate of less than $p = 0.10$ and we want the power to be 80% if $p = 0.05$. This means that $0.80 = 1-\beta$ and so $\beta = 0.20$. (Note that there are several proportions in this case. This is one example where it is easy to confuse performance rates, p, p_0 and p_a, with hypothesis testing error rates, α and β.) Here p_0 is 0.10 and we will call the value that we want to detect p_a which is 0.05 in this case. Ideally, we would like the power to be 100% but, as with all estimation, 100% accuracy is very—read infinitely—costly in terms of the number of observations that are needed. Consequently, we must trade off the desired power with its cost. Generally, a subscript of 'zero' will indicate a value to be tested and a subscript of 'a' will indicate a value inside the range of the alternative hypothesis. Having specified the desired power, the level of the test, and an alternative value for the failure rate we can turn to calculation of the appropriate sample size.

To determine the number of observations that would be required to achieve the above conditions, we start by noting that these conditions translate to the following probability statement

$$P\left(\frac{\hat{p}-p_0}{\sqrt{p_0(1-p_0)/n}} < -z_\alpha \middle| p = p_a\right) = 1-\beta, \tag{2.73}$$

where z_α is the $1-\alpha$th $\times$ 100 percentile of a $N(0,1)$ which assumes a large enough sample size so that we can use a Gaussian approximation to the sampling distribution of $\hat{p}$. We can then rewrite (2.73) as

$$P\big(\hat{p} < p_0 - z_\alpha\sqrt{p_0(1-p_0)/n} \mid p = p_a\big) = 1-\beta \tag{2.74}$$

which becomes

$$P\left(\frac{\hat{p}-p_a}{\sqrt{p_a(1-p_a)/n}} < \frac{p_0 - z_\alpha\sqrt{p_0(1-p_0)/n} - p_a}{\sqrt{p_a(1-p_a)/n}}\right) = 1-\beta. \tag{2.75}$$

If our sample size is sufficiently large, then the quantity on the left side of the inequality in (2.75) is a standard Gaussian random variable and, thus, we can equate the right hand side of the inequality to the appropriate percentile, z_β of a standard Gaussian distribution. We then have

$$z_\beta = \frac{p_0 - z_\alpha\sqrt{p_0(1-p_0)/n} - p_a}{\sqrt{p_a(1-p_a)/n}}. \tag{2.76}$$

We can rearrange (2.76) to get the following

$$z_\alpha\sqrt{p_0(1-p_0)/n} + z_\beta\sqrt{p_a(1-p_a)/n} = p_a - p_0. \tag{2.77}$$

We can then square both sides and solve for n. We then get the following result.

Result 2.29 The number of samples needed from a binomial process to achieve an α level hypothesis test of $H_0: p = p_0$ against $H_a: p < p_0$ with a power of $1-\beta$ for an alternative of p_a is

$$n = \left\lceil \frac{(z_\alpha\sqrt{p_0(1-p_0)} + z_\beta\sqrt{p_a(1-p_a)})^2}{(p_0-p_a)^2} \right\rceil. \tag{2.78}$$

Note that the difference $p_0 - p_a$ is called the 'effect size.'

Using a similar process we can get the following result for a power calculation of a process mean.

Result 2.30 For a hypothesis test of $H_0: \mu = \mu_0$ against $H_a: \mu < \mu_0$ with a significance level of α and a desired power of $1-\beta$ against an alternative of $\mu = \mu_a$, the number of samples needed from a *iid* process is

$$n = \left\lceil \sigma^2 \frac{(z_\alpha - z_\beta)^2}{(\mu_0 - \mu_a)^2} \right\rceil. \tag{2.79}$$

Comment 2.12 Result 2.30 is based upon an assumption that the standard deviation (or variance) is the same regardless of whether the population mean is μ_0 or μ_a as well as a Gaussian approximation of the sampling distribution of the sample mean. The latter assumption can be alleviated by following an iterative procedure similar to the one outlined in Comment 2.8.

Comment 2.13 An admonition on sample size and power calculations is important in this context. Often in a real world data collection, the number of individuals that agree to start the data collection process is not the same number that will complete the process. There is a certain amount of attrition that may occur. This is especially true when testing involves human beings. Consequently, in planning an evaluation, it is worthwhile to take the statistically required sample sizes or power calculations and inflate them to adjust for this attrition.

2.3.5 Resampling Methods

Resampling methods are a class of statistical methods for approximating the sampling distribution of an estimator or a group of estimators. These methods have become increasingly utilized due to the ease of implementation, the advances in computation and the wide range of disciplines to which they can be applied. The advantage of these methods is that they make few assumptions about the shape of the sampling distribution. They are especially useful when sample sizes are small and, therefore, large sample methods are not appropriate. More details on, and examples of, resampling methods can be found in Manly [62], Edgington [28], Efron and Tibshirani [29] and Lahiri [54], for example. We start this section with a discussion of randomization tests, then move to jackknife and bootstrap methodology.

Definition 2.49 A *randomization test* is one where a test statistic is computed based upon the observed data and then the data is repeatedly permuted and the same test statistic is calculated for each permuted data set that results. This is useful, for example, when comparing a statistic on two different groups. A reference distribution is then calculated based upon the calculated test statistic from the permuted data. A *p-value* is then calculated by comparing the observed test statistic from the original data to the reference distribution of the test statistic based upon the permuted data. This definition follows from Edgington [28].

We note below that there is a distinction between randomization tests and permutation tests. The basics of each test are that we want to look at the distribution of a statistic had the observations been randomly reshuffled among the groups. This permuting of observations is appropriate if the null hypothesis holds. *Note that randomization tests, as presented here, are singularly for comparing two or more groups.* A randomization test is called a permutation test if all possible permutations of the observations are obtained. Thus, the randomization test *p-value* is an empirical estimate of the *p-value* that would be obtained under a permutation test. If the number of possible permutations is small, then it is often reasonable to do all possible combinations. However, the number of permutations grows exponentially with the number of observations, in which case it is sufficient to use a randomization test to obtain an approximate *p-value* assuming that the number of permutations obtained is reasonably large, say more than 1000.

Below we outline the jackknife and bootstrap methodology. Here we will provide a short definition. The goal of both methods is to provide approximations to the sampling distributions for an estimator or statistic, say $\hat{\phi}$. Here we outline the basic approaches. Later in the book as we use these techniques, we expand out the exact methods as appropriate.

Definition 2.50 A *jackknife procedure* is one where the sampling distribution for an estimator, say $\hat{\phi} = \hat{\phi}(V_1, \ldots, V_n)$, is estimated by recalculating $\hat{\phi} n$ times, each time leaving out one of the observed values. That is, the ith element of the sampling distribution for $\hat{\phi}$ is $\hat{\phi}(V_1, \ldots, V_{i-1}, V_{i+1}, \ldots, V_n)$ which is calculated based upon $n-1$ observations.

The above is the definition for the traditional jackknife, sometimes known as the 'leave-one-out' jackknife, but there are other jackknife procedures that 'leave out' more than one observation. See Shao and Wu [90] or Friedl and Stampfer [34] for expositions of the properties of these other procedures.

Definition 2.51 A *bootstrap procedure* is one where the sampling distribution for an estimator, $\hat{\phi} = \hat{\phi}(V_1, \ldots, V_n)$ is estimated by repeatedly sampling n observations *with replacement* from the original sample of size n and calculating $\hat{\phi}$ upon each of these replicate sets. The resulting collection of estimates forms a distribution that approximates the sampling distribution of $\hat{\phi}$.

Result 2.31 We present here an algorithm for a $100 \times (1-\alpha)\%$ bootstrap confidence interval for an *iid* process mean. Denote the mean of the process by μ. The steps are as follows:

1. Calculate

$$\hat{\mu} = \bar{X} = \frac{1}{n}\sum_{i=1}^{n} X_i \tag{2.80}$$

   based upon the observed values of the sample $X_1, \ldots, X_n$.
2. Sample with replacement n integers from the list $\{1, 2, \ldots, n\}$ and call these values $b_1, b_2, \ldots, b_n$.
3. Calculate and store $e = \hat{\mu}_b - \hat{\mu}$ where

$$\hat{\mu}_b = \frac{1}{n}\sum_{i=1}^{n} X_{b_i}. \tag{2.81}$$

4. Repeat steps 2 and 3 above some large number of times, say M times.
5. Determine the $\alpha/2$th and the $1-\alpha/2$th percentiles of e and represent those by e_L and e_U, respectively.
6. Then a $100 \times (1-\alpha)\%$ confidence interval for μ can be made using

$$(\hat{\mu} - e_U, \hat{\mu} - e_L). \tag{2.82}$$

Comment 2.14 The basic bootstrap confidence interval approach we just delineated is one that we utilize in multiple ways throughout this text. The distribution of e is the distribution of differences from the statistic $\hat{\mu}$. The methodology given above differs slightly from the bootstrap confidence interval that is typically presented in biometric authentication. See, for example, Poh et al. [76] or Bolle et al. [9]. Here we made use of the so-called *Hall adjustment* which accounts for potential asymmetry in the sampling distribution of the statistic, Hall [44].

Comment 2.15 In Result 2.31, we presented a bootstrap confidence interval. In a similar manner, it is possible to create a hypothesis test for a particular process parameter. In order to ensure that the bootstrap approximation to the sampling distribution for a statistic is what would be expected when the null hypothesis is true, we have to adjust the sampling distribution so that it is centered at the value specified in the null hypothesis, for example, μ_0. To do this, we take $e = \hat{\mu}_b - \hat{\mu} + \mu_0$. In this way, we can approximate the distribution of the statistic, $\hat{\mu}$, assuming that the null hypothesis is true.

It is worthwhile mentioning that the methods above apply to a variety of data collection methodologies. That is, they can be applied to data that is not collected in a manner that would allow modeling using *iid* random variables. However, special care is needed in these circumstances to craft appropriate methodology. All of the methods in this section are quite powerful and have been underutilized in the performance evaluation of classification systems. Throughout this book, we will use these methods in a variety of circumstances.

There is an extensive amount of work that has been done on resampling methods and this continues to be an active area of research in statistics. An excellent overview of this work can be found in Manly [62]. Much of the recent interest in the statistics literature is on non-*iid* data. The interested (and *motivated*) reader is directed to the mathematically sophisticated work by Lahiri [54] for results in this area.

Part II
Primary Matching and Classification Measures

Chapter 3
False Non-Match Rate

The false non-match rate (FNMR) is the rate at which a biometric matcher miscategorizes two captures from the same individual as being from different individuals. It can be thought of as the false reject rate (FRR) for a typical classification algorithm. However, in the context of a biometric system there are other possible reasons that a system may reject an individual; therefore, biometrics differentiates between a false match and a false reject. The basic idea here is that the FRR is a system level measure while the FNMR is a classification subsystem level measure. Mansfield and Wayman have written extensively on this issue, cf. [63]. They note that there are other types of failures that can result in a false reject such as a failure to acquire an image of sufficient quality. More detail on failure to acquire can be found in Chap. 6. In Chap. 1 we discussed some additional differences between these error rates and the measurements on which estimates of these can be based.

In this chapter, we focus on statistical methods for estimation and hypothesis testing of FNMR rates. It is often the case that evaluation of a biometric device or system involves interest in how well a system can identify those in a database or on a watchlist. If the classification mechanism fails to identify an individual, then that decision is a *false non-match*. Because it is often the case that repeated biometric signals are taken from the same individual, the binomial model for classification decisions does not hold. As a consequence, we have to consider models that allow for non-zero correlations between decisions from the same individual. In the sections that follow we present statistical methods that account for these correlations. The majority of previous work has been on estimation of a single FNMR. This chapter extends and expands those works. We present a variety of methods, including large sample, bootstrap and randomization methods, meant to cover circumstances including the evaluation of multiple FNMR's. Because of the common practice of comparing results for biometric evaluations between two or more matchers or between two or more groups, we present confidence intervals and hypothesis tests that assess the equality of two or more FNMR's. These methods are based upon three approaches outlined in Sect. 2: large sample, bootstrap, and randomization. Utilization of both large sample theory and non-parametric approaches allows us to cover a broader range of biometric data collections and allows us some flexibility in the methods with which we analyze these data.

M.E. Schuckers, *Computational Methods in Biometric Authentication*,
Information Science and Statistics,
DOI 10.1007/978-1-84996-202-5_3,

There has been a good deal of attention paid to methods for false non-match rate estimation. Among the parametric approaches, Shen et al. [91] uses the binomial distribution to make confidence intervals and obtain sample size calculations, while Golfarelli et al. [38] models the features by a multivariate Gaussian distribution to obtain estimates of false non-match rates. The approach of Guyon et al. [42] is to assume that individual error rates follow a Gaussian distribution. In a similar vein, Schuckers [84] assumes a Beta distribution for the individual error rates and uses maximum likelihood estimation to create confidence interval for false non-match rates. Several non-parametric approaches have also been considered. Wayman [101] outlined exact methods for estimating the error rates for binomial data as well as for estimating the FAR when cross-comparisons are used. The 'subsets bootstrap', a resampling approach has been proposed by Bolle et al. [9]. We will make extensive use of the 'subsets bootstrap' in this chapter.

This chapter is organized in the following way. In the next section, we present the notation and the correlation structure that we will use throughout the chapter. We then present statistical methods for estimating and comparing FNMR's. This includes confidence interval and hypothesis testing for a single FNMR as well as for comparing two or more FNMR's. These methods are done using both large sample as well as non-parametric methods. A discussion of sample size and power calculations follows that section. We conclude this chapter with a section on prediction intervals and a discussion section.

3.1 Notation and Correlation Structure

The FNMR is a linear combination of decisions. As with all linear combinations, the variance depends on the covariance structure and accompanying correlation structure. In this section, we present a correlation structure for decisions made when comparing two biometric signals from the same individual. This correlation structure is necessary for estimation of the false non-match rate (FNMR) because of the possibility that multiple biometric signals will be taken from the same individual. This correlation structure has been proposed by Schuckers [85] and has been used to develop statistical methods for the FNMR, also Schuckers [87]. Since we are interested in the FNMR, our decisions will be based on comparisons from two biometric captures from the same individual. These intra-individual decisions are derived from dichotomizing the genuine distribution at some threshold, where the genuine distribution is defined as the match scores from intra-individual comparisons. Conditional on the error rate, we utilize a model that ignores—treats as zero—correlations between different individuals. We begin by introducing our correlation structure, then deriving the variance for the estimated error rate. We extend the previously mentioned work of Schuckers [85, 87], to other statistical methods and to estimation of differences between multiple FNMR's. It is important for both parametric and non-parametric methods that the correlation structure is correct so that appropriate methods are utilized. Failure to accommodate adequately and accurately for the correlation structure can lead to incorrect decisions about the equality of several process FNMR's. In particular, the type of errors that these mistakes engender is inappropriately narrow inferential intervals and *p-values* that are too small.

In Chap. 1 we introduced our notation for the general matching process of a biometric authentication system. We begin by revisiting the relevant notation for estimation of a single FNMR estimation. Let D_{iij} represent the decision for the jth pair of captures or signals collected on the ith individual, where n is the number of individuals, $i = 1, \ldots, n$ and $j = 1, \ldots, m_i$. Thus, the number of sample pairs that are compared for the ith individual is m_i, and n is the number of different individuals being compared. The use of m_i implies that we are allowing the number of comparisons made per individual to vary across individuals. We then define

$$D_{iij} = \begin{cases} 1 & \text{if } j\text{th pair of signals from individual } i \text{ is declared a non-match} \\ 0 & \text{otherwise.} \end{cases} \tag{3.1}$$

We assume for the D_{iij}'s that $E[D_{iij}] = \pi$ and $V[D_{iij}] = \pi(1 - \pi)$ represent the mean and variance, respectively. Thus, π represents the FNMR which is our quantity of interest. For the moment, we assume that we have a stationary matching process. We will later discuss relaxation of this assumption. Our estimate of π, the process FNMR, will be the total number of errors divided by the total number of decisions:

$$\hat{\pi} = \frac{\sum_{i=1}^{n} \sum_{j=1}^{m_i} D_{iij}}{\sum_{i=1}^{n} m_i}. \tag{3.2}$$

Following Schuckers [85, 87], we have the following correlation structure for the $D'_{iij}s$:

$$Corr(D_{iij}, D_{i'i'j'}) = \begin{cases} 1 & \text{if } i = i', j = j' \\ \varrho & \text{if } i = i', j \neq j' \\ 0 & \text{otherwise.} \end{cases} \tag{3.3}$$

This correlation structure for the FNMR is based upon the idea that the there will only be correlations between decisions made on signals from the same individual but not between decisions made on signals from different individuals. Thus, conditional upon the error rate, there is no correlation between decisions on the ith individual and decisions on the i'th individual, when $i \neq i'$. The degree of correlation is summarized by ϱ.

Then we can write the variance of $\hat{\pi}$, the estimated FNMR, as

$$\begin{aligned} V[\hat{\pi}] &= V\left[\frac{\sum_{i=1}^{n} \sum_{j=1}^{m_i} D_{iij}}{\sum_{i=1}^{n} m_i}\right] = V[N_\pi^{-1} \mathbf{1}^T \mathbf{D}] = N_\pi^{-2} V[\mathbf{1}^T \mathbf{D}] \\ &= N_\pi^{-2} \mathbf{1}^T \boldsymbol{\Sigma}_\pi \mathbf{1} = N_\pi^{-2} \pi(1 - \pi) \mathbf{1}^T \boldsymbol{\Phi}_\pi \mathbf{1} \\ &= N_\pi^{-2} \pi(1 - \pi) \left[N_\pi + \varrho \sum_{i=1}^{n} m_i(m_i - 1)\right] \end{aligned} \tag{3.4}$$

where $\boldsymbol{\Sigma}_\pi = V[\mathbf{D}]$, $N_\pi = \sum_{i=1}^n m_i$, $\boldsymbol{\Phi}_\pi = Corr(\mathbf{D})$, $\mathbf{1} = (1, 1, \ldots, 1)^T$ and $\mathbf{D} = (D_{111}, \ldots, D_{11m_1}, D_{221}, \ldots, D_{22m_2}, \ldots, D_{nn1}, \ldots, D_{nnm_n})^T$. The covariance matrix is $\boldsymbol{\Sigma}_\pi$ and the correlation matrix is $\boldsymbol{\Phi}_\pi$. We write the estimated FNMR, $\hat{\pi} = (N_\pi)^{-1}\mathbf{1}^T\mathbf{D}$, as a linear combination of false non-match decision here to underscore that statistical methods for linear combinations apply to this rate.

An estimator for ϱ is given by:

$$\hat{\varrho} = \frac{\sum_{i=1}^n \sum_{j=1}^{m_i} \sum_{j'=1, j' \neq j}^{m_i} (D_{iij} - \hat{\pi})(D_{iij'} - \hat{\pi})}{\hat{\pi}(1 - \hat{\pi}) \sum_{i=1}^n m_i(m_i - 1)}. \tag{3.5}$$

Note that the estimate given for ϱ in (3.5) is equivalent to averaging all of the terms in the estimated correlation matrix that correspond to ϱ to get a single estimate. Models like that found in (3.3) are known as intra-individual or intra-class models and have been studied extensively in the statistics literature, e.g. Fleiss et al. [33], Williams [103] or Ridout et al. [81]. The parameter ϱ in the models above represents the intra-class correlation. That is, it measures the degree of similarity between the decisions made on each individual. If the decisions on each individual are varying in a way that suggests that the decisions are not dependent upon the individual then ϱ is zero, meaning that the observations are uncorrelated. Negative values of ϱ are possible but such values suggest that decisions on signals from the same individual are less similar to each other than they are to all of the other decisions. This seems unlikely to be the case in the context of biometric authentication. Schuckers [83] implicitly used the model in (3.3) to create confidence intervals and to derive sample size calculations. Several authors, including Fleiss et al. [33], have suggested using the following alternative way of writing (3.4)

$$V[\hat{\pi}] = N_\pi^{-1}\pi(1 - \pi)(1 + (m_0 - 1)\varrho) \tag{3.6}$$

where

$$m_0 = \frac{\sum_{i=1}^n m_i^2}{N_\pi}. \tag{3.7}$$

If $m_i = m$ for all i, then $N_\pi = nm$ and the variance of $\hat{\pi}$ from (3.6) becomes

$$\begin{aligned} V[\hat{\pi}] &= (nm)^{-2}\pi(1 - \pi)(nm + \varrho nm(m - 1)) \\ &= (nm)^{-1}\pi(1 - \pi)(1 + (m - 1)\varrho). \end{aligned} \tag{3.8}$$

Thus (3.6) is a way to parrot the form of (3.8) that allows us to write the general case succinctly.

As mentioned above, ϱ represents the intra-class correlation. This quantity has a direct relationship with the variance of $\hat{\pi}$ found in (3.4) and the simplified version found in (3.8). As ϱ increases, the variance in both cases increases. This is a consequence of the lack of independent information from each individual. If ϱ is large, then each additional decision on a previously observed individual is providing little new information. Specifically, we note that for a single FNMR that the *effective sample size*, $N_\pi^\dagger$, is

$$N_\pi^\dagger = \frac{N_\pi}{1 + (m_0 - 1)\hat{\varrho}}. \tag{3.9}$$

Recall Definition 2.34 from Chap. 2. The effective sample size is especially important for large sample theory and will be used repeatedly in this chapter.

3.2 Statistical Methods

We now turn our attention to methods for statistical inference about a single FNMR as well as multiple FNMR's. In what follows, we describe both parametric and non-parametric confidence interval and hypothesis testing methodologies. Beginning with a single FNMR, we move to approaches for comparing two FNMR's and for evaluating simultaneously three or more FNMR's

3.2.1 Bootstrap for FNMR

The bootstrap methodology that is appropriate given the correlation structure of the FNMR is the 'subsets bootstrap' originally proposed by Bolle et al. [9]. The technique here is to sample *with replacement* the individuals and for the selected individuals we take all of the decisions. The basic algorithm is the following:

1. Sample with replacement from the n individuals. Call the selected individuals $b_1, \ldots, b_n$.
2. For each selected individual b_i, take all m_{b_i} decisions. Do this for all selected individuals $b_1, \ldots, b_n$.
3. Calculate and store the resampled FNMR to be

$$\hat{\pi}^b = \frac{\sum_{i=1}^{n} \sum_{j=1}^{m_{b_i}} D_{b_i b_i j}}{\sum_{i=1}^{n} m_{b_i}}. \tag{3.10}$$

4. Repeat the previous steps some large number of times, say M.

The result from this process is an approximation to the sampling distribution of $\hat{\pi}$. To illustrate that this methodology for bootstrapping is the appropriate one given the correlation structure that we have, we ran a simulation to compare the bootstrap estimate of $s_{\hat{\pi},boot} = \sqrt{\hat{V}[\hat{\pi}]}$ to the estimate given by using (3.4). We present results from applying both methods to matchers from the BANCA database. We considered 11 different classifiers from this database all applied to group g1 under the G protocol. The matchers that we will use as part of this illustration are listed in Table 3.1. For each classifier we ran $M = 5000$ replications of the original data. Table 3.1 illustrates how well the subsets bootstrap does in mirroring the estimated standard

Table 3.1 Subsets bootstrap standard error estimates

Modality	Matcher	Threshold	$\hat{\pi}$	$s_{\hat{\pi}}$	$s_{\hat{\pi},boot}$
Face	SURREY-SVM-MAN-0.50	0.00	0.0214	0.0136	0.0139
Face	SURREY-SVM-MAN-0.71	−0.25	0.0601	0.0222	0.0221
Face	SURREY-SVM-MAN-1.00	0.50	0.0283	0.0081	0.0080
Face	SURREY-SVM-MAN-1.41	0.30	0.0730	0.0244	0.0244
Face	SURREY-SVM-MAN-2.00	−0.10	0.0386	0.0160	0.0160
Face	SURREY-SVM-MAN-2.83	0.00	0.0257	0.0140	0.0139
Face	SURREY-SVM-MAN-4.00	0.20	0.0128	0.0093	0.0091
Face	SURREY-SVM-MAN	0.10	0.0172	0.0100	0.0098
Voice	UC3M-10-100	−0.30	0.0386	0.0191	0.0189
Voice	UC3M-10-200	0.00	0.0258	0.0152	0.0153
Voice	UC3M-10-300	−0.10	0.0043	0.0043	0.0042

deviation of $\hat{\pi}$, the estimated FNMR. $s_{\hat{\pi}} = \sqrt{\hat{V}[\hat{\nu}]}$ is the standard error of $\hat{\pi}$ following (3.4) while $s_{\hat{\pi},boot}$ is the equivalent calculated on the bootstrap replicated FNMR's, the $\hat{\pi}^{b}$'s.

3.2.2 *One Sample Methods for FNMR*

In this section, we focus on statistical methods for the false non-match rate of a single process. Assuming that we are dealing with a single stationary matching process, then we can apply the methods described below. The assumption here as noted above is that we have a single FNMR for a given biometric authentication system. It is possible to test whether a single FNMR is an appropriate representation of this process in later sections. Here we concentrate on estimation and testing of a single FNMR.

3.2.2.1 Confidence Interval for Single FNMR

Creation of a confidence interval for an FNMR assumes that some data has been collected from a biometric authentication process. That data should include individual information as well as the resulting outcome from a matching decision. Such information might appear in a format such as that give in Table 3.2. The data given there suggest that there are $n = 5$ individuals and that there are 19 total decisions. A total of 4 errors occurred from these 19 decisions. This table is meant to be an example of the format and structure of FNMR data. While it seems redundant that we include two columns for individuals, we do this for two reasons. First, we want to reinforce the idea that we are doing a comparison of two signals from the same

Table 3.2 Example of FNMR data

i	i	j	D_{iij}
1	1	1	0
1	1	2	0
1	1	3	0
1	1	4	1
2	2	1	0
2	2	2	1
2	2	3	0
3	3	1	0
3	3	2	0
4	4	1	0
4	4	2	1
4	4	3	1
4	4	4	0
5	5	1	0
5	5	2	0
5	5	3	0
5	5	4	0

individuals and, second, this notation matches more general notation that we will use in Chaps. 4 and 5.

Here we propose two approaches for making a confidence interval for an FNMR. The first is based upon large sample theory and assumes that a sufficient sample size is achieved. When the conditions for sufficient sample size are not met, a bootstrap approach is warranted and we also present a bootstrap method for making a confidence interval for a FNMR. The bootstrap approach that is appropriate here is the *subsets bootstrap* due to Bolle et al. [9]. Some practitioners prefer the lack of assumptions implicit in the bootstrap approach and, therefore, use it regardless of the situation. If the conditions for the large sample approach are met, then the two methods are likely to yield similar results.

Large Sample Approach If $N_{\pi}^{\dagger}$, the effective sample size, is large (generally $N_{\pi}^{\dagger}\hat{\pi} \geq 10$ and $N_{\pi}^{\dagger}(1-\hat{\pi}) \geq 10$), then we can use the following to make a $(1-\alpha)100\%$ confidence interval for π.

$$\hat{\pi} \pm z_{1-\frac{\alpha}{2}}\sqrt{\frac{\hat{\pi}(1-\hat{\pi})[1+(m_0-1)\hat{\varrho}]}{n\bar{m}}} \tag{3.11}$$

where

$$\hat{\pi} = \frac{\sum_{i=1}^{n}\sum_{j=1}^{m_i} D_{iij}}{\sum_{i=1}^{n} m_i}, \tag{3.12}$$

which is the estimated FNMR.

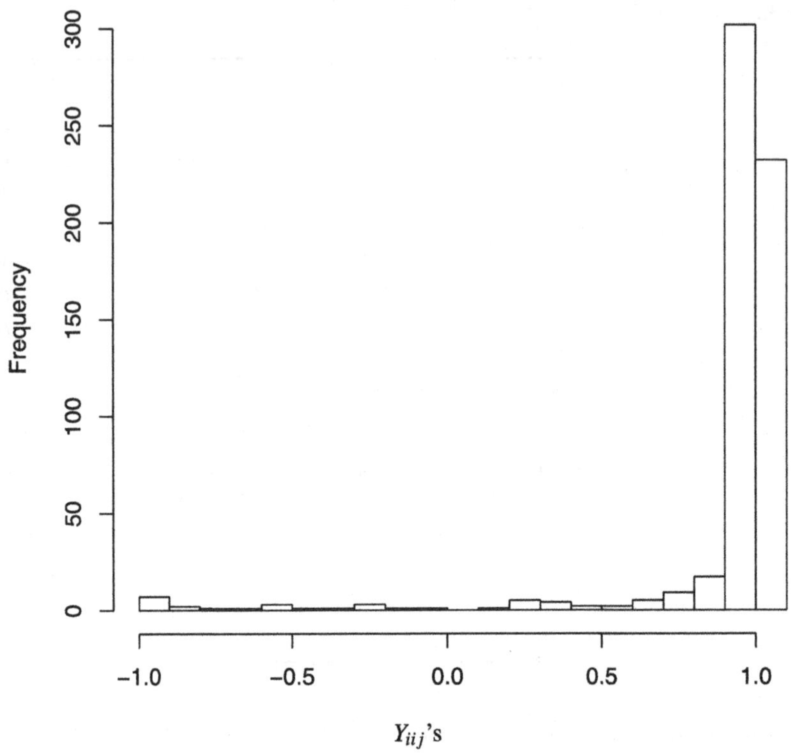

Fig. 3.1 Histogram of genuine match scores for face matcher (FH, MLP) from the XM2VTS database

Example 3.1 For this example we will use data from the XM2VTS database. See Poh et al. [74] for details. We will analyze the face matcher (FH, MLP) described in that article setting the threshold to be equal to 0.0. Note that we use a threshold for deciding if two signals are of sufficient similarity to declare them a match. To create our 95% confidence interval, we will use (3.11). We note here that in this case $n = 200$ and $m_i = 3$, for all i. Calculating some summaries from the data, our estimate of π is $\hat{\pi} = 0.0350$ and $\hat{\varrho}$ is 0.3092. Figure 3.1 has a histogram of the genuine matching scores, Y_{iij}'s, for (FH, MLP). We note that the conditions needed for use of the large sample methods do *not* require a Gaussian distribution for the match scores. Instead, we must check the conditions that the size of our sample is large enough to create a confidence interval assuming a Gaussian sampling distribution for $\hat{\pi}$. Here $N_\pi^\dagger$ is 370.76. So $N_\pi^\dagger \hat{\pi} = 370.760.0350 = 12.9765 \geq 10$ and $N_\pi^\dagger(1 - \hat{\pi}) = 370.76(1 - 0.0350) = 357.78 \geq 10$. Since these conditions are met, we can utilize this large sample approach to make a confidence interval for π, the FNMR. Our 95% confidence interval for this process FNMR is $(0.0163, 0.0537)$. Thus, we can be 95% confident that the FNMR of this process is between 1.63% and 5.37% when a threshold of 0.0 is used.

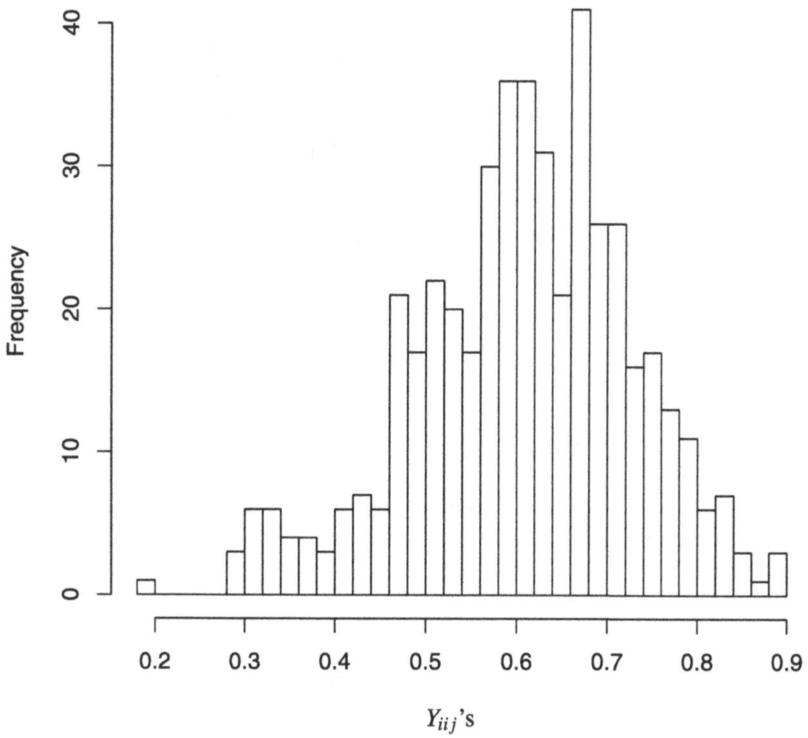

Fig. 3.2 Histogram of match scores for face matcher SURREY-NC-100 following the G protocol from the BANCA database

Example 3.2 In this example, we used decisions, D_{iij}'s, from the BANCA database. In particular, we used all decisions (both from group g1 and from group g2) for the Face Matcher SURREY_face_nc_man_scale_100 (SURREY-NC-100) under the G protocol. Figure 3.2 is a histogram of the genuine match scores for this matcher. We note that the distribution of the match scores need not be Gaussian or approximately Gaussian for us to use the methodology described above. The total number of decisions, N_π, is 467 and the number of individuals is 52. All individuals have $m_i = 9$ decisions. The estimated FNMR is 0.0407 for a threshold of 0.35 and the estimated intra-individual correlation is $\hat{\varrho} = 0.0536$. For the large sample confidence interval, it is necessary to determine if the sample size is sufficiently large. Here, the effective sample size, see Definition 2.34, is 327.01. Multiplying this number by $\hat{\pi}$ and $1 - \hat{\pi}$ yields 13.30 and 313.71, respectively. Consequently, we can use the formula given in (3.11) to create a confidence interval for the process FNMR π. Then our 99% confidence interval for π is $(0.0125, 0.0688)$. We are 99% confident that the FNMR for this process is between 1.25% and 6.88%.

Bootstrap Approach Creating a resampling-based confidence interval for a FNMR cannot be based upon the *iid* bootstrap. Instead, the bootstrap approach must account for the repeated measures nature of the FNMR data. The method we describe below was first proposed by Bolle et al. [9] who referred to it as the *subsets bootstrap*. Note that we adjust the confidence interval in a slightly different way to account for the possibility of non-symmetric sampling distribution of the bootstrapped error rate. This adjustment is due to Hall [44]. The steps for making such a confidence interval are given below.

1. Calculate $\hat{\pi}$ from the observed complete data

$$\hat{\pi} = \frac{\sum_{i=1}^{n} \sum_{j=1}^{m_i} D_{iij}}{\sum_{i=1}^{n} m_i}. \tag{3.13}$$

2. Sample n individuals with replacement from the n individuals from which there are false non-match decisions. Denote these selected individuals by $b_1, b_2, \ldots, b_n$.
3. For each selected individual, b_i, take all m_{b_i} match decisions for that individual. That is, the decision $D_{b_i b_i j}$ is the jth from resampled individual b_i.
4. Calculate $e_\pi = \hat{\pi}^b - \hat{\pi}$ where

$$\hat{\pi}^b = \frac{\sum_{i=1}^{n} \sum_{j=1}^{m_{b_i}} D_{b_i b_i j}}{\sum_{i=1}^{n} m_{b_i}}. \tag{3.14}$$

5. Repeat the three previous steps some large number of times M, each time storing the differences, e_π from the estimated FNMR, $\hat{\pi}$.
6. A $100(1-\alpha)\%$ confidence interval for the FNMR, π, is then formed by taking the interval from $\hat{\pi} - e_U$ to $\hat{\pi} - e_L$ where e_L and e_U represent the $\alpha/2$th and the $1-\alpha/2$th percentiles of the distribution of e_π.

This method of bootstrapping is the one that yields variability in $\hat{\pi}$ equivalent to that given by (3.4). Below we present two examples of the use of this methodology.

Example 3.3 For this example, we will revisit the data from the XM2VTS database considered in Example 3.1. We will analyze the Face matcher (FH, MLP) described in that article setting the threshold to be equal to 0.0. To create our 95% confidence interval, we will follow the algorithm above. The relevant summary statistics are $n = 200$, $m_i = 3$, for all $i = 1, \ldots, 200$, $\hat{\pi} = 0.0350$. Following the *subsets bootstrap* above and using $M = 1000$ bootstrap replicates, we create a 95% confidence interval for this process FNMR to be $(0.0150, 0.0517)$. The 2.5th percentile for the distribution of $\hat{\pi}^b$ is approximately 0.0183 and the 97.5th percentile is approximately 0.0550. The distribution of $\hat{\pi}^b$ can be found in Fig. 3.3. We note that this distribution is approximately Gaussian. Thus, we can be 95% confident that the FNMR of this process is between 0.0150% and 0.0517% when a threshold of 0.0 is used.

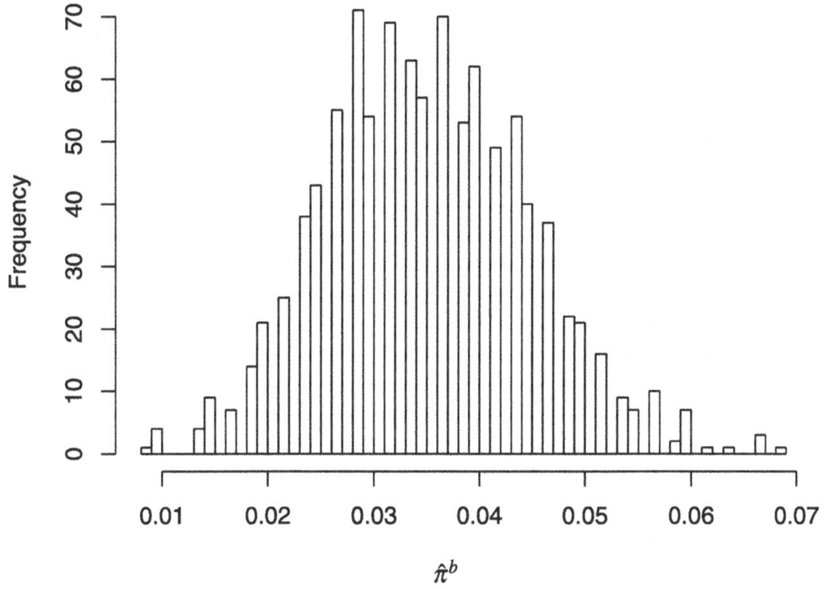

Fig. 3.3 Distribution of $\hat{\pi}^b$ for face matcher (FH, MLP) from the XM2VTS database

We note here that there are some differences between the intervals for the large sample approach in Example 3.1 and the *subsets bootstrap* approach in Example 3.3. The former method gives a symmetric interval $(0.0163, 0.0537)$ while the latter gives an asymmetric interval $(0.0150, 0.0517)$ around the estimate of the FNMR $\hat{\pi}$. These intervals are quite similar which should not come as a surprise. Both intervals are approximately the same width 0.0374 and 0.0367, respectively. The differences are due to slight asymmetry in the bootstrap distribution of $\hat{\pi}^b$ while the large sample approximation of a Gaussian distribution is symmetric and yields an confidence interval that is, consequently, symmetric. For these data either interval would be an appropriate choice for inference about the false non-match rate for the (FH, MLP) matcher.

Example 3.4 The data we will consider here is from the XM2VTS database. Our goal is to make a 90% confidence interval for the FNMR of the (LFCC, GMM) speaker matcher. The threshold here is 3.2 which gives an estimated FNMR of $\hat{\pi} = 0.005$. Here $N_\pi = 600$. Figure 3.4 shows the distribution of the $\hat{\pi}^b$'s from $M = 1000$ bootstrap replicates. The exact interval that we obtain following our algorithm above is $(-0.0017, 0.0100)$. Since a negative estimate for an FNMR is not possible, we truncate our interval to be $(0.0000, 0.0100)$. Thus, we are 90% confident that the FNMR for this matcher is between 0.00% and 1.00% based upon the data collected as part of this analysis.

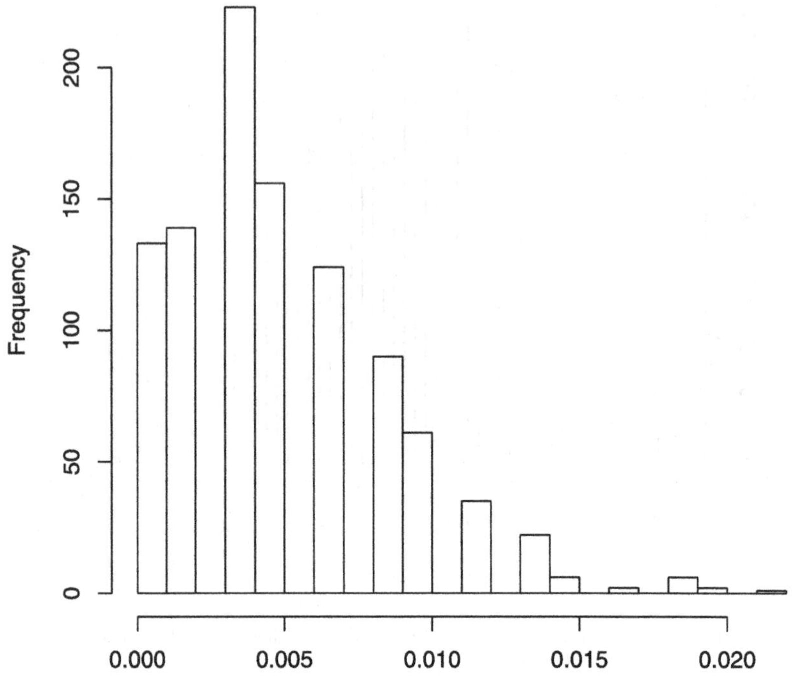

Fig. 3.4 Distribution of $\hat{\pi}^b$ for speaker matcher (LFCC, GMM) from the XM2VTS database

3.2.2.2 Hypothesis Test for Single FNMR

In this section we are testing whether a particular device or a particular process has an FNMR significantly below a set bound. We will denote this bound by π_0. This necessitates a 'less than' alternative hypothesis, H_1. The other typical alternative hypotheses, greater than and not equal to, are possible here; however, in biometric testing the focus is most often on getting devices with superior matching performance and, therefore, in the case of FNMR that means devices with low FNMR. As we did in the previous section we present two methods: one based upon large sample theory and another based upon the bootstrap.

Large Sample Approach For the large sample case, we can assume approximate Gaussianity for the sampling distribution of $\hat{\pi}$ if $N_\pi^\dagger \pi_0 \geq 10$ and if $N_\pi^\dagger(1-\pi_0) \geq 10$. If these conditions are met then the hypothesis test of interest here is given by:

$$H_0 : \pi = \pi_0$$
$$H_1 : \pi < \pi_0.$$

Test Statistic:

$$z = \frac{\hat{\pi} - \pi_0}{\sqrt{\frac{\hat{\pi}(1-\hat{\pi})(1+\hat{\varrho}(m_0-1))}{n\bar{m}}}}. \tag{3.15}$$

p-value: $p = P(Z < z)$ where Z is a standard Gaussian random variable.

We will reject the null hypothesis, H_0, if *p-value* $< \alpha$ or if the *p-value* is small.

Example 3.5 In this example, we test at a significance level of $\alpha = 0.05$ if a given FNMR is significantly below 10% or 0.10 for a FMR of 0.001. See Chap. 4 for more details on FMR's. The data here are decisions from the XM2VTS database. In particular, we are considering the speaker matcher (LFCC, GMM). Details can be found in Poh et al. [76]. To achieve a FMR of 0.001 we will use a threshold of 3.64432. For that threshold, we have an estimated FNMR of $\hat{\pi} = 0.0683$ with $N_\pi = 600$, $\hat{\varrho} = 0.3455212$ and $m_i = 3$ for all i. To verify that the large sample conditions are met, we present $N_\pi^\dagger = 354.81$ and note that $N_\pi^\dagger \pi_0 = 35.481 > 10$. The statistical question is then whether 0.0683—our estimated FNMR—is significantly less than 0.10 or that particular outcome is *not* unlikely due to chance given the variability in $\hat{\pi}$. Our test is then

$$H_0 : \pi = 0.10$$
$$H_1 : \pi < 0.10.$$

Test Statistic:

$$z = \frac{0.0683 - 0.1000}{\sqrt{\frac{0.0683(1-0.0683)(1+0.3455(3-1))}{200\times 3}}} = -2.364036. \tag{3.16}$$

Then according to Table 9.1, we have that the *p-value* is $P(Z < -2.36) = 0.0091$. Consequently, since our *p-value* is less than the significance level ($0.0091 < 0.05$), we reject the null hypothesis and we can conclude that the FNMR here is significantly less than 0.10 for the threshold that gives us an FMR of 0.001.

Bootstrap Approach The approach we take here is to follow the *subsets bootstrap* of Bolle et al. [9]. The basics of this approach are to sample with replacement individuals and take all of the decisions from those selected individuals. However, for hypothesis testing we adjust our bootstrap replicates to assume that the null hypothesis is true. The outline of this procedure is given below.

1. For the observed data, calculate
$$\hat{\pi} = \frac{\sum_{i=1}^{n} \sum_{j=1}^{m_i} D_{iij}}{\sum_{i=1}^{n} m_i}. \tag{3.17}$$
2. Randomly select with replacement n individuals from the list of individuals $\{1, 2, \ldots, n\}$ and call this list of individuals $b_1, b_2, \ldots, b_n$'s.
3. For each selected individual b_i, we will use all m_{b_i} decisions associated with that individual. We will refer to these genuine decisions as $D_{b_i b_i j}$'s where $j = 1, \ldots, m_{b_i}$.
4. Calculate and store
$$e_\pi = \hat{\pi}^b - (\hat{\pi} - \pi_0) \tag{3.18}$$

where

$$\hat{\pi}^b = \frac{\sum_{i=1}^{n} \sum_{j=1}^{m_i} D_{b_i b_i j}}{\sum_{i=1}^{n} m_{b_i}}. \tag{3.19}$$

5. Repeat the previous three steps some large number of times M and store e_π each time.
6. The *p-value* for this test is then

$$p = \frac{1 + \sum_{\varsigma=1}^{M} I_{\{e_\pi \leq \hat{\pi}\}}}{M+1}, \tag{3.20}$$

the percentage of times that the sampling distribution of $\hat{\pi}$ exceeds the observed $\hat{\pi}$ adjusting when the null hypothesis is true.

The intuition for this test is that the distribution of e_π mirrors the sampling distribution of $\hat{\pi}$ if the null hypothesis is true and hence forms a reference distribution from which to draw inference. The reasoning for adjusting e_π by subtracting the difference between $\hat{\pi}$ and π_0 is that we want to have the sampling distribution centered at the hypothesized value, π_0, which is where it should be under the null hypothesis, i.e. when the null hypothesis is true.

Example 3.6 In this example, we wish to compare the results for the bootstrap hypothesis test to those for the large sample approach. In particular, we will mirror the test found in Example 3.5. Thus, we are testing:

$$H_0 : \pi = 0.10$$
$$H_1 : \pi < 0.10.$$

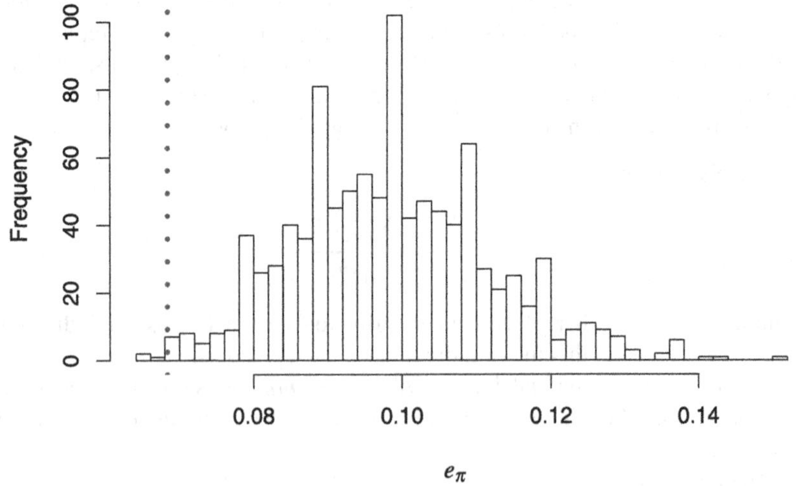

Fig. 3.5 Distribution of e_π for speaker matcher (LFCC, GMM) from the XM2VTS database

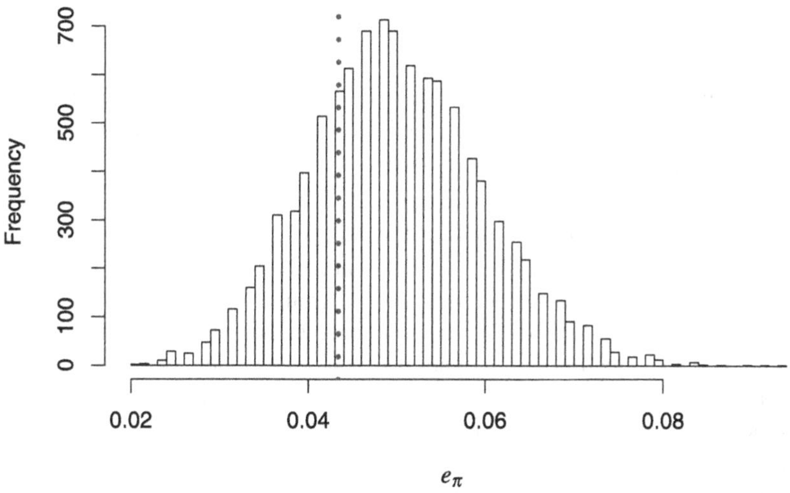

Fig. 3.6 Distribution of e_π for face matcher (DCTb, GMM) from the XM2VTS database

We are using decisions from the (LFCC, GMM) speaker matcher from the XM2VTS database. The revelant summaries are $\hat{\pi} = 0.0683$ and $\hat{\varrho} = 0.3455$ from $N_\pi = 600$ decisions made on $n = 200$ individuals and $m_i = 3$ for all i. Figure 3.5 shows the distribution of the e_b's and the dashed vertical line there is for $\hat{\pi} = 0.0683$. The *p-value* is the percent of values that are less than or equal to that from the distribution of e_b. Here with $M = 1000$, the *p-value* is 0.004 which is small and, hence, we reject the null hypothesis and conclude that the observed FNMR, $\hat{\pi}$, is significantly less than 0.1000.

Example 3.7 For this example, we will focus on the Face matcher (DCTb, GMM) from the XM2VTS database. We would like to test whether the FNMR achieved when the threshold is set to 0.29 is less than 0.05. The estimated FNMR here is 0.0433. The estimated intra-individual correlation is $\hat{\varrho} = 0.1959$ and as with all FNMR data in this database $N_\pi = 600$, $n = 200$ and $m_i = 3$ for all i. The approximate sampling distribution for e_π is summarized in Fig. 3.6. Again the dashed vertical line represents the observed value $\hat{\pi} = 0.0433$. Our hypothesis test is formally:

$$H_0 : \pi = 0.05$$
$$H_1 : \pi < 0.05.$$

Using $M = 10000$ we obtain a *p-value* of 0.2773. This *p-value* is large and, consequently, we fail to reject the null hypothesis and we conclude that the observed FNMR, 0.0433 is not significantly less than 0.05.

3.2.3 Two Sample Methods for FNMR's

In this section, we present methodology for the statistical comparison of two FNMR's. Notationally, we will let π_1 and π_2 represent the two false non match rates and we will denote the total sample sizes for each of the two groups by $N_\pi^{(g)}$ and the effective sample size will be denoted by $N_\pi^{\dagger(g)}$ for $g = 1, 2$. In general, we will differentiate between quantities for the two groups by a superscript (g). For example, we will use $m_0^{(1)}$ and $m_0^{(2)}$ to differentiate between the adjusted average number of decisions per person between the two groups.

When comparing two FNMR's, we must consider how the data were collected. Focusing on data collection that is either paired or independent, we will present inferential methods for those two cases. As a reminder, paired data is characterized by a specific association between the individual signals in both the groups being compared. It is often the case in biometric competitions, for example, that the performance of multiple classifiers is evaluated and compared on the same set of signals. This data would be paired. If the individuals in both groups are distinct then the comparison of the error rates would be an independent one. More complicated data structures are possible but are not considered in this book. We can use the methods described below to determine if a process has the same constant FNMR between two demographic groups or between two time periods, for example. Each process is assumed to be internally stationary for the group that are being analyzed.

3.2.3.1 Confidence Interval for Difference of Two Independent FNMR's

In this section, we provide a confidence interval for the case where we have two independent data collections. For example, if we have match decisions on the same biometric device for two different age groups, then we will have two independent FNMR's. The individuals in those two groups are distinct and, thus, fit the profile for an independent approach. Below we provide three methods for carrying out this hypothesis test: a large sample method, a bootstrap method and a resampling method. All of these methods attempt to approximate the sampling distribution for the estimated difference between two FNMR's, $\pi_1 - \pi_2$. The large sample approach is one that assumes a large number of decisions and uses the Gaussian distribution for the sampling distribution of $\hat{\pi}_1 - \hat{\pi}_2$. The bootstrap approach resamples individuals from each matching process to approximate the sampling distribution for $\hat{\pi}_1 - \hat{\pi}_2$. The randomization procedure aims to approximate this same sampling distribution by permuting decisions between the two groups.

Large Sample Approach The approach here is based upon an assumption that the sampling distribution of $\hat{\pi}_1 - \hat{\pi}_2$ is approximately Gaussian. This will occurs when the number of decisions in each sample is large. We define how large is sufficient below. Note that the assumption of Gaussianity is *not* an assumption about the distribution of the match scores. If $N_\pi^{\dagger(g)}$ for the gth group, is large for $g = 1, 2$

meaning that $N_\pi^{\dagger(g)}\hat{\pi}_g \geq 10$ and $N_\pi^{\dagger(g)}(1-\hat{\pi}_g) \geq 10$ for $g = 1, 2$, then we can use the following to make a $(1-\alpha)100\%$ confidence interval for $\pi_1 - \pi_2$. Recall that $N_\pi^\dagger$, following (3.9) is the effective sample size for an FNMR process.

$$\hat{\pi}_1 - \hat{\pi}_2 \pm z_{\frac{\alpha}{2}} \sqrt{\frac{\hat{\pi}_1(1-\hat{\pi}_1)[1+(m_0^{(1)}-1)\hat{\varrho}_1]}{n_1\bar{m}^{(1)}} + \frac{\hat{\pi}_2(1-\hat{\pi}_2)[1+(m_0^{(2)}-1)\hat{\varrho}_2]}{n_2\bar{m}^{(2)}}} \tag{3.21}$$

where

$$\hat{\pi}_g = \frac{\sum_{i=1}^{n_g}\sum_{j=1}^{m_i^{(g)}} D_{iij}^{(g)}}{\sum_{i=1}^{n} m_i^{(g)}}. \tag{3.22}$$

Example 3.8 For illustrating our confidence interval methodology for the difference between two independent FNMR's, we use data on two facial matchers: one from the NIST database and one from the XM2VTS database. The confidence that we use in this example will be 90%. From the former database, we use the face C matcher—process $g = 1$—and from the latter database, we use the (DCTb, GMM) face matcher—process $g = 2$. The estimated FNMR's here are 0.0367 and 0.0517 from using thresholds of 0.58 and 0.30, respectively. The number of decisions in each is the same order of magnitude, $N_\pi^{(1)} = 517$ and $N_\pi^{(2)} = 600$; however, they have distinctly different structures. For the first process, $n_1 = 517$ and $m_i^{(1)} = 1$ for all i which gives $m_0 = 1$ and implies that we are not able to estimate the intra-individual correlation since we have only a single decision per individual. In (3.21), the intra-individual correlation disappears when $m_0 = 1$, so we are still able to calculate our confidence interval. For the second process, we have $n_2 = 200$ and $m_i^{(2)} = 3$ for all i. The intra-individual correlation is $\hat{\varrho} = 0.2517$ with $m_0 = 3$. In order to use the large sample approach, we need check that our samples are of sufficient size. Recall that this means $N_\pi^{\dagger(g)}\hat{\pi}_g \geq 10$ and $N_\pi^{\dagger(g)}(1-\hat{\pi}_g) \geq 10$ for $g = 1, 2$. In this case with $N_\pi^{\dagger(1)} = 517$ and $N_\pi^{\dagger(2)} = 399.12$, these conditions are satisfied. Then our confidence interval is $(-0.0377, 0.0078)$ and we can be 90% confident that the difference in the process FNMR's for these two processes is between -3.77% and 0.78%.

Bootstrap Approach A bootstrap approach to making a confidence interval for the difference of two independent FNMR's, $\pi_1 - \pi_2$, is found by separately bootstrapping each group and then taking the difference of the two bootstrap FNMR's. We do this repeatedly to derive an approximate sampling distribution for $\hat{\pi}_1 - \hat{\pi}_2$. More formally, the procedure is:

1. Calculate $\hat{\pi}_1 - \hat{\pi}_2$ from the observed data.
2. Sample n_1 individuals with replacement from the n_1 individuals from the first group from which there are genuine match decisions. Denote these selected individuals by $b_1^{(1)}, b_2^{(1)}, \ldots, b_{n_1}^{(1)}$.

3. For each selected individual, $b_i^{(1)}$, we take all $m_{b_i^{(1)}}$ genuine decisions. Call these selected matching decisions, $D^{(1)}_{b_i^{(1)} b_i^{(1)} j}$'s, $j = 1, \ldots, m_{b_i^{(1)}}$.
4. Sample n_2 individuals with replacement from the n_2 individuals in the second group from which there are genuine match decisions. Denote these selected individuals by $b_1^{(2)}, b_2^{(2)}, \ldots, b_{n_2}^{(2)}$.
5. For each selected individual, $b_i^{(2)}$, we take all $m_{b_i^{(2)}}$ decisions on that individual. Call these selected matching attempts, $D^{(2)}_{b_i^{(2)} b_i^{(2)} j}$'s, $j = 1, \ldots, m_{b_i^{(2)}}$.
6. Calculate

$$e_\pi = \hat{\pi}_1^b - \hat{\pi}_2^b - (\hat{\pi}_1 - \hat{\pi}_2) \tag{3.23}$$

where

$$\hat{\pi}_g^b = \frac{\sum_{i=1}^{n_g} \sum_{j=1}^{m_{b_i^{(g)}}} D^{(g)}_{b_i^{(g)} b_i^{(g)} j}}{\sum_{i=1}^{n_g} m_{b_i^{(g)}}} \tag{3.24}$$

for $g = 1, 2$.
7. Repeat the five previous steps some large number of times M, each time storing the e_π's, the differences of the bootstrapped estimates, $\hat{\pi}_1^b - \hat{\pi}_2^b$, from the observed difference of FNMR's, $\hat{\pi}_1 - \hat{\pi}_2$.
8. A $100(1-\alpha)\%$ confidence interval for the difference of two FNMR's, $\pi_1 - \pi_2$ is then formed by taking the interval from $\hat{\pi}_1 - \hat{\pi}_2 - e_U$ to $\hat{\pi}_1 - \hat{\pi}_2 - e_L$ where e_L and e_U represent the $\alpha/2$th and the $1 - \alpha/2$th percentiles of the distribution of e_π.

Example 3.9 The data we analyze here is from the BANCA database [5]. We consider data from the G protocol and the voice matcher UC3M_voice_gmm_auto_scale_34_500 (UC3M-34-500). Here we compare the FNMR for the two groups g1 and g2 from that database at the threshold 0.25. Figure 3.7 shows the distribution of the e_π for this data. For both of the classifiers $n_1 = n_2 = 26$; however $N_\pi^{(1)} = 233$ and $N_\pi^{(2)} = 234$. Consequently, $m_0^{(1)} = 8.9657$ and $m_0^{(2)} = 9$ where we use $g = 1$ for the group g1 and $g = 2$ for the group g2. The estimated FNMR's are 0.0086 and 0.0214. Additionally, $N_\pi^{\dagger(1)} = 233$, $N_\pi^{\dagger(2)} = 142.44$, $\hat{\varrho}_1 = -0.0087$ which we truncate to $\hat{\varrho}_1 = 0.0000$ and $\hat{\varrho}_2 = 0.0803$. We note here that there is not sufficient decision data here for the large sample approach. The 95% confidence interval for this difference between the two FNMR's is $(-0.0384, 0.0171)$ based upon $M = 10000$ bootstrap replicates. We can then state that we are 95% confident that the difference between these two process FNMR's is between −3.84% and 1.71%.

3.2.3.2 Hypothesis Test for Difference of Two Independent FNMR's

In this section we are testing whether there are significant differences between two FNMR's. In particular, we want to know if one FNMR is significantly less than

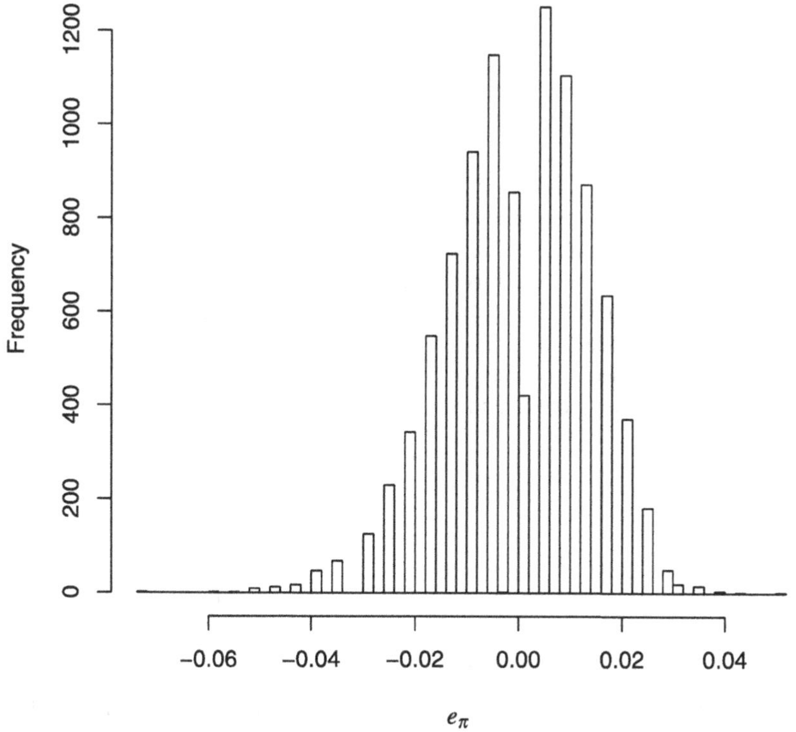

Fig. 3.7 Distribution of e_π for voice matcher (UC3M-34-500) from the BANCA database

the other. The data collection mechanism assumed here is that the individuals come from two distinct groups. If there is overlap among these groups, then the methods given below are *not* appropriate for this analysis. Below we present a large sample approach, a bootstrap approach and a randomization approach for testing the equality of two independent FNMR's.

Large Sample Approach If $N_\pi^{\dagger(1)}$ and $N_\pi^{\dagger(2)}$ are large meaning that $N_\pi^{\dagger(g)}\hat{\pi}_p > 10$ and $N_\pi^{\dagger(g)}(1-\hat{\pi}_p) > 10$ for $g = 1, 2$, then we can use the following hypothesis test for the difference of two independent FNMR's. The value, $\hat{\pi}_p$, is a pooled estimated of the FNMR which is obtained using the following equation:

$$\hat{\pi}_p = \frac{N_\pi^{(1)}\hat{\pi}_1 + N_\pi^{(2)}\hat{\pi}_2}{N_\pi^{(1)} + N_\pi^{(2)}}. \tag{3.25}$$

This pooled estimate is an estimate of the FNMR assuming that the two groups have the same FNMR which is the null hypothesis of the following hypothesis test.

$$H_0 : \pi_1 = \pi_2$$
$$H_1 : \pi_1 < \pi_2.$$

Test Statistic:

$$z = \frac{\hat{\pi}_1 - \hat{\pi}_2}{\sqrt{\hat{\pi}_p(1-\hat{\pi}_p)[\frac{1+\hat{\varrho}_1(m_0^{(1)}-1)}{n_1\bar{m}_1} + \frac{1+\hat{\varrho}_2(m_0^{(2)}-1)}{n_2\bar{m}_2}]}}. \tag{3.26}$$

p-value: $p = P(Z < z)$ where Z is a standard Gaussian random variable.

We will reject the null hypothesis, H_0, if the *p-value* is small or if the *p-value* is less than a prespecified significance level, α.

For this type of test, we need to assume that the two FNMR's are equal, since this is our null hypothesis. Consequently, we need to find the best estimate for the FNMR, if that is the case. This estimate is $\hat{\pi}_p$ and can be calculated following (3.25). The subscript 'p' is meant to denote a pooled estimator.

$\hat{\varrho}_g$ is the estimate of ϱ obtained from data in the gth group and

$$m_0^{(g)} = \frac{\sum_{i=1}^{n_g}(m_i^{(g)})^2}{\sum_{i=1}^{n_g} m_i}. \tag{3.27}$$

Example 3.10 To illustrate the approach here, we test the equality of two face matchers: face_G from the NIST database (group 1) and the (FH, MLP) matcher from the XM2VTS database (group 2). For the former, there are $n_1 = 517$ individuals each of whom contributed a single decision, i.e. $m_i^{(1)} = 1$ for all i. So for those data $N_\pi^{(1)} = 517$ and $m_0^{(1)} = 1$. For the latter matcher, $N_\pi^{(2)} = 600$ with $n_2 = 200$ and $m_0^{(2)} = 3$ based upon $m_i = 3$ for all i. The relevant FNMR summaries from these processes are $\hat{\pi}_1 = 0.0542$, $\hat{\pi}_2 = 0.0700$ and $\hat{\varrho}_2 = 0.2072$ from thresholds of 72.5 and 0.72, respectively. We note that since in the NIST database each individual contributes only a single sample that we cannot estimate ϱ_1 but it is not necessary to do so for our calculations here. In order to use the above hypothesis test appropriately, we must ensure that our sample is of sufficient size. For the face_G matcher we have that $N_\pi^{\dagger(1)} = 517$ and $N_\pi^{\dagger(2)} = 409.86$. Multiplying each of these by the respective $\hat{\pi}_g$'s and $1 - \hat{\pi}_g$'s, we obtain 28.00 and 489.00 for the face_G matcher and 28.69 and 381.17 for the (FH,MLP) matcher. Thus, our samples are sufficiently large to allow for the use of the above large sample method for testing the difference of two independent FNMR's. By substituting in the process summaries we have just given, we obtain a pooled estimator of 0.0161 and a test statistic of $z = -0.99$. Following Table 9.1, we obtain a *p-value* of 0.1611. From this we conclude that there is not a enough evidence to reject the null hypothesis. Therefore, we say that the difference between the two estimated FNMR's 0.0542 and 0.0750 is not statistically significant.

Bootstrap Approach To test the hypotheses of interest using a resampling approach, we assume the same null and alternative hypotheses,

$$H_0 : \pi_1 = \pi_2$$

$$H_1 : \pi_1 < \pi_2,$$

as in the large sample approach. Since the processes are independent, we can bootstrap the matching decisions from each group separately. We then take the difference of the bootstrapped error rates to approximate the sampling distribution of the differences. An algorithm for obtaining a *p-value* in this case is

1. Calculate the estimated difference in the FNMR's, $\hat{\pi}_1 - \hat{\pi}_2$, following (3.2).
2. Sample n_1 individuals with replacement from the n_1 individuals in the first group from which there are matching decisions. Denote these selected individuals by $b_1^{(1)}, b_2^{(1)}, \ldots, b_{n_1}^{(1)}$. For each selected individual, $b_i^{(1)}$, take all $m_{b_i^{(1)}}$ matching decisions. Call these selected decisions, $D_{b_i^{(1)} b_i^{(1)} j}$'s.
3. Sample n_2 individuals with replacement from the n_2 individuals in the second group from which there are matching decisions. Denote these selected individuals by $b_1^{(2)}, b_2^{(2)}, \ldots, b_{n_2}^{(2)}$. For each selected individual, take all $m_{b_i^{(2)}}$ matching decisions. Call these selected decisions, $D_{b_i^{(2)} b_i^{(2)} j}$'s.
4. Calculate $e_\pi = \hat{\pi}_1^b - \hat{\pi}_2^b - (\hat{\pi}_1 - \hat{\pi}_2)$ where

$$\hat{\pi}_g^b = \frac{\sum_{i=1}^{n_g} \sum_{j=1}^{m_{b_i^{(g)}}} D^{(g)}_{b_i^{(g)} b_i^{(g)} j}}{\sum_{i=1}^{n_g} m_{b_i^{(g)}}} \tag{3.28}$$

   for $g = 1, 2$. The superscript 'b' here means that the calculation is performed on the bootstrapped observations.
5. Repeat the previous three steps some large number of times, say M.
6. The *p-value* for this test is then

$$p = \frac{1 + \sum_{\varsigma=1}^{M} I_{\{e_\pi \le (\hat{\pi}_1 - \hat{\pi}_2)\}}}{M + 1}. \tag{3.29}$$

We reject the null hypothesis if the *p-value* is small or if $p < \alpha$ where α is the significance level of the test.

Example 3.11 To illustrate the use of the bootstrap approach to compare two FNMR's we consider the two groups—g1 and g2—from protocol G of the BANCA database. The classifier we consider here is the SURREY_face_nc_man_scale_50 (SURREY-NC-50) face matcher which is applied to both groups. Choosing the same threshold, 0.4, suppose we want to test the equality of these two FNMR's assuming a significance level of $\alpha = 0.05$. Treating g1 as group 1 and g2 as group 2, we have that the sample FNMR's are $\hat{\pi}_1 = 0.0300$ and $\hat{\pi}_2 = 0.0641$. For g1 we have $N_\pi = 233$ with $n_1 = 26$ and $m_0 = 8.9657$; for g2 we have $N_\pi = 234$ with $n_2 = 26$ and $m_0 = 9$. Applying the above approach, we get a *p-value* of 0.0862. Figure 3.8 shows the distribution for e_π from 4999 bootstrap replicates. The dashed vertical line there is the observed difference of $\hat{\pi}_1 - \hat{\pi}_2 = -0.0341$. Since our *p-value* is larger than our significance level, we fail to reject the null hypothesis of equality of these two FNMR's. We can say that there is not enough evidence to conclude that this matcher is not stationary for these two groups.

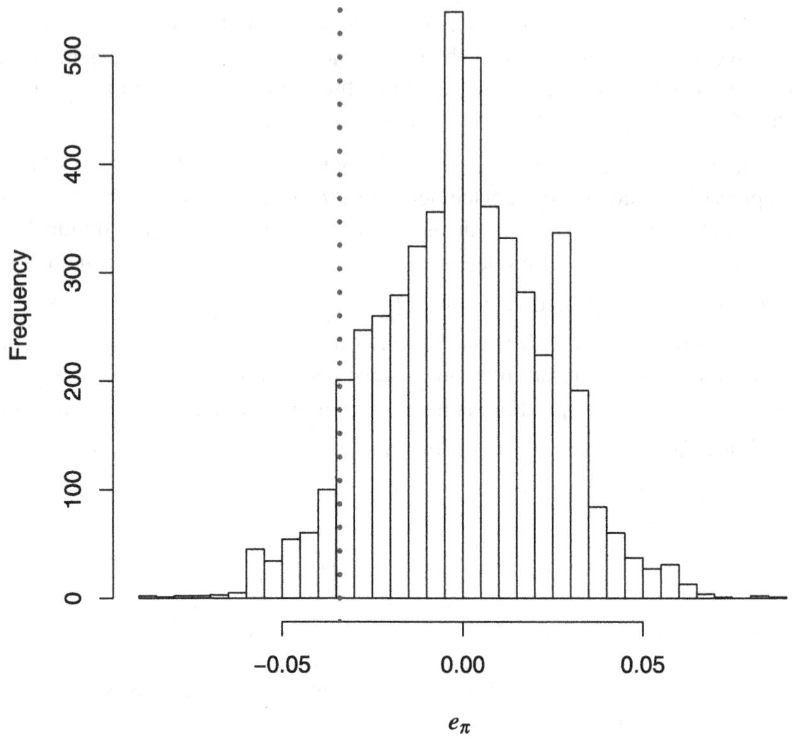

Fig. 3.8 Distribution of e_π for face matcher SURREY-NC-50 from the BANCA database comparing groups g1 and g2 for protocol G

Randomization Approach As above, we want to test whether there is a significant difference between two independent false non-match rates. To do this, we will assume the same null and the same alternative hypothesis as the bootstrap and the large sample approaches. Those hypotheses are

$$H_0 : \pi_1 = \pi_2$$
$$H_1 : \pi_1 < \pi_2.$$

The randomization approach for the difference of two independent FNMR's takes the complete list of individuals—$n_1 + n_2$ individuals—in both groups and randomly assigns n_1 of them to the first group and the remaining n_2 to the second group. We then calculate the difference in the new estimated FNMR's and repeat this process. Permuting individuals should give us an approximation to the sampling distribution of the difference between the two estimated FNMR's under the null hypothesis which is equivalent to $\pi_1 - \pi_2 = 0$. The exact procedure is given as follows:

1. Calculate $\hat{\pi}_1 - \hat{\pi}_2$ following (3.2) for each group.
2. Combine the n_1 individuals from the first group with the n_2 individuals from the second group. Then select n_1 of these individual without replacement from the

$n_1 + n_2$ total individuals. Denote the selected individuals in the first group by $b_1^{(1)}, \dots, b_{n_1}^{(1)}$ and those in the second group by $b_1^{(2)}, \dots, b_{n_2}^{(2)}$. We will refer to these groups as the resampled groups.

3. For each individual $b_i^{(g)}$ in each group take all $m_{b_i^{(g)}}$ matching decisions for that individual. Let those resampled decisions be called $D^{(g)}_{b_i^{(g)} b_i^{(g)} j}$ where $j = 1, \dots, m_{b_i^{(g)}}$.
4. Calculate and store $\Delta_\pi = \hat{\pi}_1^b - \hat{\pi}_2^b$ where

$$\hat{\pi}_g^b = \frac{\sum_{i=1}^{n_g} \sum_{j=1}^{m_{b_i^{(g)}}} D^{(g)}_{b_i^{(g)} b_i^{(g)} j}}{\sum_{i=1}^{n_g} m_{b_i^{(g)}}}. \tag{3.30}$$

5. Repeat steps 2 to 4 a large number of times, M.
6. Then the *p-value* for this test is

$$p = \frac{1 + \sum_{\varsigma=1}^{M} I_{\{\Delta_\pi \leq \hat{\pi}_1 - \hat{\pi}_2\}}}{M + 1}. \tag{3.31}$$

As with other hypothesis tests, we reject the null hypothesis of equality if the *p-value* is small and fail to reject the null hypothesis if the *p-value* is large.

Example 3.12 For this example, we reanalyze the data from Example 3.11 where we compared the same face matcher, SURREY_face _nc_man_scale_50 (SURREY-NC-50), from the BANCA database on two different groups with the same threshold to determine if there were differences between the groups. As in the previous example we treat g1 as group 1 and g2 as group 2. Then the relevant summaries for each process are $\hat{\pi}_1 = 0.0300$, $\hat{\pi}_2 = 0.0641$, $\hat{\varrho}_1 = 0.0384$ and $\hat{\varrho}_2 = 0.0797$. For g1 we have $N_\pi = 233$ with $n_1 = 26$ and $m_0 = 8.9657$; for g2 we have $N_\pi = 234$ with $n_2 = 26$ and $m_0 = 9$. From the randomization, we get a *p-value* of 0.1114. Figure 3.9 shows the distribution for e_π from 9999 randomization replicates. Since our *p-value* is larger than our significance level, we fail to reject the null hypothesis of equality of these two FNMR's. We also note that the *p-value* that we have here is approximately the same 0.1114 versus 0.0862 as the one we obtained in Example 3.11. Both tests have *p-values* that are large enough for us to conclude that we should not reject the null hypothesis, H_0.

3.2.3.3 Confidence Interval for Difference of Two Paired FNMR's

We move now to paired data and a confidence interval for the difference of two FNMR's when non-match decisions are paired. Here we will assume that it is the signals that are paired or correlated in some way. Hence, the decisions, $D_{iij}^{(1)}$ and $D_{iij}^{(2)}$, are associated or paired for all individuals i and for all decisions j. This

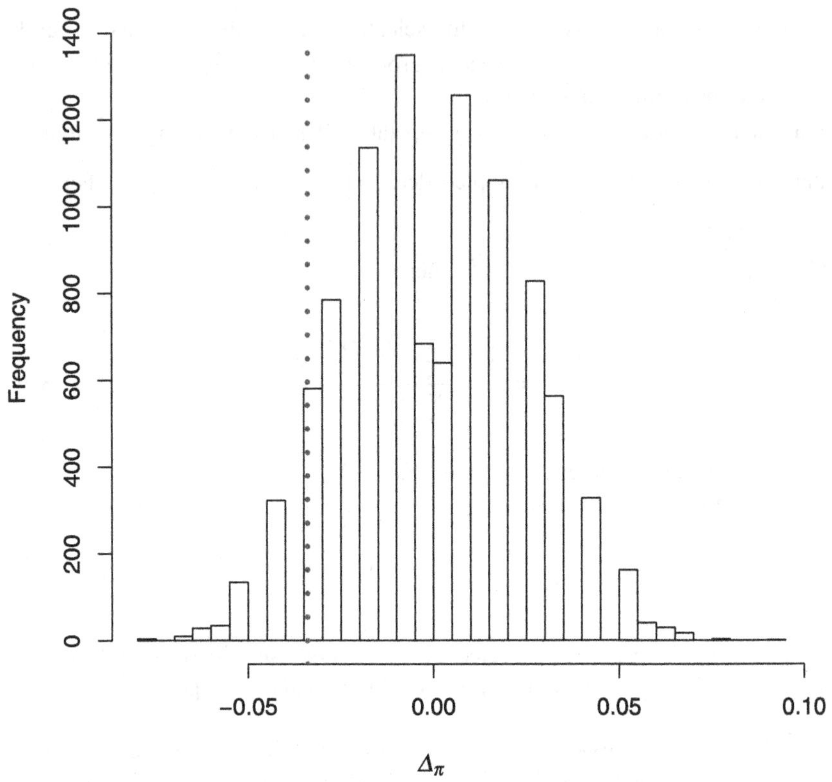

Fig. 3.9 Distribution of Δ_π, randomized differences of the FNMR's of the face matcher SURREY-NC-50 BANCA database comparing groups g1 and g2 for protocol G

sort of scenario is often present in comparing two different algorithms on the same set of biometric signals or when multiple signals are collected on a single individual by two different censors at roughly the same time. For example, we might be interested in how two different iris algorithms perform on the same database of signals/images. Since for every match decision that the first algorithm gives, the second will give a match decision; thus, we have a paired data collection. Because of pairing we can assume that the number of decisions for each individual, the m_i's, are the same in each group, i.e. $m_i^{(1)} = m_i^{(2)}$ for all i. Therefore, we will not include notation that differentiates between the n's, the N's, the m_0's and the m_i's of each group g, since these quantities are the same for both groups. However, the estimated error rates—$\hat{\pi}_1$ and $\hat{\pi}_2$—as well as the correlations—$\hat{\varrho}_1$ and $\hat{\varrho}_2$—will be different and so we distinguish between them by use of a subscript. We do not assume here that the correlations are equal but methodology based upon such an assumption is possible, though we prefer to develop methodology for the more general case.

Large Sample Approach For the large sample approach, we need to have a sample of sufficient size. In this case, we need N_π to be large meaning that

$N_{\pi}^{\dagger(g)}\hat{\pi}_g \geq 10$, and $N_{\pi}^{\dagger(g)}(1-\hat{\pi}_g) \geq 10$ for $g = 1, 2$ then we can use the following equation to make a $(1-\alpha)100\%$ confidence interval for $\pi_1 - \pi_2$.

$$\hat{\pi}_1 - \hat{\pi}_2 \pm z_{\frac{\alpha}{2}} \times \sqrt{\frac{\hat{\pi}_1(1-\hat{\pi}_1)[1+(m_0-1)\hat{\varrho}_1] + \hat{\pi}_2(1-\hat{\pi}_2)[1+(m_0-1)\hat{\varrho}_2]}{n\bar{m}} - 2Cov(\hat{\pi}_1, \hat{\pi}_2)} \tag{3.32}$$

where

$$Cov(\hat{\pi}_1, \hat{\pi}_2) = \frac{1}{n^2\bar{m}^2}\sum_{i=1}^{n}\sum_{j=1}^{m_i}(D_{iij}^{(1)} - \hat{\pi}_1)(D_{iij}^{(2)} - \hat{\pi}_2) \tag{3.33}$$

and

$$\hat{\pi}_g = \frac{\sum_{i=1}^{n}\sum_{j=1}^{m_i} D_{iij}^{(g)}}{\sum_{i=1}^{n} m_i}. \tag{3.34}$$

Example 3.13 A 90% confidence interval is the goal of this example for the difference between the FNMR for two speaker recognition systems (PAC, GMM) and (SSC, GMM) from the XM2VTS database, [74]. We use thresholds of 2.0 and 1.0 and these give estimated FNMR's of 0.0583 and 0.0267, respectively. The estimated intra-individual correlation, $\hat{\varrho}$ for the first is 0.3628 and for the second is 0.3579. Using $z_{0.05} = 1.645$, we obtain a 90% confidence interval of $(0.0091, 0.0542)$. Thus we can be 90% confident that the difference in the FNMR's for these two processes is between 0.91% and 5.42%. We note that the covariance between these two is 2.241×10^{-5}. Especially important for this confidence interval is that both endpoints of this interval are on the same side as zero. This indicates that at the 90% confidence level we can conclude that these two FNMR's are statistically different.

To illustrate the differences between paired and independent confidence intervals for the difference of two FNMR's, we note that a 90% confidence interval assuming the two data collections from Example 3.13 were independent would be slightly different $(0.0066, 0.0567)$ than the $(0.0091, 0.0542)$ that we obtained. The difference manifests itself in the widths of the two intervals since both are centered at the sample difference between the rates. The independent interval is 11% wider $0.0567 - 0.0066 = 0.0501$ than the paired interval $0.0542 - 0.0091 = 0.0451$ because of the correlations that are present in the paired data. This difference illustrates the need to ensure that the appropriate statistical method is used to analyze data.

Bootstrap Approach For the bootstrap approach, we resample the individuals and then take all genuine decisions for the selected individuals. Having done so, we take the appropriate decision from both groups. We then proceed to estimate the difference and to create a sampling distribution for $\hat{\pi}_1 - \hat{\pi}_2$ by repeating this process. Below we describe this procedure formally below:

1. Calculate

$$\hat{\pi}_1 - \hat{\pi}_2 = \frac{\sum_{i=1}^{n}\sum_{j=1}^{m_i} D_{iij}^{(1)}}{\sum_{i=1}^{n} m_i} - \frac{\sum_{i=1}^{n}\sum_{j=1}^{m_i} D_{iij}^{(2)}}{\sum_{i=1}^{n} m_i}. \tag{3.35}$$

2. Sample n individuals with replacement from the n individuals in the sample and call these individuals $b_1, \ldots, b_n$.
3. For each of the selected individuals, the b_i's, we include all m_{b_i} decisions from each group and call these decisions $D_{b_i b_i j}^{(g)}$ where $j = 1, \ldots, m_{b_i}$ for $g = 1, 2$.
4. Calculate and store $e_\pi = \hat{\pi}_1^b - \hat{\pi}_2^b - (\hat{\pi}_1 - \hat{\pi}_2)$ where

$$\hat{\pi}_g^b = \frac{\sum_{i=1}^{n_g}\sum_{j=1}^{m_{b_i^{(g)}}} D_{b_i^{(g)} b_i^{(g)} j}^{(g)}}{\sum_{i=1}^{n_g} m_{b_i^{(g)}}} \tag{3.36}$$

for $g = 1, 2$.
5. Repeat the previous three steps some large number of times, say M.
6. Find the $\alpha/2$th percentile and the $1 - \alpha/2$th percentile for the e_π's. Denote these two quantities by e_L and e_U, respectively.
7. A $100 \times (1 - \alpha)\%$ confidence interval for $\pi_1 - \pi_2$ is then the interval formed by $(\hat{\pi}_1 - \hat{\pi}_2 - e_U, \hat{\pi}_1 - \hat{\pi}_2 - e_L)$.

Example 3.14 This example creates a 90% confidence interval between two speaker matchers (PAC, GMM) and (SSC, GMM) using data from the XM2VTS database. The data here is paired since both matchers are used on the same set of matched samples. The particular focus here is the difference in the FNMR's when the FMR is 0.01. More on FMR can be found in Chap. 4. For (PAC, GMM) and (SSC, GMM), the thresholds needed are 2.7246 and 1.5049, respectively. The FNMR for (PAC, GMM) at that the specified threshold is $\hat{\pi}_1 = 0.2033$, while the FNMR for (SSC, GMM) is $\hat{\pi}_2 = 0.1367$. For these data, $N_\pi = 600$, $m_i = 3$ for all i, $\hat{\varrho}_1 = 0.4067$, and $\hat{\varrho}_2 = 0.4547$. Following the algorithm above we create a 90% confidence interval using the *subsets bootstrap* approach. Figure 3.10 gives a histogram of an approximate sampling distribution for $\hat{\pi}_1^b - \hat{\pi}_2^b$ based upon $M = 5000$ bootstrap replicates. The confidence interval that we get from the endpoints of that interval is $(0.0317, 0.1000)$. Since this interval does not contain 0, we conclude with 90% confidence that the two FNMR's here are different when the FMR is 0.01.

3.2.3.4 Hypothesis Test for Difference of Two Paired FNMR's

The focus of this section is testing for significant differences between two process FNMR's where the decisions, the D_{iij}'s, are paired in some manner. This means that there is some direct relationship between the jth decisions for the ith individual in both processes. This relationship can result, for example, if each matching attempt occurred at approximately the same time on two different devices. One characteristic

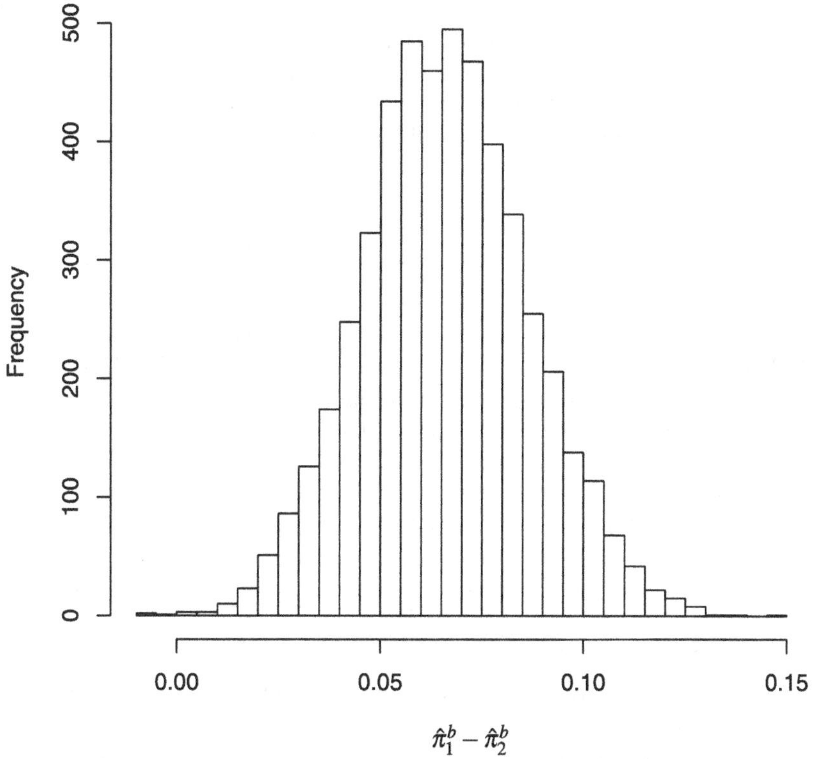

Fig. 3.10 Distribution of $\hat{\pi}_1^b - \hat{\pi}_2^b$ for the difference of two FNMR's for speaker matchers (PAC, GMM) and (SSC, GMM) from the XM2VTS database

of paired data is that the m_i's will be exactly the same for each i in each of the two groups. Data that does not have this structure cannot be paired data. We will use the same notation as above when there are two groups.

Large Sample Approach We begin by letting $\hat{\pi} = 1/2\hat{\pi}_1 + 1/2\hat{\pi}_2$. As with the confidence interval methodology above, we will assume that the decisions are paired. Thus, we need N_π to be large. This will occur if $N_\pi^{\dagger(g)}\hat{\pi}_g \geq 10$ and $N_\pi^{\dagger(g)}(1 - \hat{\pi}_g) \geq 10$ for $g = 1, 2$. If this is the case, then we can use the following hypothesis test for the difference of two paired FNMR's since the sampling distribution for $\hat{\pi}_1 - \hat{\pi}_2$ will be approximately Gaussian.

$$H_0 : \pi_1 = \pi_2$$

$$H_1 : \pi_1 < \pi_2.$$

Test Statistic:

$$z = \frac{\hat{\pi}_1 - \hat{\pi}_2}{\sqrt{\hat{\pi}(1 - \hat{\pi})[\frac{1+(m_0-1)\hat{\varrho}_1}{n\bar{m}} + \frac{1+(m_0-1)\hat{\varrho}_2}{n\bar{m}}] - 2Cov(\hat{\pi}_1, \hat{\pi}_2)}}. \tag{3.37}$$

p-value: $p = P(Z < z)$ where Z is a standard Gaussian random variable and probabilities based upon Z can be found using Table 9.1.

We will reject the null hypothesis, H_0, if the *p-value* is small, or if a significance level is specified then we will reject if $p < \alpha$.

Here $\hat{\varrho}_g$ represents the usual estimate of ϱ given in (3.5) using data from the gth group. The $Cov(\hat{\pi}_1, \hat{\pi}_2)$ can be calculated as follows:

$$Cov(\hat{\pi}_1, \hat{\pi}_2) = \frac{1}{n^2\bar{m}^2} \sum_{i=1}^{n} \sum_{j=1}^{m_i} (D_{iij}^{(1)} - \hat{\pi})(D_{iij}^{(2)} - \hat{\pi}). \tag{3.38}$$

Example 3.15 To illustrate this method we considered two face matchers from the XM2VTS database. These are (DCTb, MLP) and (DCTb, GMM) at thresholds of -0.4 and 0.4, respectively. Their sample FNMR's are $\hat{\pi}_1 = 0.0517$ and $\hat{\pi}_2 = 0.0750$, respectively. We are trying to test whether the former is significantly less than the latter. The additional summaries that we need to carry out this test include $\hat{\varrho}_1 = 0.2176$, $\hat{\varrho}_2 = 0.3273$, $N_\pi = 600$, $n = 200$, $m_i = 3$ for $i = 1, \ldots, n$, $m_0 = 3$, $N_\pi^{\dagger(1)} = 418.04$ and $N_\pi^{\dagger(2)} = 362.61$. These last two statistics are necessary for us to ensure that our sample size is large enough to permit the use of the procedure described in this section. To that end we note that we are required to check the following calculations: $N_\pi^{\dagger(1)}\hat{\pi}_1 = 21.60$, $N_\pi^{\dagger(1)}(1 - \hat{\pi}_1) = 396.44$, $N_\pi^{\dagger(2)}\hat{\pi}_2 = 27.20$, and $N_\pi^{\dagger(1)}(1 - \hat{\pi}_2) = 335.42$. All of these values are at least 10 and so we can proceed with this procedure. Substituting our summaries into the methodology outlined above, we get a test statistic of -1.47 which from Table 9.1 yields a *p-value* of approximately 0.0709. Thus, we would reject for any values of the significance level, α, which are larger than 0.0709. But with a significance level of $\alpha = 0.05$, we would fail to reject the null hypothesis, H_0, of equality between the FNMR's of these two processes. Under that scenario, we would recognize that these two sample FNMR's are different, but we would acknowledge that we do not have enough evidence in these two samples to conclude that the process FNMR's are significantly different.

Bootstrap Approach In order to test the equality of two FNMR's using a bootstrap approach, we assume the same null and alternative hypotheses as in the large sample approach. Those hypotheses are:

$$H_0: \pi_1 = \pi_2$$
$$H_1: \pi_1 < \pi_2.$$

An algorithm for obtaining a bootstrap *p-value* for these hypotheses is the following:

1. Calculate the observed difference $\hat{\pi}_1 - \hat{\pi}_2$ where

$$\hat{\pi}_g = \frac{\sum_{i=1}^{n} \sum_{j=1}^{m_i} D_{iij}^{(g)}}{\sum_{i=1}^{n} m_i} \tag{3.39}$$

for $g = 1, 2$.

2. Sample n individuals with replacement from the n individuals in our sample. Denote these selected individuals by $b_1, b_2, \ldots, b_n$.
3. For each selected individual, b_i, take all of the $2m_{b_i}$ decisions—m_{b_i} decisions from the first group and m_{b_i} decisions from the second group—that exist for that individual. Call these selected matching decisions, $D^{(g)}_{b_i b_i j}$'s where $j = 1, \ldots, m_{b_i}$ and $g = 1, 2$.
4. Calculate

$$e_\pi = (\hat{\pi}_1^b - \hat{\pi}_2^b) - (\hat{\pi}_1 - \hat{\pi}_2) \tag{3.40}$$

where

$$\hat{\pi}_g^b = \frac{\sum_{i=1}^{n} \sum_{j=1}^{m_{b_i}} D^{(g)}_{b_i b_i j}}{\sum_{i=1}^{n} m_{b_i}}. \tag{3.41}$$

5. Repeat the previous three steps some large number of times, M, storing e_π each time.
6. The *p-value* is calculated for this test is then calculated by

$$p = \frac{1 + \sum_{\varsigma=1}^{M} I_{\{e_\pi \le \hat{\pi}_1 - \hat{\pi}_2\}}}{M + 1}. \tag{3.42}$$

7. We will reject the null hypothesis, H_0, if the *p-value* is small. In a significance level, α, is specified *a priori*, then we will reject the null hypothesis, if $p < \alpha$.

Example 3.16 The data we use here is from the NIST database. We are testing the equality of the FNMR for the two face matchers, Face_C and Face_G when the FMR is 0.05. The thresholds that we are using are 0.5860 and 72.7287, respectively. For this database we have that $n = 517$ and $m_i = 1$ for all i. As a consequence, we do not have intra-individual correlation since we only have a single decision per individual. The estimated FNMR's are 0.0406 for Face_C and 0.0619 for Face_G. Figure 3.11 shows the distribution of the e_π's for $M = 1000$ bootstrap replicates. The *p-value* for this test of equality between these two FNMR's is 0.0290 which is relatively small and we would reject the null hypothesis of equality for any significance level, α, that was larger than 0.0290.

Example 3.17 For this example we draw on data from both groups (g1 and g2) of the G protocol of the BANCA database, Bailly–Baillière [5]. Here we are comparing two face matchers SURREY_face_nc_man_scale_100 (SURREY-NC-100) and SURREY_face_[2]sum_auto (SURREY-AUTO) from that database. We use a threshold of 0.33 for the former and a threshold of −0.85 for the latter. The estimated FNMR for SURREY100 is $\hat{\pi}_1 = 0.0236$ with an estimated intra-individual correlation of $\hat{\varrho}_1 = 0.0712$, while the equivalent summaries are $\hat{\pi}_2 = 0.0407$ and $\hat{\varrho}_2 = 0.0927$ for SURREY-AUTO. For this data we have $N_\pi = 467$ with $n = 52$ and $m_i = 9$ for all i except one for which $m_i = 8$. Thus, $m_0 = 8.9829$. Using the bootstrap approach given above with $M = 5000$ bootstrap replicates we get the distribution for e_π represented in Fig. 3.12. Our *p-value* is 0.0648 which is moderately

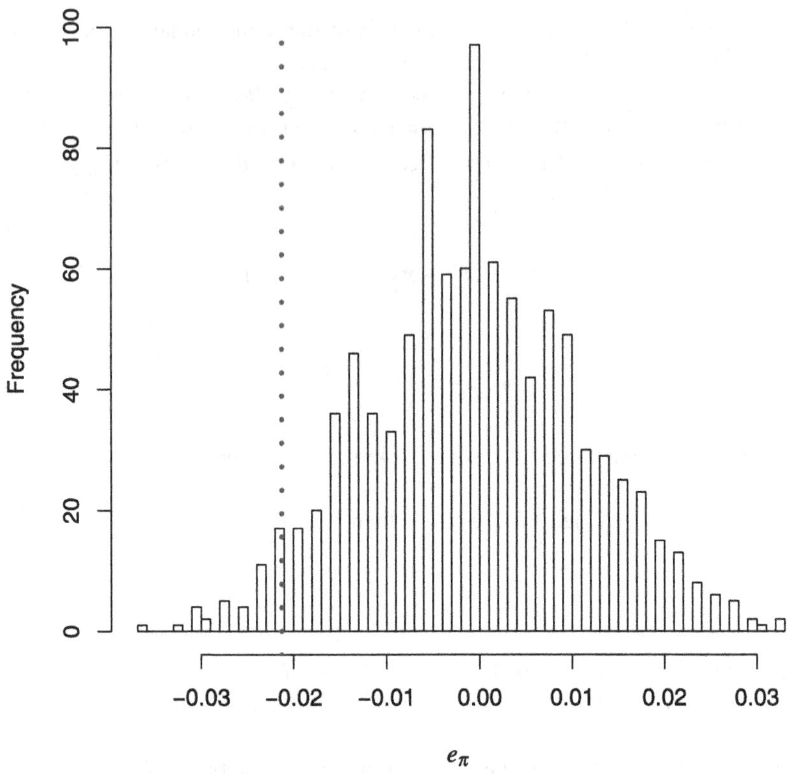

Fig. 3.11 Distribution of e_π for the bootstrapped differences of two FNMR's for face matchers Face_C and Face_G from the NIST database

large and indicates that there is not enough evidence to conclude that the process FNMR's for these two matching algorithms are different. We note this is the case even though the differences between the two FNMR's are approximately 1.7%. This serves as a reminder that we should not automatically assume that such differences automatically mean that one system *significantly* outperforms the other, particularly on a small dataset.

Randomization Approach As above, we are testing the equality of two FNMR's that come from a paired data collection. For the randomization approach here we must maintain the correlation between the 'paired' decisions. Thus we take each set of decisions associated with an individual and permute the group to which each belongs. We do this in the following way:

1. Calculate $\hat{\pi}_1 - \hat{\pi}_2$ from the observed data
2. For the ith individual, take the $2m_i$ decisions—$D^{(1)}_{ii1}, \ldots, D^{(1)}_{iim_i}, D^{(2)}_{ii1}, \ldots, D^{(2)}_{iim_i}$ —that are associated with that individual and randomly sample without replacement m_i of them. Call the selected individuals $D^{(1)*}_{ii1}, \ldots, D^{(1)*}_{iim_i}$. The remaining unsampled individuals will be denoted by $D^{(2)*}_{ii1}, \ldots, D^{(2)*}_{iim_i}$.

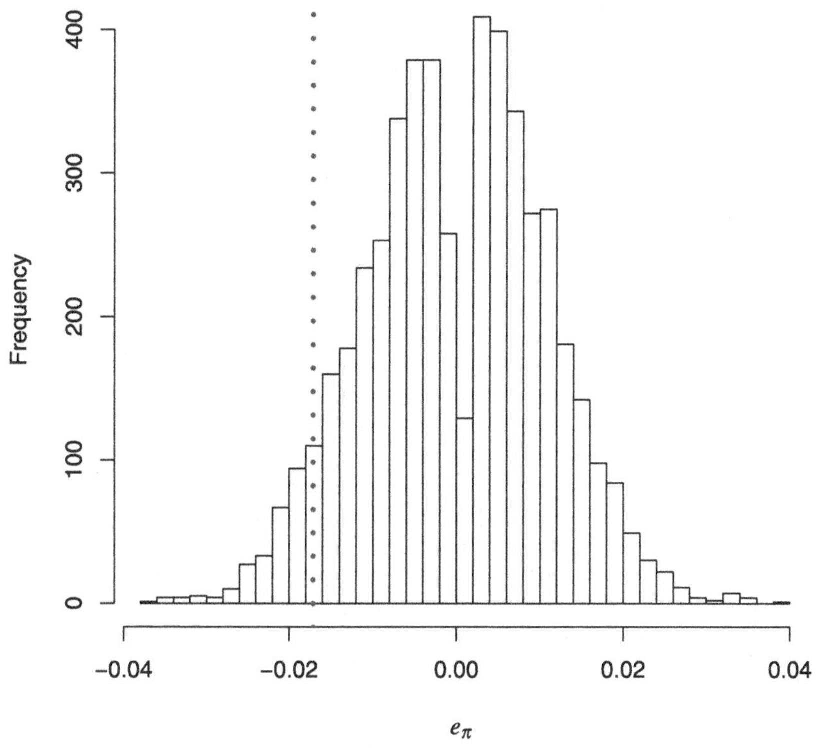

Fig. 3.12 Distribution of e_π for the bootstrapped differences of two FNMR's for face matchers SURREY-NC-100 and SURREY-AUTO from both group of the G protocol of the BANCA database

3. Repeat the previous step for all individuals $i = 1, \ldots, n$.
4. Calculate $\Delta_\pi = \hat{\pi}_1^r - \hat{\pi}_2^r$ where

$$\hat{\pi}_g^r = \frac{\sum_{i=1}^n \sum_{j=1}^{m_i} D_{iij}^{(g)*}}{\sum_{i=1}^n m_i} \tag{3.43}$$

   for $g = 1, 2$.
5. Repeat steps 2 and 3 some large number of times M.
6. Then the *p-value* for this test is given by

$$p = \frac{1 + \sum_{\varsigma=1}^M I_{\{\Delta_\pi \le \hat{\pi}_1 - \hat{\pi}_2\}}}{M+1}. \tag{3.44}$$

7. We will reject H_0 if the *p-value* is small.

Example 3.18 For this example we used the same data as were analyzed in Example 3.16. This test is a comparison of the FNMR's of two face matchers, Face_C and Face_G, from the NIST Biometrics Score Set Release 1 when their FMR's are 0.05.

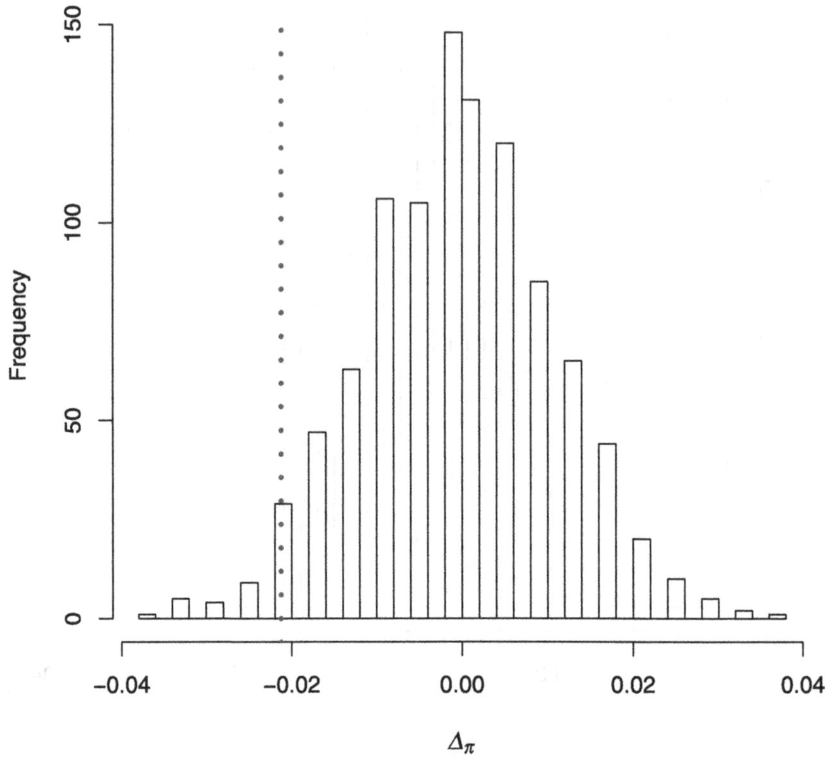

Fig. 3.13 Distribution of Δ_π for the randomized differences of two FNMR's for face matchers Face_C and Face_G from the NIST database

We ran the above randomization test on these data and found a *p-value* of 0.0190 which is quite similar to the 0.0290 found by using the bootstrap approach. Figure 3.13 has the distribution of the Δ_π's from 999 randomization replicates. Since the *p-value* is small, we reject H_0 and conclude that the differences between the two FNMR's are significant.

3.2.4 Multiple Sample Methods for FNMR's

We move now to the case where there are three or more estimated process FNMR's to consider. Confidence intervals when comparing more than two FNMR's will be beyond the scope of this text. Linear combinations of FNMRs—$\sum_{g=1}^{G} a_g \pi_g$—can be estimated using contrasts and the interested reader is directed to works on linear models, for example, Rao [78] for further details. We extend the subscript and superscript notation that we used previously for two groups, i.e. with $g = 1, 2$, to the case of G groups, i.e. with $g = 1, \ldots, G$. For comparing multiple FNMR's, we provide resampling approaches. The hypotheses that we are testing here—

$$H_0 : \pi_1 = \pi_2 = \pi_3 = \cdots = \pi_G$$
$$H_1 : \text{ not } H_0.$$

—are usually tested via an analysis of variance-type test. In general, this test takes the form of an F-test. However, because of the extravariation here, we are concerned that using a specific reference distribution, particularly one that is a ratio of two random variables may not be appropriate. Therefore, in this section we will present bootstrap methods and randomization methods for both independent processes and for paired processes.

3.2.4.1 Hypothesis Test for Multiple Independent FNMR's

The tests presented below are for evaluating the performance of three or more independent matching processes. Here the assumption of independent FNMR's means that the individuals involved are distinct for each of the G groups. A single data collection might be divided such that we want to compare different demographics. These differences could be due to age or ethnicity, for example. Or we could be comparing multiple matching processes. For instance, the data could have been collected from the same biometric authentication system in different cities. Whatever the case, in this section we will present methods for comparing the FNMR's of G independent groups. Below we begin with a bootstrap approach and that is followed by a randomization approach.

Bootstrap Approach Since the individuals and decisions are independent between groups, we bootstrap each group separately to mirror the variability in the sampling process. As with an analysis of variance (ANOVA), we use a test statistic similar to the usual F-statistic and then we compare the observed value to the bootstrapped values. Formally, our hypotheses are:

$$H_0 : \pi_1 = \pi_2 = \pi_3 = \cdots = \pi_G$$
$$H_1 : \text{ not } H_0.$$

1. Calculate

$$F = \frac{\left[\sum_{g=1}^{G} N_\pi^{(g)} (\hat{\pi}_g - \bar{\pi})^2\right]/(G-1)}{\left[\sum_{g=1}^{G} N_\pi^{(g)} \hat{\pi}_g (1-\hat{\pi}_g)(1+(m_0^{(g)}-1)\hat{\varrho}_g)\right]/(N-G)} \tag{3.45}$$

   for the observed data where

$$\bar{\pi} = \frac{\sum_{g=1}^{G} N_\pi^{(g)} \hat{\pi}_g}{\sum_{g=1}^{G} N_\pi^{(g)}}, \tag{3.46}$$

$$\hat{\pi}_g = \frac{\sum_{i=1}^{n_g} \sum_{j=1}^{m_i^{(g)}} D_{iij}^{(g)}}{\sum_{i=1}^{n_g} m_i^{(g)}} \tag{3.47}$$

and $N = \sum_{g=1}^{G} N_{\pi}^{(g)}$.

2. For each group g, sample n_g individuals with replacement from the n_g individuals in the gth group. Denote these selected individuals by $b_1^{(g)}, b_2^{(g)}, \ldots, b_{n_g}^{(g)}$. For each selected individual, $b_i^{(g)}$, in the gth group take all the $m_{b_i^{(g)}}$ non-match decisions for that individual. Call these selected decisions $D^{(g)}_{b_i^{(g)} b_i^{(g)} j}$'s with $j = 1, \ldots, m_{b_i^{(g)}}$ and calculate

$$\hat{\pi}_g^b = \frac{\sum_{i=1}^{n_g} \sum_{j=1}^{m_{b_i^{(g)}}} D^{(g)}_{b_i^{(g)} b_i^{(g)} j}}{\sum_{i=1}^{n_g} m_{b_i^{(g)}}} - \hat{\pi}_g + \bar{\pi}. \tag{3.48}$$

3. Repeat the previous two steps some large number of times, M each time calculating and storing

$$F_{\pi} = \frac{[\sum_{g=1}^{G} N_{\pi}^{(g)} (\hat{\pi}_g^b - \bar{\pi}^b)^2]/(G-1)}{[\sum_{g=1}^{G} N_{\pi}^{(g)} \hat{\pi}_g^b (1-\hat{\pi}_g^b)(1+(m_0^{(g)b}-1)\hat{\varrho}_g^b)]/(N-G)}. \tag{3.49}$$

Here $\bar{\pi}^b$ represents the calculations given above applied to the bootstrapped matching decisions,

$$\bar{\pi}^b = \frac{\sum_{g=1}^{G} N_{\pi}^{(g)b} \hat{\pi}_g^b}{\sum_{g=1}^{G} N_{\pi}^{(g)b}}, \tag{3.50}$$

where $N_{\pi}^{(g)b} = \sum_{i=1}^{n_g} m_{b_i^{(g)}}$. The values for $\hat{\varrho}_g^b$ and $m_0^{(g)b}$ are found by using the usual estimates for those quantities applied to the bootstrapped decisions from the gth group.

4. Then the *p-value* for this test is

$$p = \frac{1 + \sum_{\varsigma=1}^{M} I_{\{F_{\pi} \geq F\}}}{M+1}. \tag{3.51}$$

5. We reject the null hypothesis if the *p-value* is small. When a significance level is provided, then we will reject the null hypothesis, H_0, if $p < \alpha$.

As was the case for other bootstrap hypothesis tests, we adjust our bootstrapped sample statistic, here $\hat{\pi}^b$, to center their distributions in accordance with the null hypothesis of equality between the G FNMR's. In this case we center with respect to our estimate of the FNMR, $\bar{\pi}$, if all of the FNMR's are identical.

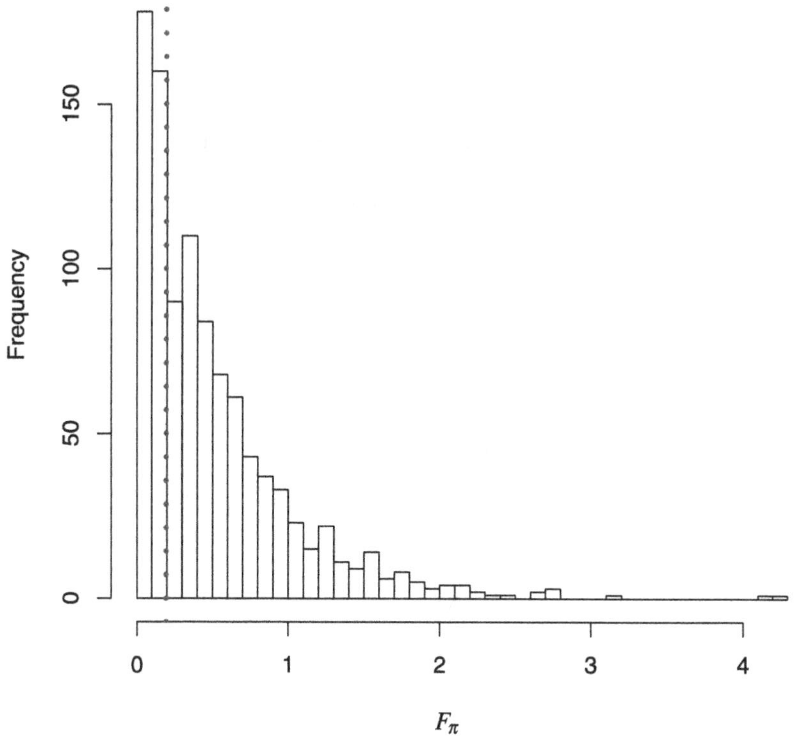

Fig. 3.14 Distribution of F_π for the randomized differences of three facial matcher FNMR's: (DCTs, GMM) from the XM2VTS database, SURREY-AUTO applied to group g1 from the BANCA database, and SURREY-AUTO applied to group g2 from the BANCA database

Table 3.3 Summary of face matcher's FNMR individual performance for Examples 3.19 and 3.20

Matcher	Threshold	$\hat{\pi}$	$\hat{\varrho}$	N_π	m_0
(DCTs, GMM)	0.00	0.0400	0.1319	600	3.0000
SURREY-AUTO (g1)	−0.75	0.0386	0.0186	233	8.9657
SURREY-AUTO (g2)	−1.00	0.0299	0.0428	234	9.0000

Example 3.19 For testing the equality of multiple FNMR's using the above bootstrap approach we will look at three implementations of face matchers. The first is the (DCTs, GMM) matcher from the XM2VTS database. The second and third are implementations of the SURREY_face_svm_auto (SURREY-AUTO) matcher to groups g1 and g2 of the G protocol from the BANCA database. Table 3.3 contains statistics based upon the sample decisions from these matchers. Applying the bootstrap approach above, we get a distribution for F_π which is summarized in Fig. 3.14. Our F is 0.1934, which is represented by the dashed vertical line in that figure. The *p-value* for this test is 0.819 using $M = 999$ bootstrapped replications

of the data. With such a large *p-value* we fail to reject the null hypothesis here of process FNMR equality for these three matchers.

Randomization Approach The approach here is one that permutes individuals among the groups in a manner similar to how we permuted individuals in the two-sample hypothesis test case. The difference here is that our measure of overall difference between the FNMR's is based upon a ratio of variability between to variability within.

$$H_0 : \pi_1 = \pi_2 = \pi_3 = \cdots = \pi_G$$

$$H_1 : \text{ not } H_0.$$

1. Calculate

$$F = \frac{\left[\sum_{g=1}^{G} N_{\pi}^{(g)} (\hat{\pi}_g - \bar{\pi})^2\right] / (G-1)}{\left[\sum_{g=1}^{G} N_{\pi}^{(g)} \hat{\pi}_g (1-\hat{\pi}_g)(1+(m_0^{(g)}-1)\hat{\varrho}_g)\right] / (N-G)} \tag{3.52}$$

   for the observed data where

$$\bar{\pi} = \frac{\sum_{g=1}^{G} N_{\pi}^{(g)} \hat{\pi}_g}{\sum_{g=1}^{G} N_{\pi}^{(g)}}, \tag{3.53}$$

$$\hat{\pi}_g = \frac{\sum_{i=1}^{n_g} \sum_{j=1}^{m_i^{(g)}} D_{iij}^{(g)}}{\sum_{i=1}^{n_g} m_i^{(g)}}, \tag{3.54}$$

   and $N = \sum_{g=1}^{G} N_{\pi}^{(g)}$.
2. Our next step is to conglomerate all of the individuals from each of the G groups into a single list and then, having permuted the list, reassign individuals to groups. Formally, let $\mathscr{I}_g$ be the collection of individuals in the gth group, i.e. $n_g = |\mathscr{I}_g|$ and let $\mathscr{I} = \bigcup_{g=1}^{G} \mathscr{I}_g$ with $n = |\mathscr{I}|$. Sample *without* replacement n_1 individuals from $\mathscr{I}$. Call these individuals $b_1^{(1)}, b_2^{(1)}, \ldots, b_{n_1}^{(1)}$ and call the collection of these individuals $\mathscr{I}_1^b$.
3. Then for the gth group where $2 \leq g \leq G-1$, sample *without* replacement n_g individuals from the set

$$\mathscr{I} \setminus \left\{ \bigcup_{t=1}^{g-1} \mathscr{I}_t^b \right\} \tag{3.55}$$

   and call these individuals $b_1^{(g)}, \ldots, b_{n_g}^{(g)}$ and the collection of them $\mathscr{I}_g^b$.
4. Assign to the Gth group the remaining n_G individuals, the set of individuals defined by

$$\mathscr{I} \setminus \left\{ \bigcup_{t=1}^{G-1} \mathscr{I}_t^b \right\} \tag{3.56}$$

and call those individuals $b_1^{(G)}, \ldots, b_{n_G}^{(G)}$ and their collection $\mathscr{I}_G^b$.

5. For every resampled individual, $b_i^{(g)}$, we take all the decisions from that individual and call those decisions, $D^{(g)}_{b_i^{(g)} b_i^{(g)} j}$ where $j = 1, \ldots, m_{b_i^{(g)}}$.
6. Calculate

$$F_\pi = \frac{\left[\sum_{g=1}^{G} N_\pi^{(g)b} (\hat{\pi}_g^b - \bar{\pi}^b)^2\right] / (G-1)}{\left[\sum_{g=1}^{G} N_\pi^{(g)b} \hat{\pi}_g^b (1 - \hat{\pi}_g^b)(1 + (m_0^{(g)b} - 1)\hat{\varrho}_g^b)\right] / (N - G)} \tag{3.57}$$

where

$$\bar{\pi}^b = \frac{\sum_{g=1}^{G} N_\pi^{(g)b} \hat{\pi}_g^b}{\sum_{g=1}^{G} N_\pi^{(g)b}}, \tag{3.58}$$

$$\hat{\pi}_g^b = \frac{\sum_{i=1}^{n_g} \sum_{j=1}^{m_{b_i^{(g)}}} D^{(g)}_{b_i^{(g)} b_i^{(g)} j}}{\sum_{i=1}^{n_g} m^{b_i^{(g)}}} \tag{3.59}$$

and $N_\pi^{(g)b} = \sum_{i=1}^{n_g} m_{b_i^{(g)}}$.

7. Repeat steps 2 through 6 some large number of times say M. Then our *p-value* for this test is

$$p = \frac{1 + \sum_{\varsigma=1}^{M} I_{\{F_\pi \geq F\}}}{M+1}. \tag{3.60}$$

8. We will reject the null hypothesis, H_0, if the *p-value* is small. If a significance level is being used, then we will reject the null hypothesis, if $p < \alpha$.

Example 3.20 To illustrate the randomization method given above, we repeated the test found in Example 3.19 using a bootstrapped approach which is a test of the equality of three different face matcher implementations. The first of these is the (DCTs, GMM) matcher from the XM2VTS database. For that matcher we used a threshold of 0.00 which gave an estimated FNMR of $\hat{\pi}_1 = 0.0400$. The second implementation we considered was the SURREY_face_svm_auto (SURREY-AUTO) matcher from the BANCA database on group g1 with a threshold of -0.75. This yielded an estimated FNMR of 0.0386. The third implementation was for that same matcher SURREY-AUTO on group g2 of the BANCA database. Since the individuals here are distinct in each data collection, we treat these collections as independent. Table 3.3 has summaries for the sample performance of these FNMR's. Figure 3.15 has the distribution of the F_π's from randomizing the individuals between the $G = 3$ groups. In that graph there is a vertical dashed line at $F = 0.1934$. Using the randomization approach above with $M = 999$ randomized versions of the data we found a *p-value* of 0.8210 which is large. Consequently, we fail to reject H_0 and say that there is not enough evidence to conclude that the process FNMR's for these three implementations are different. We note that the *p-value* we found here is similar to that found in Example 3.19 using a bootstrap approach.

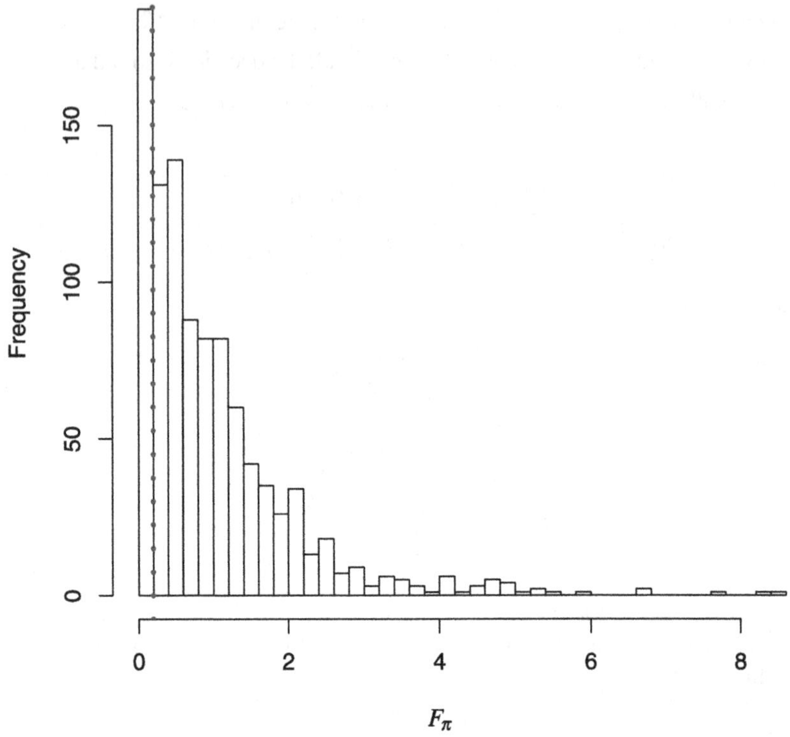

Fig. 3.15 Distribution of F_π, randomized test statistic for comparing three FNMR's—(DCTs, GMM) from the XM2VTS database, SURREY-AUTO on g1 of the BANCA database and SURREY-AUTO on g2 of the BANCA database

There is a large sample alternative to the two resampling tests above that may be worth considering, though we do not present the details here. Crowder [17] presents an analysis of variance for comparing the equality of Beta-binomial proportions using a likelihood approach. In this chapter we have not assumed a Beta-binomial distribution; however, it might be reasonable to do so in specific cases. To assess whether a Beta-binomial distribution is appropriate, one could follow the goodness-of-fit test given by Garren et al. [35]. If a Beta-binomial is appropriate, then the approach of Crowder is worthwhile to consider, since the variance-covariance structure of the non-match decisions is identical to that of a Beta-binomial distribution.

3.2.4.2 Hypothesis Test for Multiple Paired FNMR's

In this section, we present two hypothesis test approaches when we are comparing three or more FNMR's for paired decisions. Our null hypothesis here is that the G groups have equal FNMR's. As with the hypothesis tests for equality of two paired FNMR, we assume that the decisions are paired. One situation where this arises is when we are comparing three or more matching algorithms on the same database of signals. Consequently, we assume that the number of individuals is the same in each of the G groups—$n_g = n$ for all g—and the number of decisions per individual is the same for each individual—$m_i^{(g)} = m_i$ for all i and all g. (Therefore, $N_\pi^{(g)} = N_\pi$ for all g.) These conditions are necessary but not sufficient for the data to be paired. That is, if the number of individuals in each of the groups are not the same or if the number of decisions for each of those individuals differs across groups, then the methods below are not appropriate.

Next we introduce a bootstrap procedure and a randomization procedure for testing the hypothesis of equality of multiple FNMR's. The bootstrap procedure resamples with replacement individuals and then takes all the decisions for the selected individuals and all of the decisions that are paired with those decisions from other processes. The randomization procedure takes all of the decisions for a given individual and permutes the 'paired' decisions across the G groups.

Bootstrap Approach Here, we present a bootstrap method for testing whether there is a significant difference between multiple FNMR's. The hypotheses for this test are

$$H_0 : \pi_1 = \pi_2 = \pi_3 = \cdots = \pi_G$$

$$H_1 : \text{ not } H_0.$$

1. Calculate

$$F = \frac{[\sum_{g=1}^G N_\pi^{(g)}(\hat{\pi}_g - \bar{\pi})^2]/(G-1)}{[\sum_{g=1}^G \hat{\pi}_g(1-\hat{\pi}_g)(1+(m_0-1)\hat{\varrho}_g)]/(N-G)} \tag{3.61}$$

   for the observed data where

$$\bar{\pi} = \frac{\sum_{g=1}^G \hat{\pi}_g}{G}, \tag{3.62}$$

$$\hat{\pi}_g = \frac{\sum_{i=1}^n \sum_{j=1}^{m_i} D_{iij}^{(g)}}{\sum_{i=1}^n m_i} \tag{3.63}$$

   and $N = GN_\pi$.
2. Sample with replacement n individuals from the n individuals from whom data has been collected. Call these new individuals $b_1, \ldots, b_n$.
3. For each bootstrapped individual, b_i, compile all m_{b_i} decisions from the first group, $g = 1$, on that individual. For each of the selected decisions in the first group, we will take the g decisions, one from each group that is paired with a selected decisions. Denote these new decisions by $D_{b_ib_ij}^{(g)}$ where $j = 1, \ldots, m_{b_i}$.
4. Calculate and store

$$F_\pi = \frac{[\sum_{g=1}^G N_\pi(\hat{\pi}_g^b - \bar{\pi}^b)^2]/(G-1)}{[\sum_{g=1}^G \hat{\pi}_g^b(1-\hat{\pi}_g^b)(1+(m_0-1)\hat{\varrho}_g^b)]/(N-G)} \tag{3.64}$$

for the bootstrapped data where

$$\bar{\pi}^b = \frac{\sum_{g=1}^{G} \hat{\pi}_g^b}{G}, \tag{3.65}$$

$$\hat{\pi}_g^b = \frac{\sum_{i=1}^{n} \sum_{j=1}^{m_{b_i}} D_{b_i b_i j}^{(g)}}{\sum_{i=1}^{n} m_{b_i}} - \hat{\pi}_g + \bar{\pi}, \tag{3.66}$$

and $N = GN_\pi$.

5. Repeat steps 2 to 4 some large number of times, say M.
6. Then the *p-value* can be calculated as

$$p = \frac{1 + \sum_{\varsigma=1}^{M} I_{\{F_\pi \geq F\}}}{M+1}. \tag{3.67}$$

7. We will reject H_0, the null hypothesis, if the *p-value* is small or smaller than the significance level, α.

Example 3.21 For this example, we consider three different speaker matchers from the XM2VTS database. Those three matchers are (LFCC, GMM), (PAC, GMM) and (SSC, GMM). We choose thresholds of 2.8176, 2.3810 and 1.2946, respectively, which all give a false match rate (FMR) of 0.0250. See Chap. 4 for more on FMR's. Consequently, we are testing to see if there are differences between the FNMR's for these three speaker matching processes at a given FMR of 2.50%. The estimated FNMR's, $\hat{\pi}_g$'s, for these three groups are 0.0017, 0.1100 and 0.0767, respectively. $N_\pi = 600$ and since $m_i = 3$ for all $n = 200$ individuals, $m_0 = 3$. Figure 3.16 shows the distribution of F_π for these observations. The *p-value* from this test is $p = 0.001$ based upon $F = 17.7855$ with $M = 999$ bootstrap replicates. This *p-value* indicates we can reject the null hypothesis of equality between the three FNMR's and conclude that there are significant differences between these FNMR's for these three matchers.

Example 3.22 In this example, we compare 11 different face matchers from the BANCA database. A summary of those matchers is given in Table 3.4. Each of those matchers is identified by SURREY-SVM-MAN-x.xx which is short for SURREY_face_svm_man_scale_x.xx. That tables contains the estimated FNMR's, the $\hat{\pi}$'s, and the estimated intra-individual correlations, the $\hat{\varrho}$'s along with the particular threshold used in each case. There are $N_\pi = 234$ decision for each matcher and there are $n = 26$ different individuals involved in each match decision. Each individual here has 9 match decisions so that $m_i = 9$ for all i and $m_0 = 9$. F here is 0.2057 and results in a *p-value* of $p = 0.241$ with $M = 999$ replicate bootstraps. Figure 3.17 shows the distribution of the F_π's from this resampling. F in that figure is represented by the dashed vertical line. Since the *p-value* is large, we fail to reject the null hypothesis of equality between the FNMR's of these matchers and we conclude that there is not a significant difference in the FNMR's for this group of matchers at the selected thresholds.

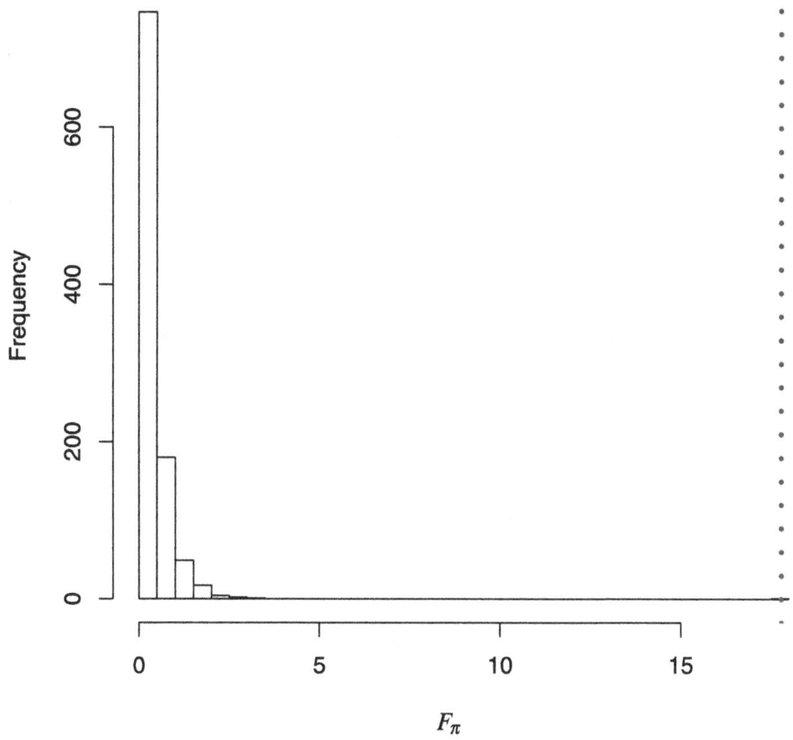

Fig. 3.16 Distribution of F_π for comparing FNMR's from three speaker matchers (LFCC, GMM), (PAC, GMM) and (SSC, GMM) from XM2VTS database

Randomization Approach In general, randomization procedures permute observations among the groups that we want to compare on some relevant statistic. Since we are here trying to assess whether we have equality of the FNMR's among G groups on observations/decisions that are paired, we must deal with the correlation induced by the pairing in our permutation scheme. Here we permute the observations that are paired by the group to which they are assigned. That is, for the G decisions associated with the jth decision from the ith individual, we randomly permute the order of the decisions, then assign the first decision from the permuted list to group 1, the second decision from the permuted list to group 2, etc. The logic here is that if FNMR's are all equal then permuting the observations should not produce test statistics that are drastically different from the value given by the test statistic calculated on the original data. Our hypothesis test here is

$$H_0: \pi_1 = \pi_2 = \pi_3 = \cdots = \pi_G$$
$$H_1: \text{ not } H_0.$$

Table 3.4 Summary of face matcher's FNMR individual performance from g2 of the G protocol from the BANCA database

Matcher	Threshold	$\hat{\pi}$	$\hat{\varrho}$
SURREY-SVM-MAN-0.13	−0.18	0.0470	0.0461
SURREY-SVM-MAN-0.18	−0.50	0.0299	0.0796
SURREY-SVM-MAN-0.25	−0.20	0.0299	0.0428
SURREY-SVM-MAN-0.35	−0.20	0.0299	0.1164
SURREY-SVM-MAN-0.50	0.00	0.0299	0.0428
SURREY-SVM-MAN-0.71	0.00	0.0427	0.0859
SURREY-SVM-MAN-1.00	0.00	0.0470	0.0699
SURREY-SVM-MAN-1.41	0.00	0.0427	0.0859
SURREY-SVM-MAN-2.00	0.00	0.0427	0.0859
SURREY-SVM-MAN-2.83	0.00	0.0427	0.0859
SURREY-SVM-MAN-4.00	0.00	0.0427	0.0859

*Indicates truncated at zero

1. Calculate

$$F = \frac{[\sum_{g=1}^{G} N_{\pi}^{(g)} (\hat{\pi}_g - \bar{\pi})^2]/(G-1)}{[\sum_{g=1}^{G} \hat{\pi}_g (1-\hat{\pi}_g)(1+(m_0-1)\hat{\varrho}_g)]/(N-G)} \tag{3.68}$$

for the observed data where

$$\bar{\pi} = \frac{\sum_{g=1}^{G} \hat{\pi}_g}{G}, \tag{3.69}$$

$$\hat{\pi}_g = \frac{\sum_{i=1}^{n} \sum_{j=1}^{m_i} D_{iij}^{(g)}}{\sum_{i=1}^{n} m_i} \tag{3.70}$$

and $N = GN_{\pi}$.

2. For the ith individual and the jth non-match decision, take the G decisions—$D_{iij}^{(1)}, D_{iij}^{(2)}, \ldots, D_{iij}^{(G)}$—randomly permute the order of those decisions.
3. Take the first non-match decision from the permuted list and assign it to the first group. Denote that decision by $D_{iij}^{(1)*}$. Take the second non-match decision from the permuted list and assign it to the second group. Denote this decisions by $D_{iij}^{(2)*}$. Continue this for the remaining groups $g = 3, \ldots, G$ assigning one of the permuted decision to each group. Repeat this process for all decisions for all individuals.
4. Calculate and store

$$F_{\pi} = \frac{[\sum_{g=1}^{G} N_{\pi}^{(g)} (\hat{\pi}_g^r - \bar{\pi})^2]/(G-1)}{[\sum_{g=1}^{G} \hat{\pi}_g^r (1-\hat{\pi}_g^r)(1+(m_0-1)\hat{\varrho}_g^r)]/(N-G)} \tag{3.71}$$

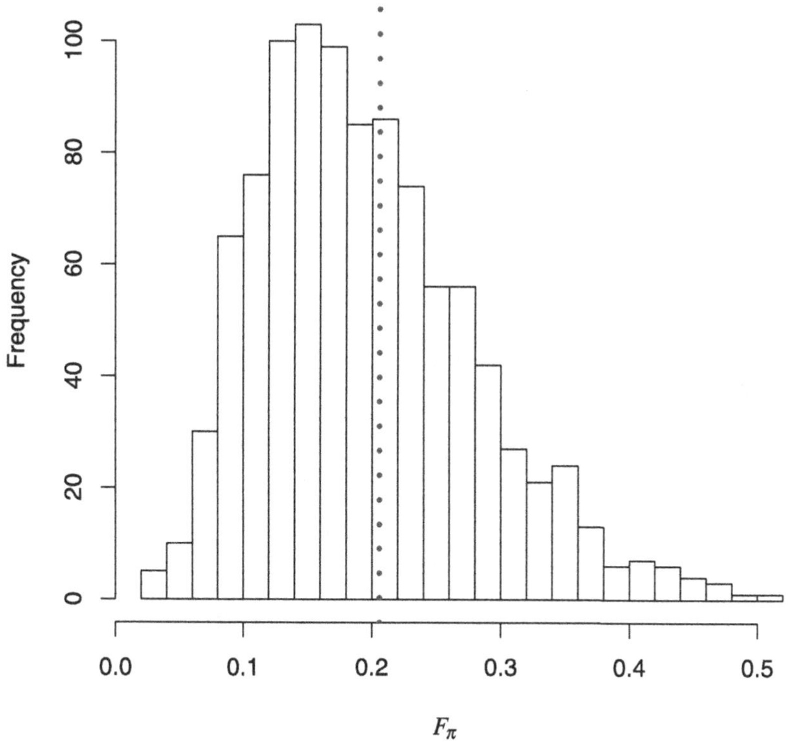

Fig. 3.17 Distribution of F_π for comparing FNMR's from 11 speaker matchers from BANCA database

where

$$\hat{\pi}_g^r = \frac{\sum_{i=1}^n \sum_{j=1}^{m_i} D_{iij}^{(g)*}}{\sum_{i=1}^n m_i} \tag{3.72}$$

and $N = GN_\pi$.

5. Repeat steps 2 and 3 some large number of times, say M. Then the *p-value* can be calculated as

$$p = \frac{1 + \sum_{\varsigma=1}^M I_{\{F_\pi \geq F\}}}{M+1}. \tag{3.73}$$

6. We will reject the null hypothesis, H_0, when the *p-value* is less than a significance level $p < \alpha$. Without a significance level, we reject the null hypothesis if p is small.

Example 3.23 In this example we reanalyze the data from Example 3.22 using the randomization approach rather than the bootstrap approach. Table 3.4 has a summary of the 11 matchers that we are testing here. We are again testing the equality

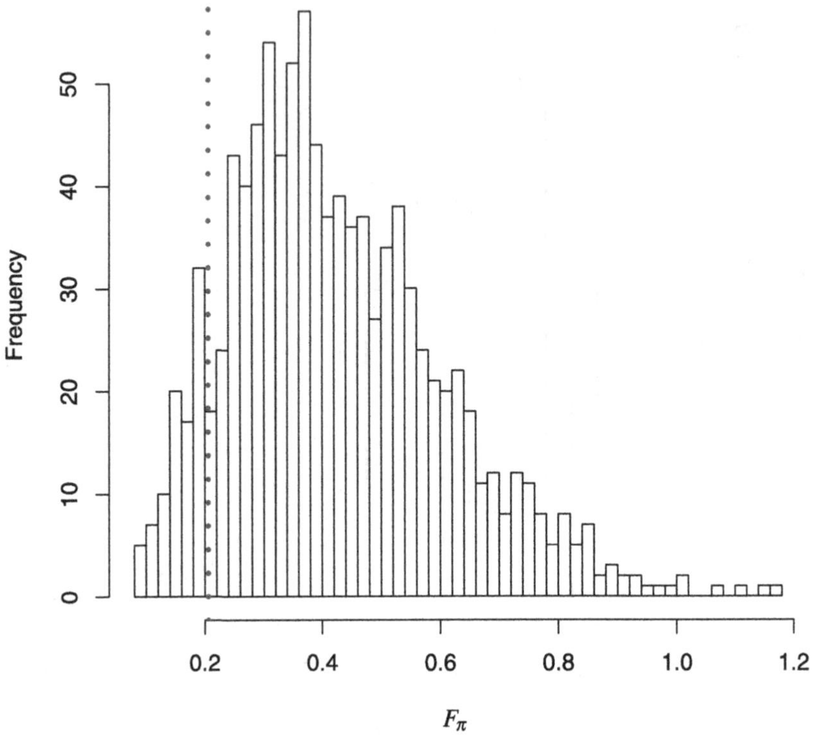

Fig. 3.18 Distribution of F_π for comparing FNMR's from 11 speaker matchers from BANCA database

of the FNMR's for these matchers from the BANCA database. The protocol used here is G and the data are on individuals from group g2. From $M = 999$ randomization replicates we find that the *p-value* is 0.393. Figure 3.18 shows the distribution of F_π here. The dashed vertical line represents $F = 0.2057$. As with the bootstrap case, we fail to reject the null hypothesis that all of the process FNMR's here are the same. Thus, there is not enough evidence to suggest that one or more of these FNMR's differed from the others.

3.3 Sample Size and Power Calculations

Below we derive calculations for determining the number of individuals needed to achieve certain statistical properties. We focus here on the case of a single FNMR rate, π. (Determining sample sizes for other estimands, for example the difference of two independent FNMR's, are not covered in this book.) Several authors have attempted to derive sample size calculations for biometric matching error rates, both FNMR and FMR. Because of misspecified the correlation structures, these methods are not appropriate. Several authors including Mansfield and Wayman [63] point to a *Rule of 30* proposed by Doddington et al. [25]. However, this rule is dependent on an assumption of independence that is not appropriate when $\varrho \neq 0$ which is often the case as many of the examples in this chapter indicate. Schuckers [87] has a discussion of some of the other attempts to develop sample size methodology. We will follows the approach of Schuckers [87] in this section. Below we develop both sample size and power calculations for a single FNMR from a matching process.

3.3.1 Sample Size Calculations

As with other sample size calculations, in order to obtain the desired sample size, we must estimate or make assumptions about some of quantities involved in the process we are considering. We previously discussed this in Sect. 2.3.4. Since our goal is a $100(1-\alpha)\%$ confidence interval for the FNMR, π, we need to specify certain parameters of the matching process. Additionally, there is the complicated manner in which FNMR decisions are collected; there are two factors in the sample sizes. This makes the determination of sample sizes more difficult than the simple case give in Result 2.28. In particular, we have the number of individuals from whom data is collected, n, and the number of decisions made on the ith individual, m_i, where $i = 1, \ldots, n$. To address this problem, Schuckers [87] proposed that we can assume some average number for the m_i's or that the values of the m_i's are the same, i.e. $m_i = m, i = 1, \ldots, n$. We follow the former convention here. As has been pointed out by several authors, including Mansfield and Wayman [63], the amount of information necessary for making a confidence interval of a given width is not a linear function of $\bar{m}$. In particular, the width depends on the value of the intra-individual correlation, ϱ. Recall that the larger the values for ϱ, the less information is gained by increasing the size of the m_i's.

Assuming that $\bar{m}$ is known, we can set the margin of error for our confidence interval for a single process FNMR equal to the desired value, say B. B might take values such as 0.01 or 0.002 which would indicate that we would like our confidence interval to be $\hat{\pi} \pm 0.01$ or $\hat{\pi} \pm 0.002$, respectively.

Then we have

$$B = z_{\alpha/2}\sqrt{\frac{\pi(1-\pi)(1+(m_0-1)\varrho)}{n\bar{m}}}. \tag{3.74}$$

Setting $m_0 \overset{SET}{=} \bar{m}$, we can solve (3.74) for n. We get the equation given below

$$n = \left\lceil \frac{z^2_{\alpha/2}\pi(1-\pi)(1+(\bar{m}-1)\varrho)}{\bar{m}B^2} \right\rceil. \tag{3.75}$$

As a practical matter, we need to specify values for the error rate π, the correlation ϱ, the confidence level $100(1-\alpha)$, the average number of decisions per person $\bar{m}$ and the margin of error B. As mentioned in Comment 2.9, we can utilize information about these quantities using pilot studies or prior similar studies. It is important to recognize the relationship that each of these plays in the calculation of n. The sample size of individuals n in (3.75) is directly related to π and ϱ meaning that as each increases *ceteris paribus*, n increases. In a similar manner α, $\bar{m}$, and B are inversely related to n all other quantities remaining constant. Understanding these relationships we can then utilize them in our estimation. Thus, we can choose to be conservative and make n larger than might be necessary. If we are unsure about any of our estimates we can inflate them—in the case of π and ϱ—or deflate them—in the case of $\bar{m}$. These adjustments can help overcome the fact that we are often dealing with estimates of the process parameters rather than known values.

We note here that α and B are often values that are determined by external factors such as oversight agencies or international standards. Finally, the selection of $\bar{m}$ warrants a brief discussion since it replaced m_0 in (3.74). If there is concern that there will be extensive variability in the number of attempts per person, the m_i's, then it is likely worthwhile to reduce the chosen value for $\bar{m}$ in (3.75) slightly below what it is likely to be.

3.3.2 Power Calculations

Power calculations are sample size calculations for hypothesis tests and they depend on a specified power for the hypothesis test. Recall that the power of a test is the probability that the test rejects the null hypothesis when that null hypothesis is false. Similar to sample size calculations, power calculations derive a given sample size but they depend on the probability of rejecting a null hypothesis, H_0, for a given alternative value of the FNMR, in our case π_a, cf. Definition 2.48. For example, suppose we would like to determine if the FNMR is less than 0.01 and we would like to reject the null hypothesis, $H_0 : \pi = 0.01$, with probability $0.80 = 1 - \beta$ if the error rate is, say, 0.005 at a significance level of $\alpha = 0.05$. Recall that the power in this case is 0.80 or 80% where β is the probability of a Type II error. As we did with the sample size calculation in the previous section, we will focus on determining the number of individuals to be tested. The power calculation given in this section will determine the number of individuals necessary given these parameters.

A power calculation is very similar to a sample size calculation in that it is necessary to specify certain parameters about the process *a priori*. In addition to those parameters specified in the sample size calculation, it is necessary to specify the power, $1 - \beta$, for the hypothesis test under consideration and specify alternative hypothesis that will be the basis for rejection. Again, $1 - \beta$ represents the probability of rejecting the null hypothesis which is our goal.

Recall that for a hypothesis of a single FNMR we have the following hypothesis test.

$$
\begin{aligned}
H_0 &: \pi = \pi_0 \\
H_1 &: \pi < \pi_0.
\end{aligned}
$$

To determine the number of individuals to test, it is necessary to specify a particular value in the alternative hypothesis. It is not possible to calculate power exactly unless an exact value is designated. We will call this value π_a. In the example above $\pi_a = 0.005$. The difference between π_0 and π_a is often known as the 'effect size' in some literature. This quantity $\pi_0 - \pi_a$ takes the place of the margin of error, B, in the sample size calculations. If the sample size is large, then we can assume that the sampling distribution of $\hat{\pi}$ is approximately Gaussian. In order to determine the number of individuals n that is required, we must first specify that

$$P\left(\frac{\hat{\pi} - \pi_0}{s_{\hat{\pi}0}} < -z_\alpha \middle| \pi = \pi_a\right) = 1 - \beta \tag{3.76}$$

where

$$s_{\hat{\pi}0} = \sqrt{\frac{\pi_0(1 - \pi_0)(1 + (m_0 - 1)\hat{\varrho})}{n\bar{m}}}. \tag{3.77}$$

This is the formal statement of our goals. We would like to reject with probability $1 - \beta$ the null hypothesis, H_0, assuming that the alternative hypothesis, $\pi = \pi_a$ is true at a significance level of α. That is, we want the probability that we reject then null hypothesis to be $1 - \beta$ when that null hypothesis is not true and the alternative $\pi = \pi_a$ is true. It then follows that

$$P(\hat{\pi} < \pi_0 - z_\alpha s_{\hat{\pi}0} \mid \pi = \pi_a) = 1 - \beta. \tag{3.78}$$

If we can assume a Gaussian distribution (due to large sample sizes), then we can write that

$$P\left(\frac{\hat{\pi} - \pi_a}{s_{\hat{\pi}a}} < \frac{\pi_0 - z_\alpha s_{\hat{\pi}0} - \pi_a}{s_{\hat{\pi}a}}\right) = 1 - \beta \tag{3.79}$$

where

$$s_{\hat{\pi}a} = \sqrt{\frac{\pi_a(1 - \pi_a)(1 + (m_0 - 1)\hat{\varrho})}{n\bar{m}}}. \tag{3.80}$$

Then the quantity on the left side of the inequality in (3.79) is a standard Gaussian random variable and, thus, we can equate it to the appropriate percentile, z_β, of that distribution. We then have

$$z_\beta = \frac{\pi_0 - z_\alpha s_{\hat{\pi}0} - \pi_a}{s_{\hat{\pi}a}} \tag{3.81}$$

which becomes

$$\frac{1}{\sqrt{n\bar{m}}}\Big(z_\beta\sqrt{\pi_a(1 - \pi_a)(1 + (m_0 - 1)\varrho_a)} + z_\alpha\sqrt{\pi_0(1 - \pi_0)(1 + (m_0 - 1)\varrho_0)}\Big) = \pi_0 - \pi_a. \tag{3.82}$$

We can then square both sides and solve for n which gives

$$n = \left\lceil \frac{(z_\alpha\sqrt{\pi_0(1-\pi_0)(1+(m_0-1)\varrho_0)} + z_\beta\sqrt{\pi_a(1-\pi_a)(1+(m_0-1)\varrho_a)})^2}{\bar{m}(\pi_0-\pi_a)^2} \right\rceil. \tag{3.83}$$

We note that the number of individuals, n, that need to be sampled as given in (3.83) is for testing a single FNMR. More sophisticated methods for determining the number of individuals needed to carry out a particular test given a particular power, $1-\beta$, are possible but not covered in this book.

3.4 Prediction Intervals

Prediction intervals are a tool that is used to create confidence intervals for statistical summaries based upon future observations. The focus of this section will be the development of prediction intervals for a future FNMR based upon decisions that were previously collected to summarize the FNMR of that same biometric matching process. We assume here that the process is sufficiently stable in that we don't expect that there have been changes in the process that would impact the FNMR. That is, the process will remain stationary. We will denote the future as yet unobserved FNMR as $\hat{\pi}^\diamond$. This is not the population FNMR but the value for the FNMR that we will get by sampling new individuals and new decisions. It needs to be reiterated that these intervals, the prediction intervals below, are for inference about $\hat{\pi}^\diamond$, a future sample FNMR, and not for inference about π, the process FNMR. Assuming that the process remains stationary allows us to utilize the estimated variability in $\hat{\pi}$ to estimate the variability in $\hat{\pi}^\diamond$. Though we note that the variability in $\hat{\pi}^\diamond$ is inherently larger because it must include the sampling variability in $\hat{\pi}$.

As above, we will assume that we have observed decisions on n individuals with m_i decisions for the ith individual. Let $n^\diamond$ be the number of individuals from whom decisions will be collected and let $m_i^\diamond$ be the number of decisions to be collected from the ith individual. It is common that these numbers will be extremely difficult to know *a priori* in practice and, therefore, we will let $\bar{m}^\diamond$ be the average number of decisions per individual, a quantity that is likely to be easier to specify *a priori*. Then we can write

$$\hat{\pi}^\diamond = \frac{\sum_{i=1}^{n^\diamond}\sum_{j=1}^{m_i^\diamond} D_{iij}^\diamond}{n^\diamond\bar{m}^\diamond} \tag{3.84}$$

where $D_{iij}^\diamond$ are the future decisions. Finally we let $m_0^\diamond$ be the value for m_0 based upon the as yet unobserved decisions, the $D_{iij}^\diamond$'s. Then the variance of $\hat{\pi}^\diamond$ is given by

$$V[\hat{\pi}^\diamond] = \hat{\pi}(1-\hat{\pi})\left[\frac{(1+(m_0-1)\hat{\varrho})}{n\bar{m}} + \frac{(1+(m_0^\diamond-1)\hat{\varrho})}{n^\diamond\bar{m}^\diamond}\right]. \tag{3.85}$$

In practice, it will be difficult to specify $m_0^\diamond$ accurately given the difficulties of predicting data collection and the uncertainties of test subject attrition. Therefore, we suggest the use of

$$m_0^\diamond = \bar{m}^\diamond \left[\frac{m_0}{\bar{m}} \right]. \tag{3.86}$$

That is, we use the anticipated mean number of decisions per person, $\bar{m}^\diamond$ for the future data collection but adjust that by the ratio of loss in the original data collection. This is a reasonable approach that is likely to give a good approximation of the type of variability in the $m_i^\diamond$'s that can be expected.

Given the variability in $\hat{\pi}^\diamond$ from (3.85), and assuming large sample size, the sampling distribution of $\hat{\pi}^\diamond$ will be approximately Gaussian. Thus we can use the following to make a $(1-\alpha)100\%$ prediction interval for $\hat{\pi}^\diamond$

$$\hat{\pi} \pm z_{\alpha/2} \sqrt{\hat{\pi}(1-\hat{\pi}) \left[\frac{(1+(m_0-1)\hat{\varrho})}{n\bar{m}} + \frac{(1+(m_0^\diamond-1)\hat{\varrho})}{n^\diamond \bar{m}^\diamond} \right]}. \tag{3.87}$$

Example 3.24 For this example, we revisit an earlier example for creation of a confidence interval for a process FNMR, Example 3.1. There we analyzed the face matcher (FH, MLP) from the XM2VTS database with a threshold of 0.0. Here we will create a 95% prediction interval for the FNMR of this process for the next $n^\diamond = 100$ individuals assuming that each of these future individuals has $\bar{m} = m_0^\diamond = m_i^\diamond = 3$ matching decisions. Recall that from the original sample there were $n = 200$ individuals and $m_i = 3$ for all of individuals $i = 1, \dots, 200$ and that the sample FNMR $\hat{\pi} = 0.0350$. The other relevant summary that we need for this procedure is $\hat{\varrho} = 0.3092$. Using that information we can calculate a 95% prediction interval for $\hat{\pi}^\diamond$ to be $(0.0026, 0.0674)$. Thus we are 95% confident that the process FNMR for the next 100 individuals will be between 0.60% and 6.74% assuming that we have three genuine matching decisions on each of those individuals. We note here that the prediction interval is, indeed, wider than the confidence interval made as part of Example 3.1 for the process FNMR, π, of $(0.0163, 0.0537)$ by about 1.4% on each side of the interval.

3.5 Discussion

Above we have presented a variety of statistical methods for false non-match rates. These techniques are for estimation and comparison of FNMR's. Both parametric methods based upon large sample theory and non-parametric methods based upon the subsets bootstrap or randomization were utilized. All of these methods are meant to provide a foundation for the numerous circumstances in which practitioners might want to evaluate the FNMR as a measure of matching system performance. Many of these methods, especially those for comparing two or more FNMR's are *new* to the bioauthentication literature. While we have provided numerous methods, these methods are not meant to be exhaustive. Instead, the intent is to proffer a general fundamental structure that can be extended and modified.

The basis for all of these methods has been a false non-match decision. Since the false non-match rate is about the rate at which a matching system makes incorrect decisions, the building block of estimation for a FNMR is a decision. Each FNMR is dependent upon a threshold for matching for determining the appropriate decision. Specific testing methodologies may allow more than one transaction before a final matching decision is rendered.

Finally, two recommendations are warranted concerning the methodology here. First, we generally prefer confidence intervals over hypothesis tests since they provide more informative output about the quantities that we are estimating. We recognize that in some circumstance such as qualified products testing that a hypothesis test may be appropriate since the goal of such testing is to determine if an FNMR is significantly below a specified value. Additionally, when we are comparing three or more FNMR's, a hypothesis test is required. Second, when planning a bootstrap or randomization analysis the number of bootstrap replicates ought to be large, say $M \geq 999$. We choose 999 rather than 1000 as our minimum, since for hypothesis testing using $M + 1 = 1000$ is an attractive choice.

Chapter 4
False Match Rate

The false match rate (FMR) is the rate at which a biometric process mismatches biometric signals from two distinct individuals as coming from the same individual. Statistical methods for that rate are the focus of this chapter. We will denote a process FMR by the Greek letter nu, ν. By a process here we are referring to the ongoing matching process from which we have taken a sample of biometric signals for matching. Biometrics differentiates between the false match rate and the false accept rate, cf. Mansfield and Wayman [63] for additional details. The latter is a measure of the entire system while the former is a measure of the matching component of such a system. Outside of bioauthentication, the methods presented here are generally applicable to the false accept rate. As was mentioned in Chap. 1, the problem of biometric matching (and more generally forensic matching) is different from other types of classification problems since it is inherently a *two-instance* problem. By two-instance we mean that we are not taking a single image and trying to classify that image into one of G groups, but rather, we are taking two instances and determining if those two instances belong to the same group. Thus two-instance matching is inherently different from other classification problems and it requires new statistical methodologies for evaluating the performance of this type of matching. In this chapter, we introduce such methodologies for the FMR. Those methods include approaches for the estimation of a single FMR and for the comparison of multiple FMR's.

Additionally, the type of matching—either identification or verification—also impacts the choice of appropriate statistical methodology. As in Chap. 3 on false non-match rates, we also differentiate between whether the data collection process is independent or paired. As in other chapters, we include both large sample and resampling approaches to these problems. *None* of the statistical approaches and algorithms in this chapter depend upon a distributional assumption for the matching scores, the $Y_{ik\ell}$'s. As before, identification is meant to refer to one to many or 1:N matching. Verification is for one to one or 1:1 matching. The focus of this chapter will be on statistical methods for verification false match rate or vFMR. For the rest of the chapter, we will simply refer to this as FMR. In Sect. 4.6, we discuss methods for identification FMR. Methods appropriate for identification FMR are

M.E. Schuckers, *Computational Methods in Biometric Authentication*, Information Science and Statistics,
DOI 10.1007/978-1-84996-202-5_4,

similar to those for FNMR which are found in Chap. 3. Because identification FMR does not depend on the second individual in the matching but rather is calculated across all individuals enrolled in the system, then it is not appropriate to model the correlation structure for identification FMR in the same way that we model the correlation structure for verification FMR. Verification FMR, unlike identification FMR, does depend on knowledge of both individuals for a given matching decision.

This chapter is structured in the following way. We begin with an introduction to false match rates and the notation that we'll use throughout this chapter. That is followed by a section on the correlation structure for the two-instance false match rate. In that section, we also discuss estimation of parameters in the general correlation structure as well as some simplifications of that general correlation structure. Section 4.2 contains a description of the *two-instance bootstrap* for estimation on an FMR. We then turn to statistical methods for a single FMR as well as for multiple FMR's. Large sample as well as bootstrap and randomization approaches to confidence intervals and hypothesis tests are given. This is followed by a section on sample size and power calculations for an FMR. Prediction intervals for the FMR are the focus on the next section. Lastly, we provide a brief discussion of the statistical methods for the FMR in this chapters.

4.1 Notation and General Correlation Structure

The basic unit of measurement for a false match rate is a false match decision. We will let $D_{ik\ell}$ represent the ℓth decision from a comparison of biometric signals from individuals i and k, $\ell = 1, \ldots, m_{ik}$, $i = 1, \ldots, n_{\mathbb{P}}$, $k = 1, \ldots, n_{\mathbb{G}}$ where $n_{\mathbb{P}}$ is the number of individuals in the probe and $n_{\mathbb{G}}$ is the number of individuals in the gallery. Match decision are typically the result of comparing a match score to a threshold. See Table 4.1 for an example of the structure of this false match decision data. Since we are dealing with false matches, we will assume that the case where $i = k$ is not present for the decisions we are considering. We note that it is possible that the probe and the gallery are not mutually exclusive. Recall that the probe is the compilation of signals which is compared against the signals in the gallery. Also, let $m_{ik} \geq 0$ represent the number of comparisons on the pair of individuals (i, k) where the order of the pair matters, i.e. that m_{ik} is not necessary equal to m_{ki}. It is possible for $m_{ik} = 0$ when no decisions for a particular ordered comparison pair are observed. It might be more accurate to use the notation $D_{ij,kw}$ which would represent the decision from comparing the jth capture from the ith individual and the wth capture from the kth individual. For notational simplicity, we have chosen to assume that the comparisons are well ordered from $\ell = 1, \ldots, m_{ik}$ and so we will use $D_{ik\ell}$ instead.

Since not all matchers are symmetric, we need to consider both the comparison pair (i, k) and the comparison pair (k, i) distinctly. This allows for possible asymmetry in the decisions; thus, $D_{ik\ell}$ is not guaranteed to equal $D_{ki\ell}$. Consequently,

Table 4.1 Example of verification FMR data

i	k	ℓ	$D_{ik\ell}$
1	2	1	0
1	2	2	1
1	2	3	1
1	2	4	0
1	3	1	0
1	3	2	0
1	3	3	0
1	4	1	0
1	4	2	1
1	5	1	0
1	5	2	0
1	5	3	1
1	5	4	0
1	5	5	0
2	1	1	0
2	1	2	0
2	1	3	0
2	1	4	0
2	1	5	0
2	1	6	0
2	3	1	0
2	3	2	0
2	3	3	0
2	4	1	0
2	4	2	0
2	5	1	0
3	1	1	1
3	1	2	0
3	2	1	1
3	2	2	0
3	2	3	0
3	2	4	0
3	5	1	1
3	5	2	0

i	k	ℓ	$D_{ik\ell}$
4	1	1	1
4	1	2	0
4	2	1	1
4	2	2	0
4	2	3	0
4	2	4	0
4	5	1	0
4	5	2	0
5	1	1	1
5	1	2	0
5	1	3	0
5	2	1	1
5	2	2	0
5	3	1	1
5	3	2	1
5	4	1	0
5	4	2	0
5	4	3	0
5	4	4	0
5	4	5	0

when we compare the ℓth decision from the comparison pair (i, k) to the ℓth decision from the pair (k, i) the resulting decision may be different. As noted in the previous paragraph, we assume that the data is well ordered in the sense that it is possible to assume that the order of the signals for a given pair of individuals (i, k)

is the same for the pair (k,i). Then, we define

$$D_{ik\ell} = \begin{cases} 1 & \text{if } \ell\text{th pair of captures from individual } i \\ & \text{and individual } k,\, i \neq k, \text{ is declared a match} \\ 0 & \text{otherwise.} \end{cases} \tag{4.1}$$

Let $E[D_{ik\ell}] = \nu$ and $V[D_{ik\ell}] = \nu(1-\nu)$ where the mean error rate or FMR is assumed to be ν. We estimate this quantity through

$$\hat{\nu} = N_\nu^{-1}\mathbf{1}^T\mathbf{D}_\nu = \frac{\sum_{i=1}^{n_{\mathbb{P}}}\sum_{k=1}^{n_{\mathbb{G}}}\sum_{\ell=1}^{m_{ik}} D_{ik\ell}}{\sum_{i=1}^{n_{\mathbb{P}}}\sum_{k=1}^{n_{\mathbb{G}}} m_{ik}} \tag{4.2}$$

where $\mathbf{1} = (1,1,\ldots,1)^T$ and

$$N_\nu = \sum_{i=1}^{n_{\mathbb{P}}}\sum_{k=1}^{n_{\mathbb{G}}} m_{ik}. \tag{4.3}$$

N_ν is the total number of decisions that go into the calculation of a given FMR. Further, let $\mathbf{D}_{ik} = (D_{ik1},\ldots,D_{ikm_{ik}})^T$ and $\mathbf{D}_\nu = (\mathbf{D}_{12}^T, \mathbf{D}_{13}^T, \ldots, \mathbf{D}_{1n}^T, \mathbf{D}_{21}^T, \mathbf{D}_{23}^T, \mathbf{D}_{24}^T, \ldots, \mathbf{D}_{2n}^T, \ldots, \mathbf{D}_{n1}^T, \mathbf{D}_{n2}^T, \ldots, \mathbf{D}_{nn-1}^T)^T$. We write the estimated FMR, $\hat{\nu}$, in (4.2) in vector notation, $N_\nu^{-1}\mathbf{1}^T\mathbf{D}_\nu$, to emphasize that it is a linear combination that fits into the framework of statistical methods for linear combinations. We will use this framework that was discussed in Chap. 2 to derive statistical methods for estimation and testing of FMR's.

We next focus on correlation models for estimation of a single stationary FMR in verification mode. As with the FNMR correlation structures in Chap. 3, these models are binary in nature because all biometric authentication decisions result in either a match or a non-match. We are concerned here with the correlation between a decision made comparing instances from two individuals and another decision made from comparing instances on two other individuals. If one or more individuals appear in both decisions, then we must account for that with our correlation structure. Thus, the amount and type of overlap in individuals from the decisions being compared will be crucial to this structure. We aim to model the correlation between inter-individual decisions. These inter-individual decisions are based on dichotomizing match scores from the imposter distribution. We follow Mansfield and Wayman [63] in using the term *imposter distribution* to refer to match scores from comparing signals from two different individuals. Below we differentiate between matching comparisons that are symmetric and those that are asymmetric. Symmetric matching algorithms are those for which the order of the individual captures does not matter and asymmetric ones are those for which the order does matter. Thus, for asymmetric matchers comparing capture 1 against capture 2 may result in a different decision than comparing capture 2 against capture 1. The decisions resulting from a symmetric matching algorithm would always be the same regardless of the order of the captures. Below, we propose a general correlation structure for the asymmetric case and show that the correlation structure for a symmetric matcher is a special case of the proposed structure.

The general correlation structure that we will use in estimation of verification FMR is one that was previously proposed by Schuckers [87]. That structure allows for the matching algorithm to be either symmetric—$D_{ik\ell} = D_{ki\ell}$—or asymmetric $D_{ik\ell}$ not necessarily equal to $D_{ki\ell}$. In the former case, the order of the individuals does not matter, while for the latter the order of the individuals does matter. See Bistarelli et al. [8] for an example of an asymmetric matching algorithm. The overall correlation structure is not dependent on the threshold from the matching process; however, the estimates for the correlation parameters are indeed threshold *dependent*. See Schuckers [87] for examples of this dependency. For a given FMR, we will suppress that dependence for notational simplicity.

The general correlation structure that was proposed by Schuckers [87] is the following:

$$Corr(D_{ik\ell}, D_{i'k'\ell'}) = \begin{cases} 1 & \text{if } i=i', k=k', \ell=\ell' \\ \eta & \text{if } i=i', k=k', \ell\neq\ell' \\ \omega_1 & \text{if } i=i', k\neq k', i\neq k, i\neq k' \\ \omega_2 & \text{if } i\neq i', k=k', i\neq k, i\neq k' \\ \omega_3 & \text{if } i=k', i'\neq k, i\neq i', i\neq k \\ \omega_3 & \text{if } i'=k, i\neq k', i'\neq i, k\neq k' \\ \xi_1 & \text{if } i=k', k=i', i\neq i', k\neq k', \ell=\ell' \\ \xi_2 & \text{if } i=k', k=i', i\neq i', k\neq k', \ell\neq\ell' \\ 0 & \text{otherwise.} \end{cases} \tag{4.4}$$

Since we are dealing with false match decisions, then $i \neq k$ for all i and k. This correlation structure is necessarily more complicated than the FNMR equivalent presented in (3.3) due to the additional individuals involved in the match decisions for calculating an FMR. Below we briefly summarize the role of each of the parameters in the model in (4.4). More details can be found in Schuckers [87]. The η here is the correlation between decisions when the same individuals appear in the same order but the captures considered are different. This is an intra-comparison pair correlation that is similar to the parameter ϱ found in the correlations of FNMR match decisions, cf. (3.3). The ω's here are the correlations between two match decisions when those match decisions share one and only one individual. ω_1 represents the case when the first individual in each decision is the same, ω_2 represents the case when the second individual is the same in each decision and ω_3 represents the case when an individual is shared between the decisions in either the 'inside' or the 'outside' positions. Recall that $Corr(D_{ik\ell}, D_{i',k',\ell'}) = Corr(D_{i',k',\ell'}, D_{ik\ell})$. The correlation parameters denoted by the ξ's serve the case where the individuals in each decision are the same but the order of those individuals to the matching algorithm is reversed. For that case the individuals involved in the two decisions must be the same but the order of those individuals must be reversed. ξ_1 denotes the case where the biometric captures involved are the same; while ξ_2 represents the case where the biometric captures differ. Lastly, we assume that if the individuals involved in the decisions are all distinct then the correlation, conditional on the FMR, is zero.

That is, for a stationary matching process we will conclude that there is no correlation between two match decisions if all four of the individuals involved in those decisions are distinct. In general, we will assume, as Schuckers [87] did, that all of the correlation parameters here are non-negative. A negative correlation for the binary decisions described above would imply that having individuals in common would make decisions less likely to be the same than those involving four distinct individuals. This seems like an unlikely result in biometrics. Additionally, it seems appropriate to note in a book on statistical methods that it is possible, indeed probable, when a particular correlation term is zero for a process that the estimate of that parameter will be non-zero for any given sample. We must be aware of this as we assess correlation parameter estimates.

Using the structure in (4.4), we calculate the variance of a linear combination to produce the following variance of the estimated FMR, $V[\hat{\nu}]$.

$$\begin{aligned}
V[\hat{\nu}] &= V[N_\nu^{-1}\mathbf{1}^T\mathbf{D}] = N_\nu^{-2}V[\mathbf{1}^T\mathbf{D}_\nu] = N_\nu^{-2}\mathbf{1}^T\boldsymbol{\Sigma}_\nu\mathbf{1} = N_\nu^{-2}\nu(1-\nu)\mathbf{1}^T\boldsymbol{\Phi}_\nu\mathbf{1} \\
&= N_\nu^{-2}\nu(1-\nu)\Bigg[N_\nu + \eta\sum_{i=1}^{n_{\mathbb{P}}}\sum_{\substack{k=1\\k\neq i}}^{n_{\mathbb{G}}} m_{ik}(m_{ik}-1) \\
&\quad + \omega_1\sum_{i=1}^{n_{\mathbb{P}}}\sum_{\substack{k=1\\k\neq i}}^{n_{\mathbb{G}}} m_{ik}\Bigg(\sum_{\substack{k'=1\\k'\neq i,k'\neq k}}^{n_{\mathbb{G}}} m_{ik'}\Bigg) + \omega_2\sum_{i=1}^{n_{\mathbb{P}}}\sum_{\substack{k=1\\k\neq i}}^{n_{\mathbb{G}}} m_{ik}\Bigg(\sum_{\substack{i'=1\\i'\neq i,i'\neq k}}^{n_{\mathbb{P}}} m_{i'k}\Bigg) \\
&\quad + \omega_3\sum_{i=1}^{n_{\mathbb{P}}}\sum_{\substack{k=1\\k\neq i}}^{n_{\mathbb{G}}} m_{ik}\Bigg(\sum_{\substack{i'=1\\i'\neq i,i'\neq k}}^{n_{\mathbb{P}}} m_{i'i} + \sum_{\substack{k'=1\\k'\neq i,k'\neq k}}^{n_{\mathbb{G}}} m_{kk'}\Bigg) \\
&\quad + \xi_1\sum_{i=1}^{n_{\mathbb{P}}}\sum_{\substack{k=1\\k\neq i}}^{n_{\mathbb{G}}} m_{ki} + \xi_2\sum_{i=1}^{n_{\mathbb{P}}}\sum_{\substack{k=1\\k\neq i}}^{n_{\mathbb{G}}} m_{ki}(m_{ki}-1)\Bigg]
\end{aligned} \tag{4.5}$$

where

$$N_\nu = \sum_{i=1}^{n_{\mathbb{P}}}\sum_{k=1}^{n_{\mathbb{G}}} m_{ik}, \tag{4.6}$$

$\mathbf{D}_\nu = (D_{121}, \ldots, D_{12m_{12}}, D_{131}, \ldots, D_{13m_{13}}, \ldots, D_{n_{\mathbb{P}}n_{\mathbb{G}}1}, \ldots, D_{n_{\mathbb{P}}n_{\mathbb{G}}m_{n_{\mathbb{P}}n_{\mathbb{G}}}})^T$, $\Sigma_\nu = V[\mathbf{D}_\nu]$ and $\Phi_\nu = Corr(\mathbf{D}_\nu)$. N_ν is the total number of match decisions for a particular sample. Here we have let the covariance matrix be Σ_ν and we let the correlation matrix be Φ_ν. We use the matrix and vector notation to emphasize that the estimated FMR is a linear combination and; hence, statistical methods for linear combinations apply.

We next derive parameter estimators based upon the setting the variance of $\mathbf{D}$, $(\mathbf{D}_\nu - \hat{\nu}\mathbf{1})(\mathbf{D}_\nu - \hat{\nu}\mathbf{1})^T$, equal to the model variance, Σ_ν, whose correlation structure is defined in (4.4) and solving for the correlation parameters. We can estimate the variance of the sample FMR by substituting estimates of the parameters into (4.5).

This gives us $\hat{V}[\hat{\nu}]$, the estimated variance of a sample FMR. Our approach is equivalent to averaging the individual components of the sample correlation matrix that correspond to each correlation parameter. The estimator for η is then

$$\hat{\eta} = \left(\hat{\nu}(1-\hat{\nu}) \sum_{i=1}^{n_{\mathbb{P}}} \sum_{\substack{k=1 \\ k \neq i}}^{n_{\mathbb{G}}} m_{ik}(m_{ik}-1) \right)^{-1} \times \sum_{i=1}^{n_{\mathbb{P}}} \sum_{\substack{k=1 \\ k \neq i}}^{n_{\mathbb{G}}} \sum_{\ell=1}^{m_{ik}} \sum_{\substack{\ell'=1 \\ \ell' \neq \ell}}^{m_{ik}} (D_{ik\ell}-\hat{\nu})(D_{ik\ell'}-\hat{\nu}). \tag{4.7}$$

Note the similarities in this estimator to the estimator for ϱ in Chap. 3. The other estimators for the correlation parameters are

$$\hat{\omega}_1 = \left(\hat{\nu}(1-\hat{\nu}) \sum_{i=1}^{n_{\mathbb{P}}} \sum_{\substack{k=1 \\ k \neq i}}^{n_{\mathbb{G}}} m_{ik} \left(\sum_{\substack{k'=1 \\ k' \neq i, k' \neq k}}^{n_{\mathbb{G}}} m_{ik'} \right) \right)^{-1} \times \left(\sum_{i=1}^{n_{\mathbb{P}}} \sum_{\substack{k=1 \\ k \neq i}}^{n_{\mathbb{G}}} \sum_{\substack{k'=1 \\ k' \neq i, k' \neq k}}^{n_{\mathbb{G}}} \sum_{\ell=1}^{m_{ik}} \sum_{\ell'=1}^{m_{ik'}} (D_{ik\ell}-\hat{\nu})(D_{ik'\ell'}-\hat{\nu}) \right), \tag{4.8}$$

$$\hat{\omega}_2 = \left(\hat{\nu}(1-\hat{\nu}) \sum_{i=1}^{n_{\mathbb{P}}} \sum_{\substack{k=1 \\ k \neq i}}^{n_{\mathbb{G}}} m_{ik} \left(\sum_{\substack{i'=1 \\ i' \neq i, i' \neq k}}^{n_{\mathbb{P}}} m_{i'k} \right) \right)^{-1} \times \left(\sum_{i=1}^{n_{\mathbb{P}}} \sum_{\substack{k=1 \\ k \neq i}}^{n_{\mathbb{G}}} \sum_{\substack{i'=1 \\ i' \neq i, i' \neq k}}^{n_{\mathbb{P}}} \sum_{\ell=1}^{m_{ik}} \sum_{\ell'=1}^{m_{i'k}} (D_{ik\ell}-\hat{\nu})(D_{i'k\ell'}-\hat{\nu}) \right), \tag{4.9}$$

$$\hat{\omega}_3 = \left(\hat{\nu}(1-\hat{\nu}) \sum_{i=1}^{n_{\mathbb{P}}} \sum_{\substack{k=1 \\ k \neq i}}^{n_{\mathbb{G}}} m_{ik} \left(\sum_{\substack{i'=1 \\ i' \neq i, i' \neq k}}^{n_{\mathbb{P}}} m_{i'i} + \sum_{\substack{k'=1 \\ k' \neq k, k' \neq i}}^{n_{\mathbb{G}}} m_{kk'} \right) \right)^{-1} \times \left(\sum_{i=1}^{n_{\mathbb{P}}} \sum_{\substack{k=1 \\ k \neq i}}^{n_{\mathbb{G}}} \sum_{\substack{i'=1 \\ i' \neq i, i' \neq k}}^{n_{\mathbb{P}}} \sum_{\ell=1}^{m_{ik}} \sum_{\ell'=1}^{m_{i'i}} (D_{ik\ell}-\hat{\nu})(D_{i'i\ell'}-\hat{\nu}) + \sum_{i=1}^{n_{\mathbb{P}}} \sum_{\substack{k=1 \\ k \neq i}}^{n_{\mathbb{G}}} \sum_{\substack{k'=1 \\ k' \neq i, k' \neq k}}^{n_{\mathbb{G}}} \sum_{\ell=1}^{m_{ik}} \sum_{\ell'=1}^{m_{kk'}} (D_{ik\ell}-\hat{\nu})(D_{kk'\ell'}-\hat{\nu}) \right), \tag{4.10}$$

$$\hat{\xi}_1 = \left(\hat{\nu}(1-\hat{\nu})\sum_{i=1}^{n_{\mathbb{P}}}\sum_{\substack{k=1\\k\neq i}}^{n_{\mathbb{G}}} \min(m_{ik}, m_{ki})\right)^{-1}$$

$$\times \sum_{i=1}^{n_{\mathbb{P}}}\sum_{\substack{k=1\\k\neq i}}^{n_{\mathbb{G}}}\sum_{\ell=1}^{m_{ki}} (D_{ik\ell}-\hat{\nu})(D_{ki\ell}-\hat{\nu}), \tag{4.11}$$

and

$$\hat{\xi}_2 = \left(\hat{\nu}(1-\hat{\nu})\sum_{i=1}^{n_{\mathbb{P}}}\sum_{\substack{k=1\\k\neq i}}^{n_{\mathbb{G}}} m_{ik}m_{ki} - \min(m_{ik}, m_{ki})\right)^{-1}$$

$$\times \sum_{i=1}^{n_{\mathbb{P}}}\sum_{\substack{k=1\\k\neq i}}^{n_{\mathbb{G}}}\sum_{\ell=1}^{m_{ki}}\sum_{\substack{\ell'=1\\\ell'\neq\ell}}^{m_{ki}} (D_{ik\ell}-\hat{\nu})(D_{ki\ell'}-\hat{\nu}). \tag{4.12}$$

Not all data collections will allow for estimation for all parameters since certain combinations of individuals may not be present. In those cases there will be no need to estimate those parameters since they will not appear in the sample variance matrix.

4.1.1 Alternative Correlation Structures

As mentioned above, the correlation structure for the FMR is necessarily more complicated than the structure for the FNMR. It is possible to simplify the FMR correlation structure found in (4.4) in several ways. Below we provide some possible approaches to this simplification. One straightforward simplification occurs when the matcher is symmetric. As a consequence, $\xi_1 = 1$ and $\xi_2 = \eta$. This outcome is because the symmetric case yields identical results regardless of the order of the captures. Consequently, with a symmetric matcher the correlation ξ_1 is 1 and the correlation ξ_2 is the same as η. This can be written as

$$Corr(D_{ik\ell}, D_{i'k'\ell'}) = \begin{cases} 1 & \text{if } i=i', k=k', \ell=\ell', \\ 1 & \text{if } i=k', i'=k, \ell=\ell', \\ \eta & \text{if } i=i', k=k', \ell\neq\ell', \\ \eta & \text{if } i=k', i'=k, \ell\neq\ell', \\ \omega_1 & \text{if } i=i', k\neq k', i\neq k', \\ \omega_2 & \text{if } k=k', i\neq i', k\neq k', \\ \omega_3 & \text{if } i'=k, i\neq k', i\neq i', k\neq k' \\ \omega_3 & \text{if } i=k', i'\neq k, i\neq i', k\neq k' \\ 0 & \text{otherwise.} \end{cases} \tag{4.13}$$

The variance and the estimators for the parameters in this model would follow the same approach that we took above for the general correlation structure, though we have different estimators for each parameter. More on this model is found in an appendix to this chapter in Sect. 4.7.1.

The next model we present we refer to as the *equal cross correlation model* since we let $\omega = \omega_1 = \omega_2 = \omega_3$. The idea behind this model is to suggest that the correlation between decisions is the same when two and *only* two of the four individuals involved in the decisions are the same. This correlation structure is

$$Corr(D_{ik\ell}, D_{i'k'\ell'}) = \begin{cases} 1 & \text{if } i = i', k = k', \ell = \ell' \\ \eta & \text{if } i = i', k = k', \ell \neq \ell' \\ \omega & \text{if } i = i', k \neq k', i \neq k, i \neq k' \\ \omega & \text{if } i \neq i', k = k', k \neq i, k \neq i' \\ \omega & \text{if } i = k', i' \neq k, i \neq i', i \neq k \\ \omega & \text{if } i' = k, i \neq k', i' \neq i, i' \neq k' \\ \xi_1 & \text{if } i = k', k = i', i \neq i', k \neq k', \ell = \ell' \\ \xi_2 & \text{if } i = k', k = i', i \neq i', k \neq k', \ell \neq \ell' \\ 0 & \text{otherwise.} \end{cases} \quad (4.14)$$

We can use the same approach to estimation that was used above. The variance and the estimators for the parameters in this model can be found in Sect. 4.7.2 as (4.96) and (4.97). The estimator given in (4.7) for η remains the same in this model.

Other simplified models—for example, $\omega_1 = \omega_2 = \omega_3$, $\xi_1 = \xi_2$—may be reasonable for a given data collection or with some additional knowledge about a particular matching algorithm or data collection. For a given simplified model, deriving the estimates for the correlation parameters should be straightforward. When reporting results on estimation of the correlation parameters it is important to identify the specific correlation structure used in order to make readers fully aware of the exact methodology used.

4.2 Bootstrap for Two-Instance FMR

In this section we introduce a *new* method for bootstrapping the FMR. Because of the unique correlation structure of the false match decisions, it is necessary to develop a resampling methodology that will maintain that correlation structure. The method we will use below is a bootstrap approach that is done in two stages. Note that this method is *different* from the 'two sample' bootstrap given by Wu [105], which is aimed at ROC curves and assumes *iid* data collection. We will refer to the bootstrap described here as a *two-instance bootstrap*. The basic approach is to draw a sample of individuals with replacement from the probe and, then, to draw a separate sample of individuals with replacement from the gallery for each selected individual in the probe. That is, for each individual in the resampled probe, we will create a different bootstrapped gallery. Taking all of the possible matching decisions for the selected pair of individuals, we concatenate them into a bootstrapped version of the original data. All of these steps are done while ensuring that all of the decisions are imposter decisions, i.e. $i \neq k$. Below, we formally enumerate the algorithm for this process.

The key to resampling the decisions that encompass an FMR is to recreate the structure of the full set of data while trying to mimic the correlation structure for an FMR. To that end, the methods given below assume the correlation structure found in (4.4) with $\omega_1 = \omega_2 = \omega_3$.

1. Sample $n_{\mathbb{P}}$ individuals from the list of all $n_{\mathbb{P}}$ individuals in the probe with replacement and call those individuals $b_1, b_2, \ldots, b_{n_{\mathbb{P}}}$.
2. For the b_ith individual selected in the previous step we then sample with replacement from the $n_{\mathbb{G}}$ individuals in the gallery and call those individuals $h_{i,k}$'s for $k = 1, \ldots, n_{\mathbb{G}}$.
3. Take all of the decisions $D_{b_i h_{i,k} 1}, \ldots, D_{b_i h_{i,k} m_{b_i h_{i,k}}}$ from each of the resampled pairs of individuals, $(b_i, h_{i,k})$, found in the two previous steps.
4. The replicated value for the estimated FMR, $\hat{\nu}$, is then given by

$$\hat{\nu}^b = \frac{\sum_{i=1}^{n_{\mathbb{P}}} \sum_{k=1}^{n_{\mathbb{G}}} \sum_{\ell=1}^{m_{b_i h_{i,k}}} D_{b_i h_{i,k} \ell}}{\sum_{i=1}^{n_{\mathbb{P}}} \sum_{k=1}^{n_{\mathbb{G}}} m_{b_i h_{i,k}}}. \tag{4.15}$$

Some commentary is warranted regarding the bootstrap approach given above. First, this is a *new* approach with a goal of matching the correlation structure for estimation of ν. This approach yields estimated variances, $\hat{V}[\hat{\nu}]$'s, that are approximately the same as the parametric equivalent calculated in (4.5). The reason that we use this *two-instance bootstrap* is that the resulting variance of the estimated FMR's using this approach seems to consistently approximate the variance as calculated by the general correlation structure. We considered many alternative bootstrap approaches and this methodology was the only one that worked consistently across all datasets that we considered. The key to this approach is the second step. It is necessary that for a given individual in the probe, we resample the individuals in the gallery against whom a comparison could be made. And we do this *separately* for *every* individual. Consequently, we will refer to this algorithm as the *two-instance bootstrap*. Note that for some data collections, we may choose a combination of individuals such that $m_{b_i h_{i,k}}$ is 0. In that case, we move to the next pair of selected individuals that have been selected via the process above. If there is variability in the m_{ik}'s, then there will be variability in the number of decisions that are summed as part of (4.15). However, the average number of terms in the denominator of (4.15) should be the same as the denominator of the original data given in (4.2).

We illustrate the utility of the *two-instance bootstrap* by applying it to each of the matchers in the XM2VTS database. There are $N_\nu = 40000$ match decisions that comprise each estimated FMR, $\hat{\nu}$, from this database. A variety of different estimated FMR values were chosen by specifying different thresholds to assess the performance of this algorithm. We choose here $M = 1000$ bootstrap replications. A summary of the results for this evaluation are given in Table 4.2. In that table, $s_{\hat{\nu}}$ is the standard error of $\hat{\nu}$ based upon the estimated variance $\hat{V}[\hat{\nu}]$ from (4.5) and

Table 4.2 Two-instance bootstrap standard error comparison

Modality	Matcher	Threshold	$\hat{\nu}$	$s_{\hat{\nu}}$	$s_{\hat{\nu},boot}$
Face	(FH, MLP)	0.0	0.0038	0.0007	0.0009
Face	(DCTs, GMM)	0.0	0.0582	0.0067	0.0066
Face	(DCTb, GMM)	0.2	0.0041	0.0007	0.0008
Face	(DCTs, MLP)	−0.8	0.1057	0.0086	0.0091
Face	(DCTb, MLP)	−0.5	0.0580	0.0072	0.0074
Speaker	(LFCC, GMM)	3.0	0.0142	0.0039	0.0038
Speaker	(PAC, GMM)	2.0	0.0570	0.0090	0.0090
Speaker	(SSC, GMM)	1.0	0.0692	0.0105	0.0099

$s_{\hat{\nu},boot}$ is the standard error of $\hat{\nu}$ obtained by taking the standard deviation of the bootstrap replicates, the $\hat{\nu}$'s. We can readily see in Table 4.2 that the bootstrapped standard error using the *two-instance bootstrap* performs quite well in matching the parametric standard errors across all of the matchers.

4.3 Statistical Methods

We move now to statistical methods about the false match rate, ν. Starting with a single error rate, we present confidence interval and hypothesis testing methodology. Inferential tools are then proffered in subsequent sections for comparing two or more FMR's. Bootstrap and randomization methods will be presented along with large sample approaches. There have been previous attempts to create methods for a single FMR including Mansfield and Wayman [63], Schuckers [84], Poh et al. [76] among others. None of these approaches fully accounted for the correlation structure as we do here, though the 'so-called' Bickel equations given in Mansfield and Wayman [63] may be a special case of the general approach here.

Because of the complicated nature of the parametric variance structure for the FMR, cf. (4.5), and in the interest of space, we will denote the estimated variance for a sample FMR by $\hat{V}[\hat{\nu}]$ rather than provide the full variance structure for each large sample procedure. For these approaches, we assume that the number of decisions is sufficiently large for their use. We will denote the effective sample size for an FMR to be

$$N_{\nu}^{\dagger} = N_{\nu}\left[\frac{\hat{\nu}(1-\hat{\nu})/N_{\nu}}{\hat{V}[\hat{\nu}]}\right] = \left[\frac{\hat{\nu}(1-\hat{\nu})}{\hat{V}[\hat{\nu}]}\right] \tag{4.16}$$

following Definition 2.34. Similarly, it is necessary to deal with the complicated FMR correlation structure appropriately when resampling. We will make extensive use of the *two-instance bootstrap* methodology given in the previous section since that bootstrap methodology well approximates the standard error from the general parametric correlation structure for a false match process. This approach is inspired by work done by Lahiri [54] on resampling for spatial data. We will use this methodology extensively both for the estimation and comparison of FMR's.

As we do in other chapters, we present statistical methods in increasing complexity. Starting with a single FMR, we present both confidence interval and hypothesis test methodology. We then do the same for a comparison of two FMR's whether they are from a paired data collection or from an independent data collection. The last set of statistical methods that we present are for comparing the equality of three or more FMR's. Both paired and independent data collections are considered in our comparisons of multiple FMR's.

4.3.1 One Sample Methods for FMR

The focus of this section is on making inference for a single stationary false match rate, ν. We begin with confidence interval methodology and move to hypothesis test methodology. In both cases we use large sample approaches as well as a bootstrap approach based upon the *two-instance bootstrap* from the previous section. Examples are provided for each approach.

4.3.1.1 Confidence Interval for Single FMR

We develop two approaches to making a confidence interval for a process FMR based upon a large sample approach and a bootstrap approach. Implicit to these methods is an assumption that the matching process begin analyzed is stationary in the sense of Definition 2.23. If that is not the case, then these methods are not appropriate. As we will see, the assumption of a stationary process is testable using methods that are given in this chapter for comparing two or more FMR's.

Large Sample Approach We derive a confidence interval for the false match rate, ν, based upon the estimated variance of our estimator. Thus, a $100(1-\alpha)\%$ CI for ν is

$$\hat{\nu} \pm z_{\alpha/2}\sqrt{\hat{V}[\hat{\nu}]} \qquad (4.17)$$

where $\hat{\nu}$ is calculated following (4.2) and $z_{\alpha/2}$ is the $1-\alpha/2$th percentile from a Gaussian distribution. We get the estimated variance, $\hat{V}[\hat{\nu}]$, by substituting our parameter estimates into (4.5). This confidence interval is appropriate to use if $N_\nu^\dagger \hat{\nu} \geq 10$ and $N_\nu^\dagger(1-\hat{\nu}) \geq 10$. Thus, for smaller FMR's, a larger number of decisions will be required for using this method than for larger FMR's. The assumption that for large $N_\nu^\dagger$ the sampling distribution of $\hat{\nu}$ is approximately Gaussian draws upon recent work by Schuckers and Dietz [23].

Example 4.1 To illustrate the methodology above, we start by considering data from the G protocol of the BANCA database. In particular, we use decisions from the

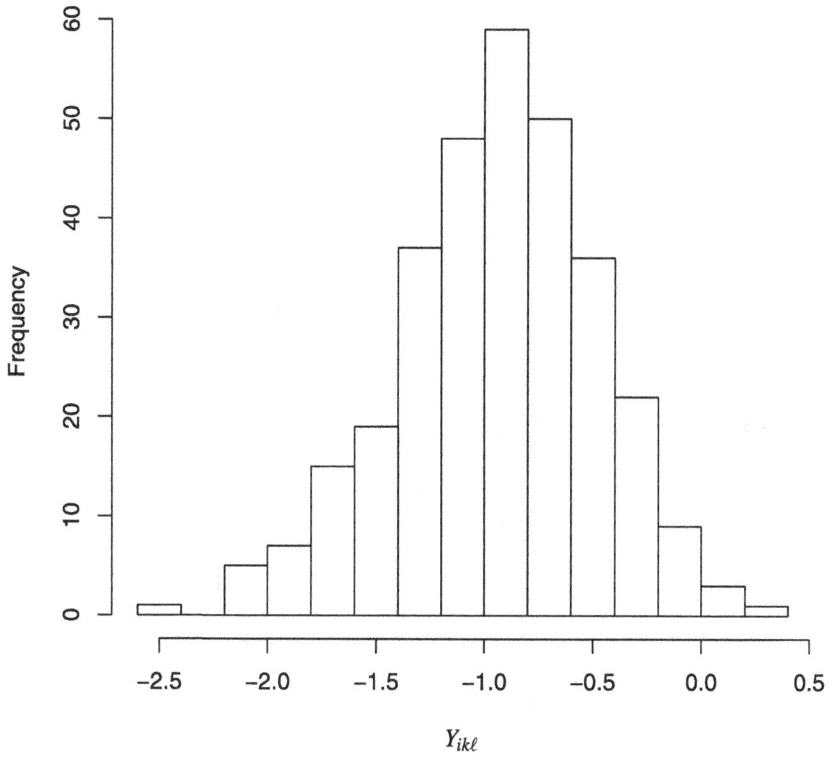

Fig. 4.1 Distribution of $Y_{ik\ell}$, the imposter match scores, for the SURREY-AUTO classifier applied to group g1 of the BANCA database following protocol G

SURREY_face_svm_auto (SURREY-AUTO) classifier on group g1 with a threshold of -0.25. Our goal will be a 99% confidence interval for the process FMR at that threshold. Figure 4.1 shows the distribution of the imposter match scores for this application. Note that this histogram is *not* that of a Gaussian distribution; however, that is not required for application of the large sample method. Instead, we must ensure that $N_\nu^\dagger$ is of sufficient size. For these data, $N_\nu = 312$ and $N_\nu^\dagger = 240.34$. Our estimate of the FMR at this threshold is $\hat{\nu} = 0.0737$. Then $N_\nu^\dagger \hat{\nu} = 17.72$ and $N_\nu^\dagger(1 - \hat{\nu}) = 222.62$. Consequently, our sample size is large enough for us to proceed. For this example, we do not have multiple decisions from a given pair of individuals and so it is not possible to estimate η or ξ_2. Our estimates for ω_1, ω_2, ω_3, and ξ_1 are 0.0058, 0.0000, 0.0100 and 0.0143, respectively. The calculated estimate for ω_2 is $\hat{\omega}_2 = -0.0284$, but we truncate it to zero for reasons outlined above. With $\hat{V}[\hat{\nu}] = 0.0169$, we obtain a 99% confidence interval for ν of $(0.0303, 0.1171)$. Thus we can be 99% confident that the FMR for this matching process is between 3.03% and 11.7%.

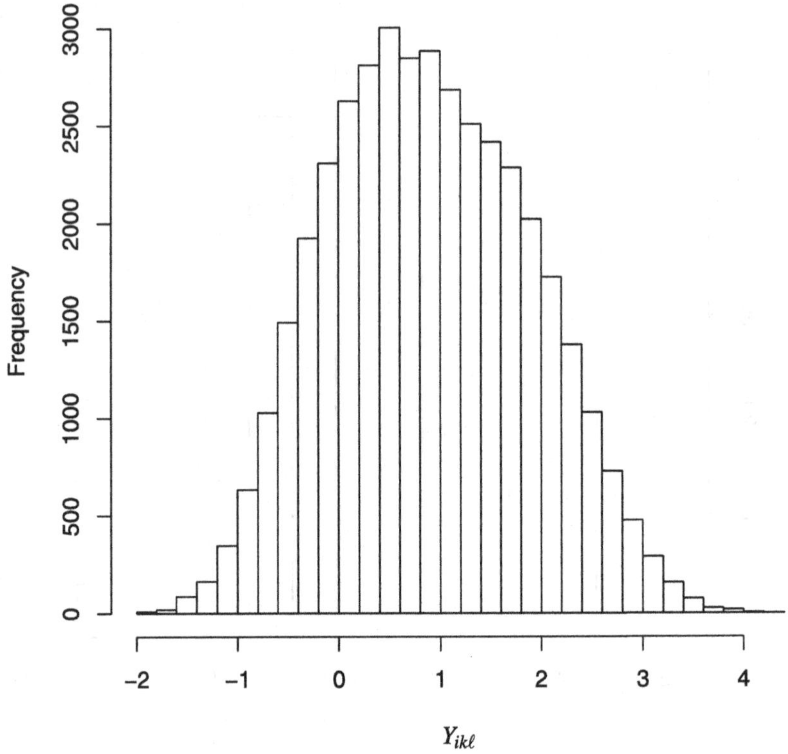

Fig. 4.2 Distribution of $Y_{ik\ell}$, the imposter match scores, for the speaker matcher (LFCC, GMM) from the XM2VTS database

Example 4.2 We want to create a 99% confidence interval for the FMR of the (LFCC, GMM) speaker matcher from the XM2VTS database. A histogram of the distribution of matching scores from this matcher is given in Fig. 4.2. Using a threshold of 3.0, we have an estimated process FMR of $\hat{\upsilon} = 0.0142$. This estimated FMR is based upon $N_\upsilon = 40000$ decisions from $n_{\mathbb{G}} = 25$ individuals in the probe and $n_{\mathbb{P}} = 200$ individuals in the gallery. Due to the combinations present in these data, we have the following correlation parameter estimates $\hat{\eta} = 0.3328$, $\hat{\omega}_1 = 0.0226$ and $\hat{\omega}_2 = 0.0184$. The other correlations parameters for the general correlation model, (4.4), are not estimable in this case. With these terms we have that $N_\upsilon^\dagger = 933.02$ which results from the fact that $\sqrt{\hat{V}[\hat{\upsilon}]} = 0.0039$. Then, $N_\upsilon^\dagger \hat{\upsilon} = 13.30$ and $N_\upsilon^\dagger(1 - \hat{\upsilon}) = 919.73$. Both of those quantities are greater than 10 and, consequently, we can use the large sample methodology given above. Thus, our 99% confidence interval for the FMR is $(0.0043, 0.0242)$. We are 99% confident that the process FMR for the (LFCC, GMM) speaker matcher is between 0.43% and 2.42%.

As is illustrated by the previous examples, some data collections do not allow for estimation of all of the correlation parameters. Since we have no data upon which to base estimates, we should *not* simply treat those estimates as 0's. Another data collection from this same process could yield data from which it would be possible to estimate these parameters.

Bootstrap Approach We next present a bootstrap approach for creating a confidence interval for a single process FMR. We will use the *two-instance bootstrap* methodology presented in Sect. 4.2 to obtain a sampling distribution of $\hat{\nu}$, the estimated FMR.

1. Calculate $\hat{\nu}$ following

$$\hat{\nu} = \frac{\sum_{i=1}^{n_{\mathbb{P}}} \sum_{k=1}^{n_{\mathbb{G}}} \sum_{\ell=1}^{m_{ik}} D_{ik\ell}}{\sum_{i=1}^{n_{\mathbb{P}}} \sum_{k=1}^{n_{\mathbb{G}}} m_{ik}}. \tag{4.18}$$

2. Bootstrap the decisions following the algorithm outlined in Sect. 4.2 to generate a $\hat{\nu}_b$. This calculation is

$$\hat{\nu}_b = \frac{\sum_{i=1}^{n_{\mathbb{P}}} \sum_{k=1}^{n_{\mathbb{G}}} \sum_{\ell=1}^{m_{b_i h_{i,k}}} D_{b_i h_{i,k}\ell}}{\sum_{i=1}^{n_{\mathbb{P}}} \sum_{k=1}^{n_{\mathbb{G}}} m_{b_i h_{i,k}}}. \tag{4.19}$$

3. Calculate and store $e_\nu = \hat{\nu}_b - \hat{\nu}$.
4. Repeat Steps 2 and 3 above M times where M is large.
5. Find the $\alpha/2$th and $1-\alpha/2$th percentile for the e_ν's and call these e_L and e_U, respectively.
6. Then a $100(1-\alpha)\%$ confidence for ν is given by the interval

$$(\hat{\nu} - e_U, \hat{\nu} - e_L). \tag{4.20}$$

The distribution of e_ν is meant to approximate the sampling distribution of the differences between $\hat{\nu}_b - \hat{\nu}$. We follow Hall [44] in creating our bootstrap confidence interval based upon subtracting the upper and lower endpoints of that distribution.

Example 4.3 For this example we reanalyze the decisions from the SURREY-AUTO classifier applied to group g1 that were collected under the G protocol in the BANCA database in order to obtain a 99% confidence interval. This data was previously analyzed in Example 4.1 using a large sample approach. Since we are using the bootstrap approach here we need not concern ourselves with verifying that we have a sample of sufficient size. Rather we generate $M = 5000$ e_ν's following the protocol above. These bootstrapped replicate errors are summarized in the histogram given in Fig. 4.3. For a 99% confidence interval we need the 0.5^{th} percentile and the 99.5^{th} percentile of the distribution of the e_b's. From our analysis we found that these were -0.0489 and 0.0573, respectively. Subtracting both of these from our estimated FMR, $\hat{\nu} = 0.0737$, we obtain a confidence interval of $(0.0164, 0.1226)$. Thus, at the 99% level we are confident that the process FMR for this classifier is between 0.16% and 10.69%. Comparing this interval to the interval from the large sample approach found in Example 4.1 of $(0.0303, 0.1171)$, we

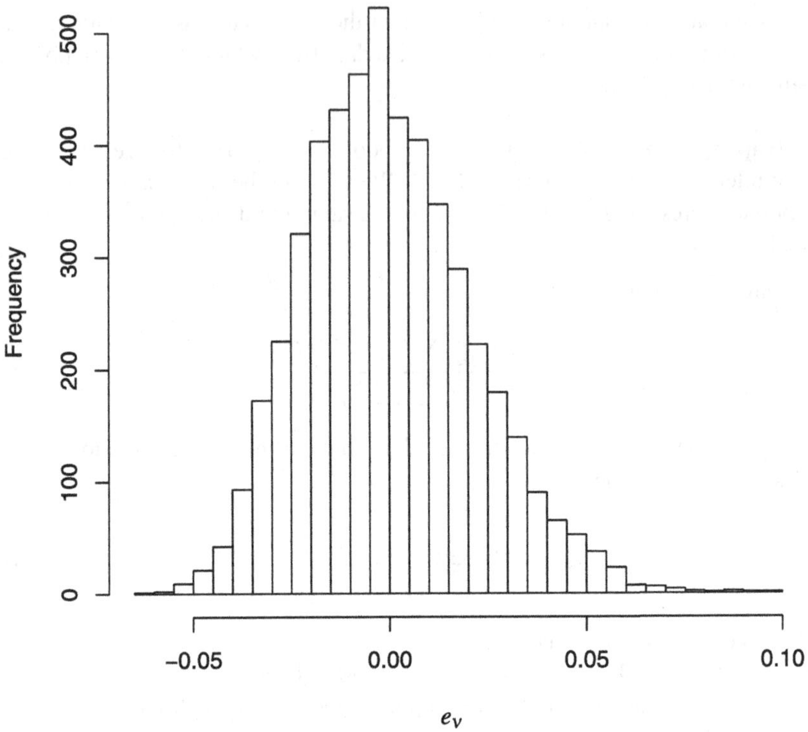

Fig. 4.3 Distribution of the e_ν, the bootstrapped differences from the estimated FMR for the SURREY-AUTO matcher applied to group g1 of the BANCA database following protocol G

see that they are quite similar, though the lower end of the bootstrapped interval is slightly lower, the upper bounds here are roughly the same. The difference in these two intervals is due to the asymmetric nature of the distribution of the bootstrapped replicated errors, the e_ν's. Both intervals are appropriate to use for decisions and inference about this process.

Note that it is possible to derive a confidence interval with a lower bound that is negative using the methodology here. In that case, it is appropriate to truncate the lower bound to 0.

4.3.1.2 Hypothesis Test for Single FMR

The goal of the hypothesis test we give in this section is to determine if the sample FMR, $\hat{\nu}$, is significantly less than a specified value ν_0. Formally, the test we are doing uses the following hypotheses:

$$H_0 : \nu = \nu_0$$
$$H_1 : \nu < \nu_0.$$

For example, we might be interested in testing whether an FMR is less than 0.01. In that case $\nu_0 = 0.01$ and our hypotheses would be:

$$H_0 : \nu = 0.01$$
$$H_1 : \nu < 0.01.$$

Below we provide two methods for the creation of such a hypothesis test. The first is a large sample approach and the second is a bootstrap approach.

Large Sample Approach We can assume approximate Gaussianity for the sampling distribution of $\hat{\nu}$ if the number of match decisions in our sample is sufficiently large. Here we mean that $N_\nu^\dagger \hat{\nu} \geq 10$ and $N_\nu^\dagger (1 - \hat{\nu}) \geq 10$. If these conditions are met then the hypothesis test is given by the following algorithm.

Test Statistic:

$$z = \frac{\hat{\nu} - \nu_0}{\sqrt{\hat{V}[\hat{\nu}]}} \tag{4.21}$$

where $\hat{V}[\hat{\nu}]$ is the estimated variance for $\hat{\nu}$ as calculated following (4.5).

p-value: Our *p-value* can be calculated using $p = P(Z < z)$ where Z is a standard Gaussian random variable. We can use Table 9.1 to calculate the *p-value* explicitly.

We will reject H_0 if the *p-value* is small, say less than an *a priori* specified α where α is the level of the test.

Example 4.4 In this example, we want to test whether the voice matcher IDIAP_voice_gmm_auto_scale_33_300_pca (IDIAP-33-300-PCA) has a process FMR significantly less than 0.10 at a threshold of −0.05 when applied to group g1 using the G protocol of the BANCA database. Figure 4.4 has a histogram of the matching scores for IDIAP-33-300-PCA. From our sample we have estimated that the FMR is $\hat{\nu} = 0.0801$. We will use a significance level of $\alpha = 0.05$ to test this hypothesis. Before we can proceed to test the hypotheses

$$H_0 : \nu = 0.10$$
$$H_1 : \nu < 0.10,$$

we must determine if our sample size is sufficiently large. Here N_ν is 312 with $n = 26$. Our effective sample size is $N_\nu^\dagger = 248.07$ and $N_\nu^\dagger \hat{\nu} = 19.87$ which is certainly larger than 10. Similarly $N_\nu^\dagger (1 - \hat{\nu}) = 228.20 \geq 10$, so it is appropriate to proceed

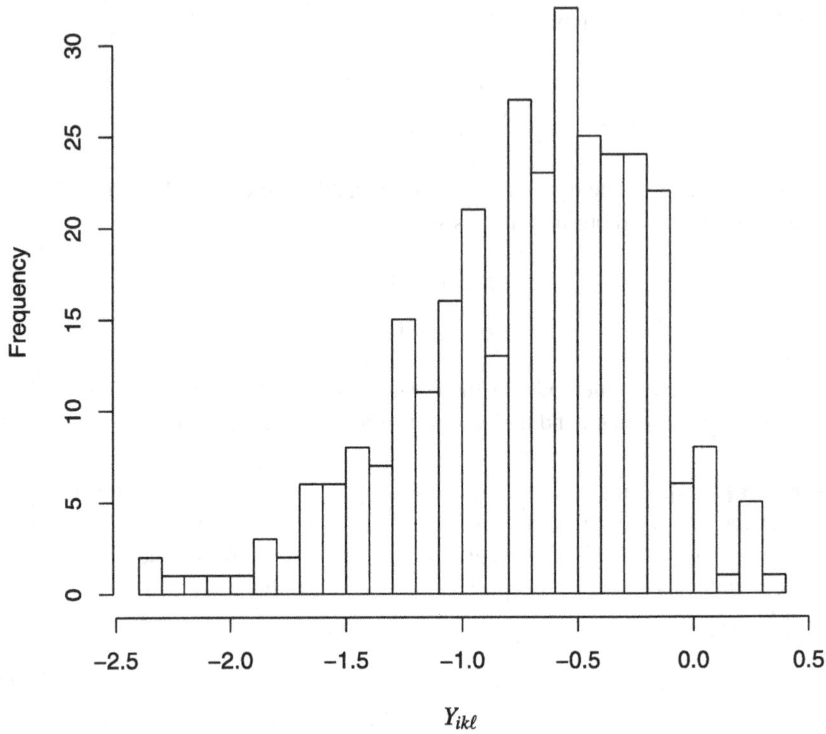

Fig. 4.4 Distribution of $Y_{ik\ell}$, the imposter match scores, for the voice matcher IDIAP-33-300-PCA applied to group g1 of the BANCA database following protocol G

with this test. Here $\sqrt{V[\hat{\nu}]} = 0.0172$ and the test statistic is

$$z = \frac{0.0801 - 0.1000}{0.0172} = -1.15. \tag{4.22}$$

From Table 9.1 we have a *p-value* of $p = P(Z < -1.15) = 0.1251$. Since this *p-value* is larger than our significance level $\alpha = 0.05$, we fail to reject the null hypothesis that the process FMR is 10% or 0.1000. Consequently, we conclude that the observed FMR here for IDIAP-33-300-PCA is *not* significantly less than 0.1000.

Bootstrap Approach The bootstrap approach for testing a single FMR assumes for the purposes of *p-value* calculation that the null hypothesis, $\nu = \nu_0$, is true. This assumption is also made by the large sample approach given above. As a consequence, the bootstrapped FMR's are adjusted to center the sampling distribution of the bootstrapped FMR's, $\hat{\nu}^b$ at ν_0. As with all bootstrapped hypothesis tests, the bootstrapped sampling distribution replaces the relevant appropriate reference distribution for calculating p-values. The hypotheses used here are the same as those give for the large sample approach above. We give this approach below.

1. Calculate the estimated FMR $\hat{\nu}$ following (4.2). That is,

$$\hat{\nu} = \frac{\sum_{i=1}^{n_{\mathbb{P}}} \sum_{k=1}^{n_{\mathbb{G}}} \sum_{\ell=1}^{m_{ik}} D_{ik\ell}}{\sum_{i=1}^{n_{\mathbb{P}}} \sum_{k=1}^{n_{\mathbb{G}}} m_{ik}}. \tag{4.23}$$

2. Bootstrap the decisions based upon the algorithm given in Sect. 4.2 to get the replicated FMR given by

$$\hat{\nu}_b = \frac{\sum_{i=1}^{n_{\mathbb{P}}} \sum_{k=1}^{n_{\mathbb{G}}} \sum_{\ell=1}^{m_{b_i h_{i,k}}} D_{b_i h_{i,k} \ell}}{\sum_{i=1}^{n_{\mathbb{P}}} \sum_{k=1}^{n_{\mathbb{G}}} m_{b_i h_{i,k}}}. \tag{4.24}$$

3. Calculate and store $e_\nu = \hat{\nu}_b - \hat{\nu} + \nu_0$. In this way, we obtain a sampling distribution for $\hat{\nu}$ under the assumption that the null hypothesis holds, i.e. $\nu = \nu_0$.
4. Repeat Steps 2 and 3 above some M times where M is large.
5. Then the *p-value* for this test is given by

$$p = \frac{1 + \sum_{\varsigma=1}^{M} I_{\{e_\nu \leq \hat{\nu}\}}}{M+1}. \tag{4.25}$$

6. We reject the null hypothesis H_0 if the *p-value* calculated above is small which for a given significance level α means that $p < \alpha$.

Example 4.5 For this example, we would like to test whether the UC3M_voice_gmm_auto_scale_10_100 (UC3M-10-100) matcher has an FMR of less than 0.10 when the FNMR is 0.10. We use data from the BANCA database following the G protocol applied to group g2. Formally, our hypothesis test is:

$$H_0 : \nu = 0.10$$
$$H_1 : \nu < 0.10.$$

The threshold at which we achieve an FNMR of 0.10 is 0.0903. At that value, the estimated FMR is $\hat{\nu} = 0.0577$. Another way to describe this test is to say that we are testing if the estimated FMR of 0.0577 for this process is significantly less than 0.10. We note that our answer to that question will depend upon how much variability there is in the estimated sampling distribution generated by our bootstrap approach. The distribution of the e_ν's is summarized in Fig. 4.5. This figure also includes a dashed vertical line at $\hat{\nu} = 0.0577$. Our *p-value* from generating $M = 4999$ bootstrap replicates is *p-value* $= 0.0016$ suggesting that we should reject the null hypothesis and conclude that the process FMR for this matcher applied to group g2 is significantly less than 0.10.

Example 4.6 In this example, we repeat the hypothesis test from Example 4.4 using the bootstrap approach. The classifier under consideration is the voice matcher IDIAP_voice_gmm_auto_scale_33_300_pca (IDIAP-33-300-PCA). These data come from group g1 of the BANCA database and were collected under the G

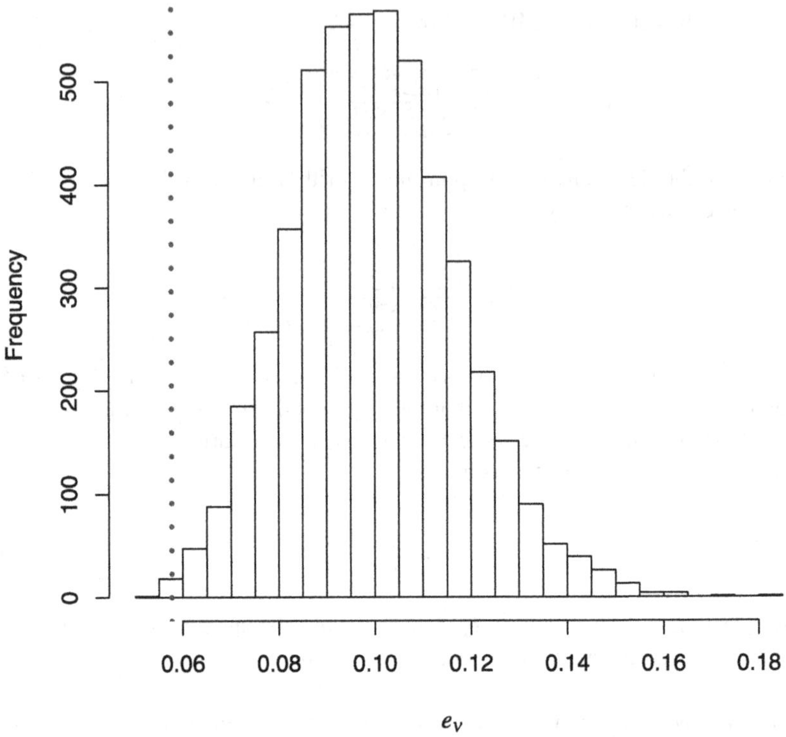

Fig. 4.5 Distribution of e_ν, bootstrapped differences from the estimated FMR for the voice matcher UC3M-10-100 from the G protocol of the BANCA database applied to group g2

protocol. Recall from that previous example that the threshold of -0.05 yields an estimated FMR of 0.0801. As we did before, we will test the hypotheses

$$H_0 : \nu = 0.10$$
$$H_1 : \nu < 0.10$$

with a significance level of $\alpha = 0.05$. There are $N_\nu = 312$ decisions that are available for estimation of this process FMR. These data come from matching comparisons on a total of $n = 26$ individuals. We created $M = 999$ bootstrap replicates following the two-instance bootstrap given above. Figure 4.6 gives a summary of the distribution of the e_ν's following this approach. The dashed vertical line given there is at $\hat{\nu} = 0.0801$. The e_ν's at or to the left of that line give the *p-value* which is 0.1110. Since this *p-value* is larger than our significance level, we fail to reject the null hypothesis and conclude that the estimated FMR of 0.0769 is not significantly less than 0.1000. This conclusion is the same as the one we came to in Example 4.4 where the *p-value* was 0.0853.

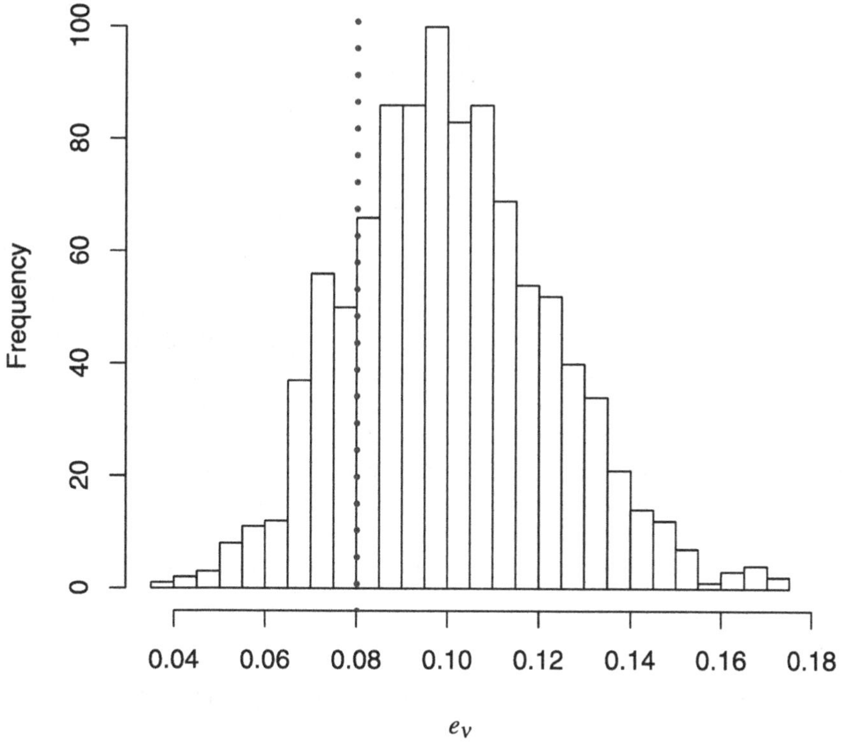

Fig. 4.6 Distribution of e_ν, bootstrapped differences from the estimated FMR, for the voice matcher IDIAP-33-300-PCA from the G protocol of the BANCA database applied to group g2

It is worth noting that there is a difference between the *p-values* quoted in the previous example. This discrepancy is one of the reasons that statisticians and other researchers often prefer to report the *p-value* along with the decision. Another important consideration is M, the number of bootstrap replicates obtained. Concerned about the variability in the *p-value* can be alleviated by choosing a value for M that is particularly large, say $M > 10000$.It is not, however, ethical to repeat a resampling procedure until your *p-value* is on the desired side of α. The *p-value* that is reported should be calculated based upon all bootstrap replications that have been calculated.

4.3.2 Two Sample Methods for FMR's

We move now to the comparison of two FMR's. Denoting these quantities by ν_1 and ν_2, we will focus on whether these rates differ significantly. That is, we want to determine whether the differences in the sample FMR's from these two processes vary more than we would suspect by chance alone. Methods are given below for both independent and paired data collections that result in two samples. To differentiate between these two samples, we will use the letter $g = 1, 2$ as both a subscript and a superscript. For the total number of decisions in each sample, we will use $N_\nu^{(g)}$. The effective sample size in each group will be $N_\nu^{\dagger(g)}$. For most other quantities a superscript of (g) will be used to denote the respective group from which a summary is derived. These statistics will be based upon decisions from each group which will be represented by $D_{ik\ell}^{(g)}$. One important application of the methods below is to look at potential changes in a process and whether those differences are significant. Thus, we can use the tests below to determine if two processes have the same FMR or if two groups of a given matching process have significantly different FMR's. Similarly, we can check if a process is stationary of over two groups of decisions. For the two groups we are comparing in this chapter, we will assume that each process is stationary in the sense of Definition 2.23. For example, we might want to know if there are differences in the false match rate for a bioauthentication matching system that is located in an airport in Orlando, Florida, USA and the same system located in Ottawa, Ontario, Canada. Similarly we might want to know if there are differences in the FMR between two matching algorithms run on the same database of signals.

4.3.2.1 Confidence Interval for Difference of Two Independent FMR's

The aim in this section is to create a $100(1-\alpha)\%$ confidence interval for the difference of two FMR's from distinct or independent processes. The focus of our inference will be the difference between the two FMR's, $\nu_1 - \nu_2$. One situation where we would have independent FMR's is the aforementioned example of comparing FMR's from the same bioauthentication system at two different locations. Because the individuals from whom signals are gathered will be distinct, these data collections are independent. A bootstrap approach for making a confidence interval for the difference of two independent FMR's follows the large sample approach that is shown below.

Large Sample Approach The large sample approach depends on having collected a large number of decisions from each process. Specifically, the number of decisions collected must be large enough that the sampling distribution of $\hat{\nu}_1 - \hat{\nu}_2$ approximately follows a Gaussian distribution. That will be the case here if $N_\nu^{\dagger(1)}$ and $N_\nu^{\dagger(2)}$, the effective sample sizes, are large meaning that $N_\nu^{\dagger(g)}\hat{\nu}_g \geq 10$ and $N_\nu^{\dagger(g)}(1-\hat{\nu}_g) \geq 10$ for $g = 1, 2$. With those conditions met, we can use the following to make a $(1-\alpha)100\%$ confidence interval for the difference of the two FMR's, $\nu_1 - \nu_2$:

$$\hat{\nu}_1 - \hat{\nu}_2 \pm z_{\alpha/2}\sqrt{\hat{V}[\hat{\nu}_1] + \hat{V}[\hat{\nu}_2]}, \tag{4.26}$$

where

$$\hat{\nu}_g = \frac{\sum_{i=1}^{n_{\mathbb{P}}^{(g)}} \sum_{k=1}^{n_{\mathbb{G}}^{(g)}} \sum_{\ell=1}^{m_{ik}^{(g)}} D_{ik\ell}^{(g)}}{\sum_{i=1}^{n_{\mathbb{P}}^{(g)}} \sum_{k=1}^{n_{\mathbb{P}}^{(g)}} m_{ik}^{(g)}} \tag{4.27}$$

which is the fraction of observed false match errors to total number of matching decisions in the sample for group g. The estimated variances, $\hat{V}[\hat{\nu}_g]$'s, are calculated based solely on the data in each group, g, and following (4.5) or the equivalent using a different correlation structure.

Example 4.7 Suppose that we would like to estimate the difference between two process FMR's from a classifier applied to two distinct groups. To illustrate the method above, we consider the IDIAP_voice_gmm_auto_scale_33_300_ pca (IDIAP-33-300-PCA) classifier applied to groups g1 and g2 from the G protocol of the BANCA database at the same threshold -0.05. We propose making at 95% confidence interval for the difference of these two FMR's. The estimated FMR's are $\hat{\nu}_1 = 0.0801$ and $\hat{\nu}_2 = 0.0577$, respectively, for groups g1 and g2. In order to ensure that our methodology is appropriate here, we need to verify that the sample sizes here are sufficiently large. We first note that the number of decisions is the same in both groups $N_\nu^{(1)} = N_\nu^{(2)} = 312$ and the number of individuals is the same in both groups $n_1 = n_2 = 26$. The effective sample sizes are, however, different. The effective sample size for the first group is $N_\nu^{\dagger(1)} = 248.07$ which is large enough since $N_\nu^{\dagger(1)}\hat{\nu}_1 = 19.88$ and $N_\nu^{\dagger(1)}(1-\hat{\nu}_1) = 228.19$. Similarly $N_\nu^{\dagger(2)} = 258.15$ is sufficiently large because $N_\nu^{\dagger(2)}\hat{\nu}_2 = 14.89$ and $N_\nu^{\dagger(2)}(1-\hat{\nu}_2) = 243.26$. Following the methodology above we can calculate that $\sqrt{\hat{V}[\hat{\nu}_1] + \hat{V}[\hat{\nu}_2]} = 0.0225$. From that we get a 95% confidence interval that is $0.0801 - 0.0577 \pm 1.96(0.0225)$ which is $(-0.0218, 0.0666)$. Consequently, we are then 95% confident that the difference between these two process FMR's applied to these groups is between -2.18% and 6.66%. Given that *zero*, 0, is inside the above interval, we could conclude that there is not enough evidence to say that these two FMR's differ.

Bootstrap Approach As we did with the large sample approach to a confidence interval for $\nu_1 - \nu_2$, the bootstrap approach aims to approximate the sampling distribution for the estimator $\hat{\nu}_1 - \hat{\nu}_2$. The basic concept of this approach is that we bootstrap each of the two groups separately to get bootstrapped estimates of the difference between the FMR's. As above, we use the *two-instance bootstrap* to obtain our bootstrap replicates of estimated FMR's. We call the bootstrapped difference in the two FMR's $\hat{\nu}_1^b - \hat{\nu}_2^b$. Our methodology is given below.

1. Calculate $\hat{\nu}_1 - \hat{\nu}_2$ following (4.2) for each estimated FMR where

$$\hat{\nu}_g = \frac{\sum_{i=1}^{n_{\mathbb{P}}^{(g)}} \sum_{k=1}^{n_{\mathbb{G}}^{(g)}} \sum_{\ell=1}^{m_{ik}^{(g)}} D_{ik\ell}^{(g)}}{\sum_{i=1}^{n_{\mathbb{P}}^{(g)}} \sum_{k=1}^{n_{\mathbb{G}}^{(g)}} m_{ik}^{(g)}}. \tag{4.28}$$

2. Bootstrap the decisions in the first group and denote the resulting replicated FMR by $\hat{\nu}_1^b$ where

$$\hat{\nu}_{1b} = \frac{\sum_{i=1}^{n_{\mathbb{P}}^{(1)}} \sum_{k=1}^{n_{\mathbb{G}}^{(1)}} \sum_{\ell=1}^{m_{b_i h_{i,k}}^{(1)}} D_{b_i h_{i,k} \ell}^{(1)}}{\sum_{i=1}^{n_{\mathbb{P}}^{(1)}} \sum_{k=1}^{n_{\mathbb{G}}^{(1)}} m_{b_i h_{i,k}}^{(1)}}. \tag{4.29}$$

3. Bootstrap the decisions in the second group and denote the resulting replicated FMR by $\hat{\nu}_2^b$ where

$$\hat{\nu}_{2b} = \frac{\sum_{i=1}^{n_{\mathbb{P}}^{(2)}} \sum_{k=1}^{n_{\mathbb{G}}^{(2)}} \sum_{\ell=1}^{m_{b_i h_{i,k}}^{(2)}} D_{b_i h_{i,k} \ell}^{(2)}}{\sum_{i=1}^{n_{\mathbb{P}}^{(2)}} \sum_{k=1}^{n_{\mathbb{G}}^{(2)}} m_{b_i h_{i,k}}^{(2)}}. \tag{4.30}$$

4. Calculate and store $e_\nu = \hat{\nu}_1^b - \hat{\nu}_2^b - (\hat{\nu}_1 - \hat{\nu}_2)$.
5. Repeat Steps 2, 3 and 4 above some M times where M is large.
6. Find the $\alpha/2$th and $1 - \alpha/2$th percentile for the e_ν's and call these e_L and e_U, respectively.
7. Then a $100(1 - \alpha)\%$ confidence for $\nu_1 - \nu_2$ is given by the interval

$$(\hat{\nu}_1 - \hat{\nu}_2 - e_U, \hat{\nu}_1 - \hat{\nu}_2 - e_L). \tag{4.31}$$

Example 4.8 In the BANCA database there were two distinct groups, g1 and g2 for each protocol. Since the individuals in those groups are different, we can use the above bootstrap methodology to made a 95% confidence interval for the difference of FMR's from the IDIAP_voice_gmm_auto_scale _33_300_pca (IDIAP-33-300-PCA) classifier at the same threshold. Above, in Example 4.7, we created a similar interval using a large sample approach. For group g1, the estimated FMR is $\hat{\nu}_1 = 0.0577$ and for group g2 that estimate is $\hat{\nu} = 0.0481$. Although the groups are different under the G protocol, the number of decisions is the same $N_\nu^{(1)} = N_\nu^{(2)} = 312$ for each group and the number of individuals is the same $n_1 = n_2 = 26$ for each group. Figure 4.7 gives a summary of the distribution of the e_ν's in this case. It is worth noting that this distribution is approximately Gaussian. Using $M = 1000$ bootstrapped replicates we find the values of e_L and e_U are -0.0470 and 0.0487, respectively. We can then create our interval which is $(-0.0391, 0.0567)$. Thus we can be 95% confident that the difference between these two FMR's is between -3.91% and 5.67%. We note that this interval is slightly wider than the interval from the large sample approach of $(-0.0312, 0.0504)$. Both are valid and appropriate approaches to estimation in this situation.

4.3.2.2 Hypothesis Test for Difference of Two Independent FMR's

Here we are testing whether one process FMR is significantly less than another. The methodology presented in this section focuses on the case where the FMR decisions

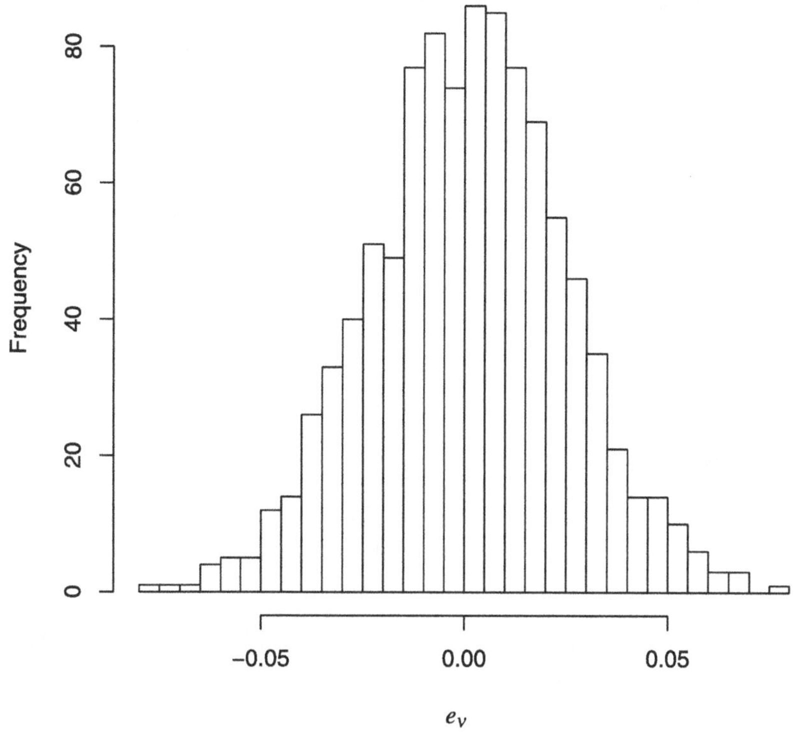

Fig. 4.7 Distribution of e_ν, the bootstrapped difference from the estimated difference in the FMR's for comparing the voice matcher IDIAP-33-300-PCA applied to groups g1 and g2 from the G protocol of the BANCA database

are independent meaning that the data collections for the two processes are distinct. Formally, the hypotheses that we will be testing are:

$$H_0 : \nu_1 = \nu_2$$

$$H_1 : \nu_1 < \nu_2.$$

One application of this methodology would be to compare a single bioauthentication system's FMR's from two different demographic groups, for example, men and women. To test these hypotheses, we present procedures based both on large sample methods and on bootstrapping methods.

Large Sample Approach Using the hypotheses above, we assume that the sampling distribution follows an approximately Gaussian distribution. This assumption is reasonable if the effective sample sizes for each group are large, i.e. if $N_\nu^{\dagger(g)}\hat{\nu}_g \geq 10$ and $N_\nu^{\dagger(g)}(1 - \hat{\nu}_g) \geq 10$ for $g = 1, 2$. Our large sample test is then the following:

Table 4.3 Correlation parameter estimates for Example 4.9

	g2	g1
g	1	2
$N_{\nu}^{\dagger(g)}$	142.92	150.29
$\hat{\eta}$	–	–
$\hat{\omega}_1$	0.0042	0.0039
$\hat{\omega}_2$	0.0397	0.0369
$\hat{\omega}_3$	0.0264	0.0242
$\hat{\xi}_1$	0.1197	0.1152
$\hat{\xi}_2$	–	–
$\sqrt{\hat{V}[\hat{\nu}_g]}$	0.0214	0.0221

Test Statistic:

$$z = \frac{\hat{\nu}_1 - \hat{\nu}_2}{\sqrt{\hat{V}[\hat{\nu}_1] + \hat{V}[\hat{\nu}_2]}} \tag{4.32}$$

where $\hat{V}[\hat{\nu}_g]$ is the estimated variance for $\hat{\nu}_g$.

p-value: $p = P(Z < z)$ where Z is a standard Gaussian random variable. We can calculate the *p-value* by using Table 9.1

Our decision rule is then to reject H_0 if *p-value* $< \alpha$ or if the *p-value* is small.

Example 4.9 The first example of the large sample approach given above is for testing the difference between a matcher applied to different groups in the BANCA database. The data used here is from the G protocol. The matcher of interest here is UCL_face_lda_man (UCL-LDA-MAN) and it is applied to groups g1 and g2 from the BANCA database. Table 4.3 has some summaries for these two groups. Although the data collection is independent, both groups had the same number of decisions $N_{\nu}^{(1)} = 312$ and $N_{\nu}^{(2)} = 312$. We evaluated this matcher at the same threshold -66 for both groups. The first group, g2, had an estimated FMR of $\hat{\nu}_1 = 0.0705$ and for the second group, g1, had an estimated FMR that was $\hat{\nu}_2 = 0.0801$. We are putting the group g2 first here since it has a lower FMR and our test above is one of determining if the smaller FMR is significantly less than the larger FMR. As an aside, it is possible to carry out the test

$$H_0 : \nu_1 = \nu_2$$

$$H_1 : \nu_1 > \nu_2$$

using the same test statistic given above and by calculating $p = P(Z > z)$. In order to use the large sample approach given above, it is necessary that our samples be of sufficient size. To that end, we note that $N_{\nu}^{\dagger(1)}\hat{\nu}_1 = 142.92 \times 0.0705 = 10.08$, $N_{\nu}^{\dagger(1)}(1 - \hat{\nu}_1) = 142.92 \times (1 - 0.0705) = 132.84$ $N_{\nu}^{\dagger(2)}\hat{\nu}_2 = 150.29 \times 0.0801 = 12.04$, and $N_{\nu}^{\dagger(2)}(1 - \hat{\nu}_2) = 142.92 \times (1 - 0.0801) = 138.25$. All of these quantities

Table 4.4 Correlation parameter estimates for Example 4.10

	g2	g1
g	1	2
$N_\nu^{\dagger(g)}$	8188.7	167.02
$\hat{\eta}$	0.2271	–
$\hat{\omega}_1$	0.0014	−0.0013
$\hat{\omega}_2$	−0.0006	0.0234
$\hat{\omega}_3$	–	0.0234
$\hat{\xi}_1$	–	0.0972
$\hat{\xi}_2$	–	–
$\sqrt{\hat{V}[\hat{\nu}_g]}$	0.0007	0.0206

are at least 10 and, thus, the large sample approach is appropriate here. Evaluating the test statistic, we find that $z = -0.34$ and that the *p-value* is $p \approx 0.3745$ from Table 9.1. This is a large *p-value* and so there is not enough evidence to reject the null hypothesis, H_0, that the FMR of the (UCL-LDA-MAN) matcher applied to these two groups is the same. It is then reasonable to say that the FMR of this matcher at a threshold of −66 is stationary over both groups.

Example 4.10 In this example, we compare two face matchers. The first matcher is the (FH, MLP) matcher from the XM2VTS database. The second matcher is the UCL_face_lda_man (UCL-LDA-MAN) matcher from the BANCA database. We use decisions from applying the second matcher to group g2 of that database under the G protocol. A threshold of 0.0 was used for the former matcher and a threshold of −67 was used for the latter matcher. These values produced estimated FMR's of 0.0038 and 0.0769, respectively. Since we are carrying out a large sample procedure here, we need to consider if the sample sizes are sufficiently large. For this test this means that both $N_\nu^{\dagger(g)}\hat{\nu}_g \geq 10$ and $N_\nu^{\dagger(g)}\hat{\nu}_g \geq 10$ for both groups $g = 1, 2$. Here we have $N_\nu^{\dagger(1)} = 8188.7$ from $N_\nu^{(1)} = 40000$ decisions and $N_\nu^{\dagger(2)} = 167.02$ from $N_\nu^{(2)} = 312$. Given the estimated FMR's, we can conclude that our sample sizes are sufficiently large since $8188.7 \times 0.0038 = 31.53 \geq 10$, $8188.7 \times (1 - 0.0038) = 8157.6 \geq 10$, $167.02 \times 0.0769 = 12.85$, and $167.02 \times (1 - 0.0769) = 154.18$. Table 4.4 contains a summary of the estimated parameters for these data. Based upon these values, we get a test statistic of $z = -3.54$ and, consequently, a *p-value* that is quite small, less than 0.0002. Therefore, we can conclude that the FMR's here are significantly different.

Bootstrap Approach For the estimation of a single FMR, we bootstrapped the sample decisions that are used to calculate that FMR following the *two-instance bootstrap* outlined in Sect. 4.2. For the case of testing the difference in two independent FMR's, we bootstrap each sample separately and take the difference of the bootstrapped FMR's. The hypotheses used here are the same as those give for the large sample approach above. We give a bootstrap approach to this hypothesis test below.

1. Calculate $\hat{\nu}_1 - \hat{\nu}_2$ where $\hat{\nu}_g$ is calculated following (4.2) which is

$$\hat{\nu}_g = \frac{\sum_{i=1}^{n_{\mathbb{P}}^{(g)}} \sum_{k=1}^{n_{\mathbb{G}}^{(g)}} \sum_{\ell=1}^{m_{ik}} D_{ik\ell}^{(g)}}{\sum_{i=1}^{n_{\mathbb{P}}^{(g)}} \sum_{k=1}^{n_{\mathbb{G}}^{(g)}} m_{ik}}. \tag{4.33}$$

2. Bootstrap the decisions from each group separately following the *two-instance bootstrap* algorithm outlined in Sect. 4.2 to generate $D_{b_i h_{i,k}\ell}^{(g)}$'s where $g = 1, 2$ to denote the group.
3. Calculate and store $e_\nu = (\hat{\nu}_1^b - \hat{\nu}_2^b) - (\hat{\nu}_1 - \hat{\nu}_2)$ where

$$\hat{\nu}_g^b = \frac{\sum_{i=1}^{n_{\mathbb{P}}^{(g)}} \sum_{k=1}^{n_{\mathbb{G}}^{(g)}} \sum_{\ell=1}^{m_{b_i h_{i,k}}^{(g)}} D_{b_i h_{i,k}\ell}^{(g)}}{\sum_{i=1}^{n_{\mathbb{P}}^{(g)}} \sum_{k=1}^{n_{\mathbb{G}}^{(g)}} m_{b_i h_{i,k}}^{(g)}}. \tag{4.34}$$

4. Repeat Steps 2 and 3 above a large number of times, say M times.
5. Then the *p-value* for this test is given by

$$p = \frac{1 + \sum_{\varsigma=1}^{M} I_{\{e_\nu \le \hat{\nu}_1 - \hat{\nu}_2\}}}{M+1}. \tag{4.35}$$

6. We will reject the null hypothesis that the two FMR's are equal if the *p-value* is small which may be smaller than some specified significance level α.

Example 4.11 For this example we consider the face matcher UCL_lda_man (UCL-LDA-MAN) applied to two different groups: g1 and g2 from the BANCA database. This data was previously considered in Example 4.9 and the FMR's were compared using a large sample approach. The threshold that we used in each case is -66 which gives $\hat{\nu}_1 = 0.0801$ for group g1 and $\hat{\nu}_2 = 0.0705$ for group g2. Here we will test

$$H_0 : \nu_1 = \nu_2$$

$$H_1 : \nu_1 > \nu_2$$

since for these estimated FMR's it is unlikely we would prefer an alternative of $\nu_1 < \nu_2$ when $\hat{\nu}_1 > \hat{\nu}_2$. This hypothesis test has the same form as the one above except that the *p-value* is based upon looking for evidence to reject the null hypothesis at the opposite tail of the distribution. This *p-value* is given by

$$p = \frac{1 + \sum_{\varsigma=1}^{M} I_{\{e_\nu \ge \hat{\nu}_1 - \hat{\nu}_2\}}}{M+1}. \tag{4.36}$$

An analysis of these data indicates that the $\sqrt{V[\hat{\nu}_1]} = 0.0221$ and $\sqrt{V[\hat{\nu}_2]} = 0.0214$. Using the bootstrap algorithm we tested the equality of these two independent FMR's. The resulting distribution of e_ν's is given in Fig. 4.8. The *p-value* for this test is 0.371 which is visually represented in that figure by the area to the right of

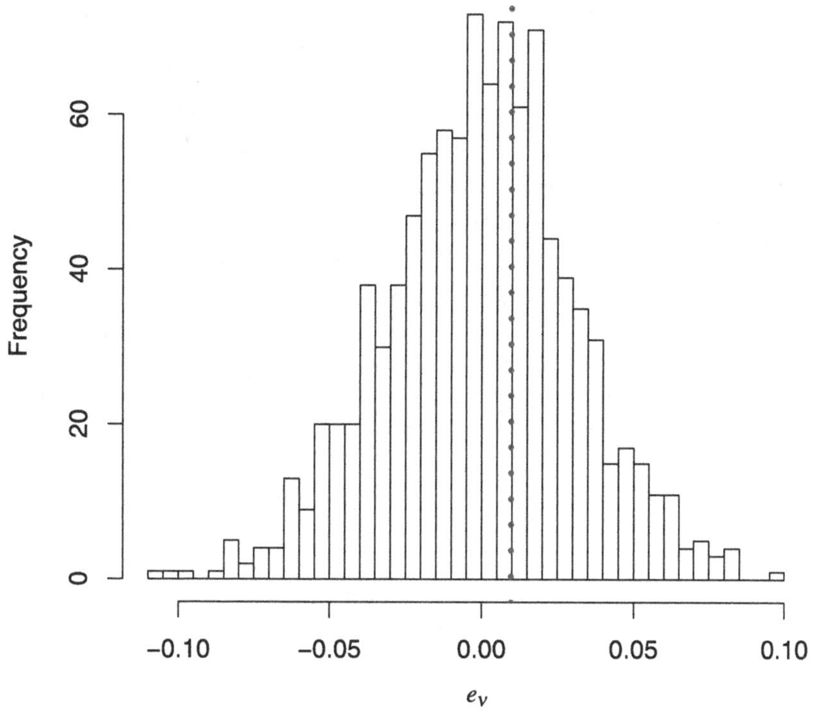

Fig. 4.8 Distribution of e_ν for comparing the facial matcher UCL-LDA-MAN applied to two groups from the BANCA database

the vertical dashed line at 0.0096. To obtain that value we used $M = 999$. Since the *p-value* is large, here we conclude as we did in Example 4.9 that the FMR is not significantly different between the two groups. Further we note that the *p-values* for the two approaches is relative similar: 0.371 and 0.4524. Both of these *p-values* result in the same conclusion and would for any reasonable choice of a significance level α.

4.3.2.3 Confidence Interval for Difference of Two Paired FMR's

We next present a confidence interval for comparing the difference of two FMR's when the data collection process is paired. As in the previous section we would like to make a $100(1-\alpha)\%$ confidence interval for $\nu_1 - \nu_2$ which is the difference of two process FMR's. One application of the paired methodology below would be to compare two the false match rate performance of two matching systems applied to the same database of captures. We produce both a large sample and a bootstrap approach below.

Large Sample Approach Any large sample approach depends on having sample sizes large enough that the sampling distribution of the statistic, $\hat{\nu}_1 - \hat{\nu}_2$, approximately follows a Gaussian distribution. We are now considering the case where the data collection process has yielded paired data. That is, we have the same number of decisions per comparison pair, i.e. $m_{ik}^{(1)} = m_{ik}^{(2)}$ for all i and all k. Our conditions for the sampling distribution approximating a Gaussian distribution are that $N_\nu^{\dagger(1)}$ and $N_\nu^{\dagger(2)}$, the effective samples sizes, are large. Accordingly, both $N_\nu^{\dagger(g)}\hat{\nu}_g \geq 10$ and $N_\nu^{\dagger(g)}(1 - \hat{\nu}_g) \geq 10$ for $g = 1, 2$. Worth noting here is that, although $N_\nu^{(1)}$ and $N_\nu^{(2)}$ are guaranteed to be the same by the paired nature of our data collection, this does not guarantee that $N_\nu^{\dagger(1)}$ and $N_\nu^{\dagger(2)}$ are equal. This is because these quantities depend upon the decisions in each group and the intra-group correlations within each. If these conditions are met, we can use the following to make a $(1-\alpha)100\%$ confidence interval for the difference of the two FMR's, $\nu_1 - \nu_2$:

$$\hat{\nu}_1 - \hat{\nu}_2 \pm z_{\alpha/2}\sqrt{\hat{V}[\hat{\nu}_1] + \hat{V}[\hat{\nu}_2] - 2Cov(\hat{\nu}_1, \hat{\nu}_2)}, \tag{4.37}$$

where

$$\hat{\nu}_g = \frac{\sum_{i=1}^{n_\mathbb{P}} \sum_{k=1}^{n_\mathbb{G}} \sum_{\ell=1}^{m_{ik}} D_{ik\ell}^{(g)}}{\sum_{i=1}^{n_\mathbb{P}} \sum_{k=1}^{n_\mathbb{G}} m_{ik}} \tag{4.38}$$

and

$$Cov(\hat{\nu}_1, \hat{\nu}_2) = \left(\sum_{i=1}^{n_\mathbb{P}} \sum_{k=1}^{n_\mathbb{G}} m_{ik} \right)^{-2} \sum_{i=1}^{n_\mathbb{P}} \sum_{k=1}^{n_\mathbb{G}} \sum_{\ell=1}^{m_{ik}} (D_{ik\ell}^{(1)} - \hat{\nu}_1)(D_{ik\ell}^{(2)} - \hat{\nu}_2). \tag{4.39}$$

Example 4.12 To illustrate this methodology we consider two classifiers from the BANCA database: the SURREY_face_svm_auto (SURREY-AUTO) classifier and the SURREY_face_svm_man (SURREY-SVM-MAN). We use data from g1 of the G protocol for this comparison to make a 95% confidence interval with the thresholds of -0.3 and -0.2, respectively. For SURREY-AUTO our estimated FMR is $\hat{\nu}_1 = 0.0641$ and for SURREY-SVMMAN our estimated FMR is $\hat{\nu}_2 = 0.0833$. Our data are paired here since we have aligned comparisons and hence aligned decisions for each of the two classifiers. For this data we have $n = 26$ and $N_\nu = 312$. The correlation estimates here for both matchers are given in Table 4.5. For this data collection we are not able to find estimates for η, ξ_1 or ξ_2 because this particular data collection does not allow for the calculation of estimates of those parameters. Additionally, the estimates found in Table 4.5 are the raw estimates for the other correlation parameters. We truncate those values that are negative to 0.0000 for calculation of our estimated standard deviation of $\hat{\nu}_g$, the $\sqrt{\hat{V}[\hat{\nu}_g]}$. We also note that the $Cov(\hat{\nu}_1, \hat{\nu}_2) = 0.0001$. In order to be able to use the methodology presented above, we must verify that the sample sizes are sufficiently large. The effective sample sizes, the $N_\nu^{\dagger(g)}$'s, needed for these calculations are also given in Table 4.5. Assessing this, we find that $N_\nu^{\dagger(1)}\hat{\nu}_1 = 16.63$, $N_\nu^{\dagger(1)}(1 - \hat{\nu}_1) = 242.78$, $N_\nu^{\dagger(2)}\hat{\nu}_2 = 15.36$,

Table 4.5 Correlation parameter estimates for Example 4.12

Measure	SURREY-AUTO	SURREY-SVM-MAN
g	1	2
$N_{\nu}^{\dagger(g)}$	259.41	184.36
$\hat{\eta}$	–	–
$\hat{\omega}_1$	−0.0005	0.0006
$\hat{\omega}_2$	−0.0296	0.0083
$\hat{\omega}_3$	0.0092	0.0197
$\hat{\xi}_1$	−0.0685	0.0161
$\hat{\xi}_2$	–	–
$\sqrt{\hat{V}[\hat{\nu}_g]}$	0.0152	0.0204

and $N_{\nu}^{\dagger(2)}(1 - \hat{\nu}_2) = 169.00$. We can then be confident that our samples sizes are sufficient to use the method introduced above. Thus, we obtain a 95% confidence interval for the difference of these paired FMR's to be $(-0.0637, 0.0253)$. From this result, we can conclude that the two process FMR's are not different since zero is contained inside our interval.

Bootstrap Approach A bootstrap approach to making a confidence interval for the difference of two paired FMR's is presented here. The difference between this bootstrap approach and the bootstrap approach for the difference of two independent FMR's is that for each bootstrap replication we only bootstrap a single time to obtain the bootstrapped individuals from which we obtain the decisions that comprise our estimates. Since all of the decisions are paired, we take the decisions from both groups for the selected individuals. This approach is compared to the need to bootstrap once for each of the groups in the case of comparing two independent FMR's. We then take both decisions that are connected to those indices due to the paired nature of the decisions.

1. Calculate $\hat{\nu}_1 - \hat{\nu}_2$ where $\hat{\nu}_g$ is calculated following

$$\hat{\nu}_g = \frac{\sum_{i=1}^{n_{\mathbb{P}}^{(g)}} \sum_{k=1}^{n_{\mathbb{G}}^{(g)}} \sum_{\ell=1}^{m_{ik}} D_{ik\ell}^{(g)}}{\sum_{i=1}^{n_{\mathbb{P}}^{(g)}} \sum_{k=1}^{n_{\mathbb{G}}^{(g)}} m_{ik}}. \tag{4.40}$$

2. Bootstrap the decisions for the first group following the algorithm outlined in Sect. 4.2 to generate $\hat{\nu}_1^b$. Since each decision from the first group is paired with a decision from the second, we use the exact same set of bootstrapped indices for the second group that we used for the first. From those we calculate $\hat{\nu}_2^b$. That is,

$$\hat{\nu}_1^b - \hat{\nu}_2^b = \frac{\sum_{i=1}^{n_{\mathbb{P}}} \sum_{k=1}^{n_{\mathbb{G}}} \sum_{\ell=1}^{m_{b_i h_{i,k}}} (D_{b_i h_{i,k}\ell}^{(1)} - D_{b_i h_{i,k}\ell}^{(2)})}{\sum_{i=1}^{n_{\mathbb{P}}} \sum_{k=1}^{n_{\mathbb{G}}} m_{b_i h_{i,k}}}$$

$$= \frac{\sum_{i=1}^{n_{\mathbb{P}}} \sum_{k=1}^{n_{\mathbb{G}}} \sum_{\ell=1}^{m_{b_i h_{i,k}}} D_{b_i h_{i,k} \ell}^{(1)}}{\sum_{i=1}^{n_{\mathbb{P}}} \sum_{k=1}^{n_{\mathbb{G}}} m_{b_i h_{i,k}}} - \frac{\sum_{i=1}^{n_{\mathbb{P}}} \sum_{k=1}^{n_{\mathbb{G}}} \sum_{\ell=1}^{m_{b_i h_{i,k}}} D_{b_i h_{i,k} \ell}^{(2)}}{\sum_{i=1}^{n_{\mathbb{P}}} \sum_{k=1}^{n_{\mathbb{G}}} m_{b_i h_{i,k}}}. \tag{4.41}$$

3. Calculate and store $e_\nu = \hat{\nu}_1^b - \hat{\nu}_2^b - (\hat{\nu}_1 - \hat{\nu}_2)$.
4. Repeat Steps 2 and 3 above a large number of times, say M times.
5. Find the $\alpha/2$th and $1 - \alpha/2$th percentile for the e_ν's and call these e_L and e_U, respectively.
6. Then a $100(1 - \alpha)\%$ confidence for $\nu_1 - \nu_2$ is given by the interval

$$(\hat{\nu}_1 - \hat{\nu}_2 - e_U, \hat{\nu}_1 - \hat{\nu}_2 - e_L). \tag{4.42}$$

Example 4.13 Two facial matchers (FH, MLP) and (DCTb, GMM) from the XM2VTS database are used to illustrate this bootstrap approach to making a confidence interval for the difference of two paired FMR's. We use a threshold of -0.2 for (FH, MLP) and a threshold of 0.2 for (DCTb, GMM). These values give estimated FMR's of $\hat{\nu}_1 = 0.0055$ and $\hat{\nu}_2 = 0.0041$. This gives a difference of 0.0014. We then run the above algorithm to obtain $M = 1000$ values for e_ν. Figure 4.9 has a histogram of these values. For a 99% confidence interval we find e_U and e_L to be 0.0029 and -0.0027, respectively. From this we can obtain a 99% confidence interval of $(-0.0015, 0.0041)$. Thus between -0.15% and 0.41% is the range within which we can be 99% confident that the difference between these two process FMR's lies.

4.3.2.4 Hypothesis Test for Difference of Two Paired FMR's

Our goal in this section is to test the hypotheses

$$H_0 : \nu_1 = \nu_2$$
$$H_1 : \nu_1 < \nu_2$$

assuming that the data collection for both groups is paired. This is likely to happen, for example, when we are comparing two matchers applied to the same database of images or biometric signals. We also note at the beginning of this section that the number of decisions per pair of individuals must be the same in both of the groups, $m_{ik}^{(1)} = m_{ik}^{(2)}$ for all i and k. Likewise, the total number of decisions calculated on each group must be identical, $N_\nu^{(1)} = N_\nu^{(2)}$, by virtue of the paired nature of the data collection. As a consequence of this, we will drop the superscript that distinguishes these quantities.

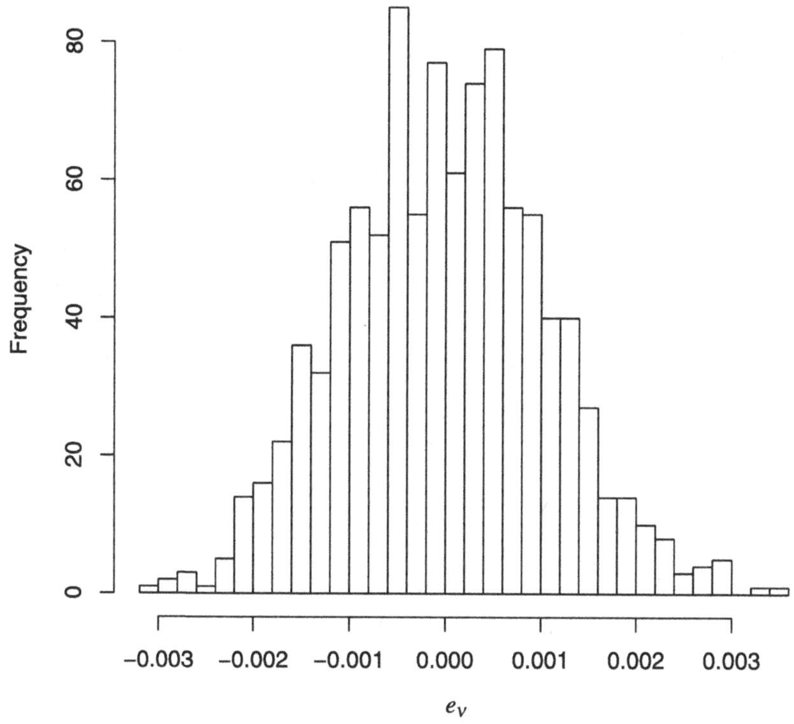

Fig. 4.9 Distribution of e_ν, the bootstrapped differences from the estimated difference of the sample FMR's, for comparing the facial matchers (FH, MLP) and (DCTb, GMM) from the XM2VTS database

Large Sample Approach Using the hypotheses

$$H_0 : \nu_1 = \nu_2$$

$$H_1 : \nu_1 < \nu_2,$$

we assume that the sampling distribution approximately follows a Gaussian distribution. This assumption is reasonable if $N_\nu^{\dagger(g)} \hat{\nu}_g \geq 10$ and $N_\nu^{\dagger(g)}(1 - \hat{\nu}_g) \geq 10$ for $g = 1, 2$. Again, as with all large sample procedures, we do not need to have a distribution of scores, $Y_{ik\ell}$'s, that is approximately Gaussian. Instead, what we need is that the number of observations is sufficiently large, so that the sampling distribution of $\hat{\nu}_1 - \hat{\nu}_2$ is approximately Gaussian. The resulting Gaussianity is due to the fact that we are averaging a large number of decisions. If both samples are sufficiently large, then the methodology for this test is the following:

Test Statistic:

$$z = \frac{\hat{\nu}_1 - \hat{\nu}_2}{\sqrt{\hat{V}[\hat{\nu}_1] + \hat{V}[\hat{\nu}_2] - 2Cov(\hat{\nu}_1, \hat{\nu}_2)}} \tag{4.43}$$

Table 4.6 Correlation parameter estimates for Example 4.14

Measure	IDIAP-33-300	IDIAP-25-300
g	1	2
$N_\nu^{\dagger(g)}$	312.00	174.12
$\hat{\eta}$	–	–
$\hat{\omega}_1$	−0.0198	0.0042
$\hat{\omega}_2$	−0.0198	0.0486
$\hat{\omega}_3$	−0.0062	0.0086
$\hat{\xi}_1$	−0.0470	0.0219
$\hat{\xi}_2$	–	–
$\sqrt{\hat{V}[\hat{\nu}_g]}$	0.0117	0.0194

where

$$\hat{\nu}_g = \frac{\sum_{i=1}^{n_{\mathbb{P}}} \sum_{k=1}^{n_{\mathbb{G}}} \sum_{\ell=1}^{m_{ik}} D_{ik\ell}^{(g)}}{\sum_{i=1}^{n_{\mathbb{P}}} \sum_{k=1}^{n_{\mathbb{G}}} m_{ik}}, \tag{4.44}$$

$$Cov(\hat{\nu}_1, \hat{\nu}_2) = (N_\nu)^{-2} \sum_{i=1}^{n_{\mathbb{P}}} \sum_{k=1}^{n_{\mathbb{G}}} \sum_{\ell=1}^{m_{ik}} (D_{ik\ell}^{(1)} - \hat{\nu}_p)(D_{ik\ell}^{(2)} - \hat{\nu}_p) \tag{4.45}$$

and

$$\hat{\nu}_p = \frac{1}{2}\hat{\nu}_1 + \frac{1}{2}\hat{\nu}_2. \tag{4.46}$$

p-value: $p = P(Z < z)$ where Z is a standard Gaussian random variable, z is the calculated test statistic and we can calculate this probability using Table 9.1.

Thus we will reject H_0 if $p < \alpha$ or, more generally, if the *p-value* is small.

Example 4.14 To illustrate the method above, we consider two paired FMR's. In this case we will use two different matchers applied to group g1 of the BANCA database. These matchers are IDIAP_voice_gmm_auto_scale_33_300 (IDIAP-33-300) and IDIAP_voice_gmm_auto_scale_25_300 (IDIAP-25-300). Both are voice classifiers. Table 4.6 gives the correlation estimates here for both matchers. These correlations are derived using thresholds of −0.05 and 0.02, respectively. Noting that $\hat{\nu}_1 = 0.0449$ and $\hat{\nu}_2 = 0.0705$, it is clear that the sample sizes are sufficiently large to use the above method to carry out the proposed hypothesis test. We note that the covariance here $Cov(\hat{\nu}_1, \hat{\nu}_2)$ is 0.0001. Thus, we obtain a value for our test statistics of $z = -1.24$ and we get a *p-value* from Table 9.1 of 0.1075. Since this *p-value* is large, we conclude that there is not enough evidence to suggest that the process FMR's are significantly different.

Bootstrap Approach We now turn to a bootstrap approach to testing whether two independent process FMR's are significantly different. The hypotheses used here are the same as those give for the large sample approach above. Due to the nature of the paired comparisons, it is necessary to keep intact the link between the decisions in the two groups. Thus, there is a single bootstrapping process here that yields both groups. As above, we note that since the data is paired $m_{ik}^{(1)} = m_{ik}^{(2)}$ for all i and k. Thus, for each decision in the first group, we have a decision in the second group that is connected with the decision in the first. We give this bootstrap approach below.

1. Calculate $\hat{\nu}_1 - \hat{\nu}_2$ where $\hat{\nu}_g$ is calculated following (4.2) which is

$$\hat{\nu}_g = \frac{\sum_{i=1}^{n_{\mathbb{P}}} \sum_{k=1}^{n_{\mathbb{G}}} \sum_{\ell=1}^{m_{ik}} D_{ik\ell}^{(g)}}{\sum_{i=1}^{n_{\mathbb{P}}} \sum_{k=1}^{n_{\mathbb{G}}} m_{ik}}. \tag{4.47}$$

2. Bootstrap the decisions for the first group following the algorithm outlined in Sect. 4.2 to generate $D_{b_i h_{i,k} \ell}^{(1)}$'s. Since each decision from the first group is paired with a decision from the second, we use the exact same set of bootstrapped indices, b_i's and $h_{i,k}$'s for the second group that we used for the first. For each pair of indices selected we take all $m_{b_i h_{i,k}}$ decisions from that selected pair of individuals.
3. Calculate and store $e_\nu = \hat{\nu}_1^b - \hat{\nu}_2^b - (\hat{\nu}_1 - \hat{\nu}_2)$ where

$$\begin{aligned} \hat{\nu}_1^b - \hat{\nu}_2^b &= \frac{\sum_{i=1}^{n_{\mathbb{P}}} \sum_{k=1}^{n_{\mathbb{G}}} \sum_{\ell=1}^{m_{b_i h_{i,k}}} (D_{b_i h_{i,k} \ell}^{(1)} - D_{b_i h_{i,k} \ell}^{(2)})}{\sum_{i=1}^{n_{\mathbb{P}}} \sum_{k=1}^{n_{\mathbb{G}}} m_{b_i h_{i,k}}} \\ &= \frac{\sum_{i=1}^{n_{\mathbb{P}}} \sum_{k=1}^{n_{\mathbb{G}}} \sum_{\ell=1}^{m_{b_i h_{i,k}}} D_{b_i h_{i,k} \ell}^{(1)}}{\sum_{i=1}^{n_{\mathbb{P}}} \sum_{k=1}^{n_{\mathbb{G}}} m_{b_i h_{i,k}}} \\ &\quad - \frac{\sum_{i=1}^{n_{\mathbb{P}}} \sum_{k=1}^{n_{\mathbb{G}}} \sum_{\ell=1}^{m_{b_i h_{i,k}}} D_{b_i h_{i,k} \ell}^{(2)}}{\sum_{i=1}^{n_{\mathbb{P}}} \sum_{k=1}^{n_{\mathbb{G}}} m_{b_i h_{i,k}}}. \end{aligned} \tag{4.48}$$

4. Repeat Steps 2 and 3 above a large number of times, say M times.
5. Then the *p-value* for this test is given by

$$p = \frac{1 + \sum_{\varsigma=1}^{M} I_{\{e_\nu \le \hat{\nu}_1 - \hat{\nu}_2\}}}{M + 1}. \tag{4.49}$$

6. Our decision will be to reject the null hypothesis if the *p-value* is less than some specified significance level α or, more generally, if the *p-value* is small. Otherwise, we will fail to reject that same null hypothesis.

Example 4.15 In this example, we consider two matchers from the XM2VTS database. Both of these are facial matches: (DCTs, MLP) and (DCTs, GMM). To assess these matchers, we use thresholds of 0.4 and −0.3, respectively. We implement the bootstrap algorithm above to test the following hypotheses:

$$H_0 : \nu_1 = \nu_2$$

$$H_1 : \nu_1 < \nu_2.$$

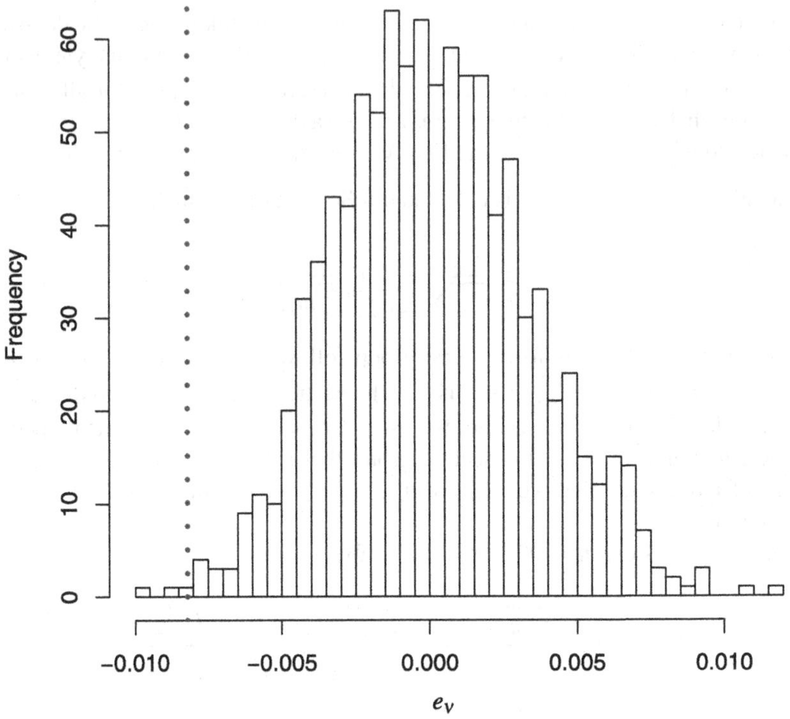

Fig. 4.10 Distribution of e_ν, bootstrapped differences from the sample difference in the FMR's, for comparing the facial matchers (DCTs, MLP) and (DCTs, GMM) from the XM2VTS database

The estimated FMR's are $\hat{\nu}_1 = 0.0123$ and $\hat{\nu}_2 = 0.0204$, respectively. Each of these FMR's is based upon $N_\nu = 40000$ decisions. The distribution of the e_ν's is summarized in Fig. 4.10. The observed difference between the two error rates $\hat{\nu}_1 - \hat{\nu}_2$ is -0.0082 and this value is represented by the dashed vertical line in that graph. We can calculate the *p-value* to be 0.003 based upon $M = 999$ bootstrap replicates of the difference in the two FMR's. Since this *p-value* is small we can conclude that the two error rates here are significantly different. In particular, we conclude that the error rate for (DCTs, MLP) is significantly less than that for (DCTs, GMM).

Randomization Approach The next approach that we will present is a randomization approach to comparing two paired FMR's. As above, we are testing the equality of two FMR's, ν_1 and ν_2, based upon decisions collected in samples of each process against a one-sided alternative hypothesis that one is less than the other. We assume that the sampling process has produced decisions, $D_{ik\ell}$'s, that are paired. Due to the nature of the paired collection, it is necessary to keep intact the 'connections' between the decisions, $D_{ik\ell}^{(1)}$ and $D_{ik\ell}^{(2)}$ in the two groups. With the randomization approach, for each decision in each comparison pair we randomize to which group each decision belongs. That is, for each set of decision indices we will collect the $2m_{ik}$ decisions and randomly divide them between the two groups.

1. Calculate $\hat{\nu}_1 - \hat{\nu}_2$ where $\hat{\nu}_g$ is calculated following

$$\hat{\nu}_g = \frac{\sum_{i=1}^{n_{\mathbb{P}}} \sum_{k=1}^{n_{\mathbb{G}}} \sum_{\ell=1}^{m_{ik}} D_{ik\ell}^{(g)}}{\sum_{i=1}^{n_{\mathbb{P}}} \sum_{k=1}^{n_{\mathbb{G}}} m_{ik}}. \tag{4.50}$$

2. For the each pair of individuals i and k, take the m_{ik} decisions from the first group and the m_{ik} decisions from the second group and combine them. Take these $2m_{ik}$ decisions and randomly permute their order. Then reassign the first m_{ik} decisions from the permuted list to the first group denoted by $D_{ik1}^{(1)*}, \dots, D_{ikm_{ik}}^{(1)*}$. The remaining m_{ik} decisions are assigned to the second group and denoted by $D_{ik1}^{(2)*}, \dots, D_{ikm_{ik}}^{(2)*}$ Repeat this for every pair of individual i and k.
3. Calculate $\hat{\nu}_1^r$ from the N_ν decisions randomized to group 1 and $\hat{\nu}_2^r$ from the N_ν decisions randomized to group 2 and store

$$\Delta_\nu = \hat{\nu}_1^r - \hat{\nu}_2^r \tag{4.51}$$

where

$$\hat{\nu}_g^r = \frac{\sum_{i=1}^{n_{\mathbb{P}}} \sum_{k=1}^{n_{\mathbb{G}}} \sum_{\ell=1}^{m_{ik}} D_{ik\ell}^{(g)*}}{\sum_{i=1}^{n_{\mathbb{P}}} \sum_{k=1}^{n_{\mathbb{G}}} m_{ik}}. \tag{4.52}$$

4. Repeat Steps 2 and 3 above a large number of times, say M times.
5. Then the *p-value* for this test is given by

$$p = \frac{1 + \sum_{\varsigma=1}^{M} I_{\{\Delta_\nu \leq \hat{\nu}_1 - \hat{\nu}_2\}}}{M+1}. \tag{4.53}$$

6. We will reject the null hypothesis if this *p-value* is small.

Example 4.16 From the BANCA database, we focus on two face matchers following the G protocol—the SURREY_face_svm_auto (SURREY-AUTO) and SURREY_face_svm_man (SURREY-SVM-MAN)—for this example. These matchers were previously considered in Example 4.12 to illustrate a confidence interval for paired FMR's. Here our focus is on a different set of thresholds and a hypothesis test for paired FMR's rather than a confidence interval. Note that having different thresholds—we use −0.3 and −0.2—implies that the correlation parameters would be very likely to change if we were to estimate them at these thresholds. (Schuckers [87] has a demonstration of this for other data.) Since we are doing a randomization test, it is not necessary to estimate the correlation parameters. The sample FMR's from these two processes are $\hat{\nu}_1 = 0.0641$ and $\hat{\nu}_2 = 0.0833$. The difference of −0.0192 seems large relative to the size of the FMR's. Applying this randomization test approach, we calculated $M = 999$ randomized differences Δ_ν's. Figure 4.11 summarizes the distribution of the randomized differences in these two FMR's. The dashed vertical line represents the difference in the sample FMR's, $\hat{\nu}_1 - \hat{\nu}_2 = -0.0192$. The *p-value* was found to be 0.146. This *p-value* is large and

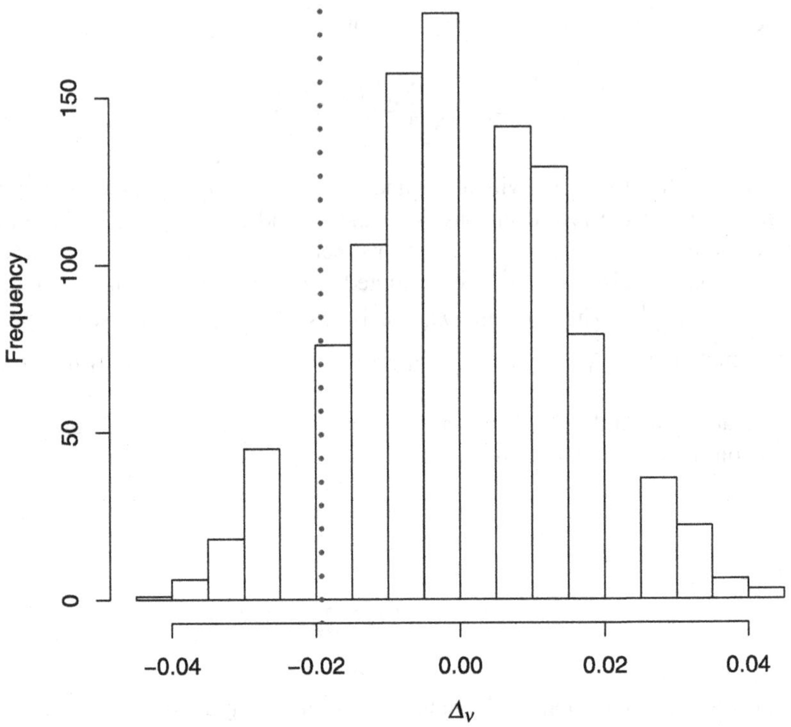

Fig. 4.11 Distribution of Δ_ν, randomized differences in the FMR's from the sample difference in the FMR's, for comparing the facial matches SURREY-AUTO and SURREY-SVM-MAN from the BANCA database

this suggests that there is not a significant difference between the two process FMR's at these thresholds.

Example 4.17 In this example, we apply the randomization test methodology to two facial matchers, (DCTs, MLP) and (DCTs, GMM), from the XM2VTS database. For the former we use a threshold of -0.3 to get an estimated process FMR of $\hat{\nu}_1 = 0.0123$ and for the latter we use a threshold of 0.4 to get an estimated process FMR of $\hat{\nu}_2 = 0.0204$. We apply the randomization approach to these decisions and a summary of the resulting Δ_ν's is given in Fig. 4.12. The observed difference $\hat{\pi}_1 - \hat{\pi}_2 = -0.0082$ is represented by a dashed vertical line. The *p-value* is calculated to be 0.001 based upon $M = 999$ randomized differences. From this we can conclude that there is enough evidence to reject the null hypothesis that these two FMR's are equal. Consequently, we can conclude that the process FMR for (DCTs, MLP) is significantly less than the process FMR for (DCTs, GMM) at these thresholds.

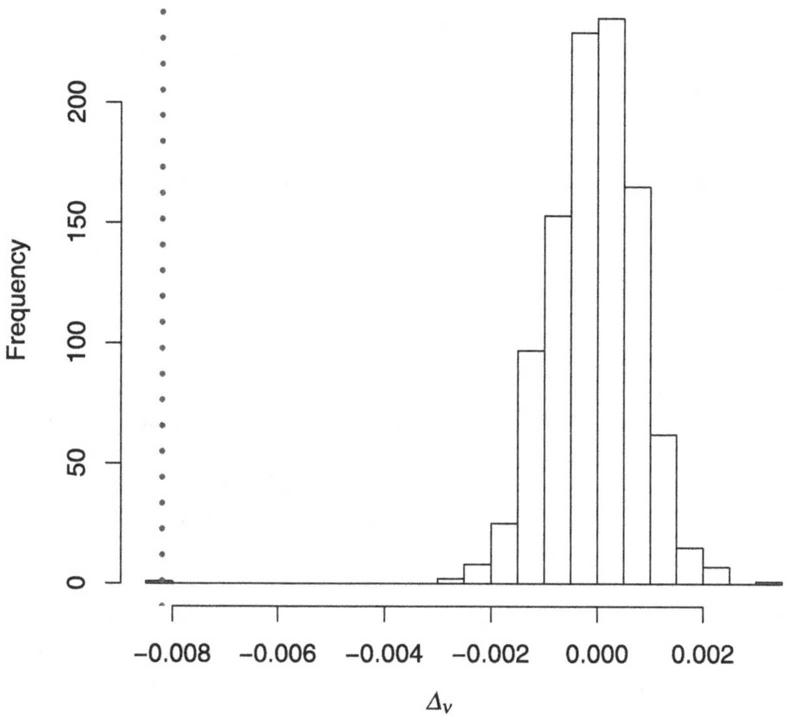

Fig. 4.12 Distribution of Δ_ν, the randomized differences in the FMR's from the observed differences in the FMR's, for comparing the facial matchers (DCTs, MLP) and (DCTs, GMM) from the XM2VTS database

4.3.3 Multiple Sample Methods for FMR's

The methods we develop and present in this section are focused on testing the equality of $G \geq 3$ FMR's. Formally, we are testing the following hypotheses:

$$H_0 : \nu_1 = \nu_2 = \cdots = \nu_G$$
$$H_1 : \text{ not } H_0.$$

All of the approaches in this section do not depend upon the shape of the distribution of the match scores or the shape of the distribution of the decisions. They are dependent upon the type of data collection. Thus, methods are given for testing the above hypothesis for both independent data collections and paired data collections.

4.3.3.1 Hypothesis Test for Multiple Independent FMR's

As mentioned above, we are testing the equality of G different FMR's. In this section, our focus in on process FMR's that have been collected independently. One circumstance where this is applicable is the case where a single system might be deployed in five locations and we could compare the FMR's from those five distinct locations. Below we present a bootstrap approach to this hypothesis test for the case of testing multiple independent FMR's. We will denote differences among our groups $g = 1, \ldots, G$ by a superscript (g) except for the group FMR's which will be denoted by a subscript g, i.e. ν_g. In this section we have only a single approach, the bootstrap approach, for carrying out this hypothesis test. As with the other bootstrap approaches in this chapter, our bootstrap methodology is based upon the *two-instance bootstrap*.

Bootstrap Approach Since we have independent FMR's, the approach we take here is to bootstrap each of the groups separately as was done for the two-sample test for independent FMR's above. We will use the *two-instance bootstrap* described in Sect. 4.2. The test statistics that we will use here is similar to a traditional one that is used as part of an analysis of variance test; however, we will only use the numerator from that test. Thus, our test statistic, F, is effectively the sum of squares between each groups individual FMR. Compare (4.54) to (3.45) to observe the differences between these two types of test statistics. Our reason for using this form of the test is the computational efficiency gained by not recalculating the variance of $\hat{\nu}_g^b$ for each bootstrapped replicate of the data. Additionally, the bootstrapped values for F are calculated assuming that H_0 is true. Consequently, we use the difference of each bootstrapped FMR from the original estimated FMR for that group to approximate the distribution of F when H_0 holds.

1. Calculate

$$F = \sum_{g=1}^{G} N_\nu^{(g)} (\hat{\nu}_g - \bar{\nu})^2 \tag{4.54}$$

   where $\hat{\nu}_g$ is calculated following (4.2) and

$$\bar{\nu} = \frac{\sum_{g=1}^{G} N_\nu^{(g)} \hat{\nu}_g}{\sum_{g=1}^{G} N_\nu^{(g)}}. \tag{4.55}$$

   The value $\bar{\nu}$ is a weighted average of the G sample FMR's and is the estimate of the FMR for all G processes if they all share the same FMR. That is, if the null hypothesis, $\nu_1 = \nu_2 = \cdots = \nu_G$, is true.
2. Bootstrap the decisions for the each of the G groups following the algorithm outlined in Sect. 4.2 to generate $D_{b_i h_{i,k\ell}}^{(g)}$'s.
3. Calculate and store

$$F_\nu = \sum_{g=1}^{G} N_\nu^{(g)b} (\hat{\nu}_g^b - \hat{\nu}_g)^2 \tag{4.56}$$

where

$$\hat{\nu}_g^b = \frac{\sum_{i=1}^{n_{\mathbb{P}}^{(g)}} \sum_{k=1}^{n_{\mathbb{G}}^{(g)}} \sum_{\ell=1}^{m_{b_i h_{i,k}}^{(g)}} D_{b_i h_{i,k} \ell}^{(g)}}{\sum_{i=1}^{n_{\mathbb{P}}^{(g)}} \sum_{k=1}^{n_{\mathbb{G}}^{(g)}} m_{b_i h_{i,k}}^{(g)}} \tag{4.57}$$

and

$$N_\nu^{(g)b} = \sum_{i=1}^{n_{\mathbb{P}}^{(g)}} \sum_{k=1}^{n_{\mathbb{G}}^{(g)}} m_{b_i h_{i,k}}^{(g)}. \tag{4.58}$$

4. Repeat Steps 2 and 3 above a large number of times, say M times.
5. Then the *p-value* for this test is given by

$$p = \frac{1 + \sum_{\varsigma=1}^{M} I_{\{F_\nu \geq F\}}}{M+1}. \tag{4.59}$$

6. We reject the null hypothesis H_0 if the *p-value* is small. For a specified significance level α, we will reject if $p < \alpha$.

Note that when the null hypothesis $H_0 : \nu_1 = \cdots = \nu_G$ is true, the variability in the resampled FMR should be obtained via the recentered

$$\hat{\nu}_g^b - \hat{\nu}_g + \bar{\nu}. \tag{4.60}$$

Using the above recentered bootstrapped FMR, we can derive the appropriate F_ν to be

$$F_\nu = \sum_{g=1}^{G} N_\nu^{(g)} ((\hat{\nu}_g^b - \hat{\nu}_g + \bar{\nu}) - \bar{\nu})^2 = \sum_{g=1}^{G} N_\nu^{(g)} (\hat{\nu}_g^b - \hat{\nu}_g)^2. \tag{4.61}$$

Example 4.18 Here we are interested in testing whether three independent FMR's are equal or whether at least one of the FMR's is significantly different from the rest. We are testing three speaker/voice matchers. The first is a voice matcher, UC3M_voice_gmm_auto_scale_34_300 (UC3M-34-300), from the BANCA database following protocol G. The data analyzed in this example come from group g1 of that study. The second is also a voice matcher from the same database and following the same protocol but on group g2. That matcher is IDIAP_voice_gmm_auto_scale _33_300 (IDIAP-33-300). The last matcher to be compared is a speaker matcher, (PAC, GMM) from the XM2VTS database. For these matchers we use thresholds of 0.0, 0.0 and 2.0, which, in turn, yield estimated FMR's of 0.0673, 0.0321 and 0.0570, respectively. Figure 4.13 has the distribution of F_ν based upon $M = 999$. The vertical dashed line here represents $F = 0.2268$ which is the test statistic from the original data. The *p-value* here is 0.873 and is large. Therefore, we conclude that there is not enough evidence to reject the null hypothesis of equality of these three FMR's based upon the decisions in these three samples.

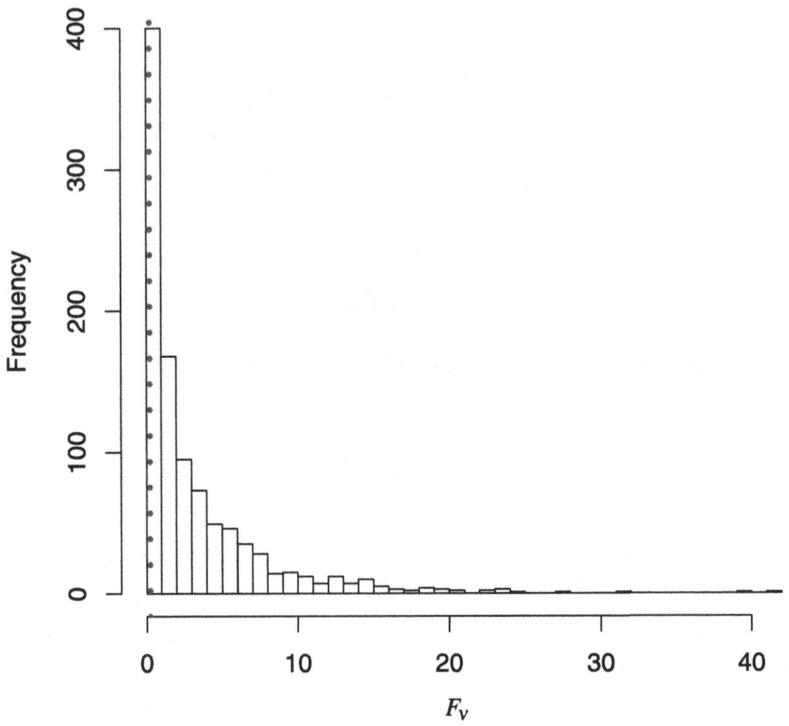

Fig. 4.13 Distribution of F_ν, the bootstrapped values of the test statistic for comparing three independent facial matchers—UC3M-34-300, IDIAP-33-300 and (PAC, GMM)—found in Example 4.18

4.3.3.2 Hypothesis Test for Multiple Paired FMR's

Paired data are often the result of multiple classifiers or matchers being applied to the biometric captures of a single database. One possible focus of such an application is how well these matchers perform on that database. We might be interested in the false match rate of seven different facial algorithms applied to a single database of signals/images, for example. Since each decision on a pair of faces would have six other 'associated' decisions on that same pair of faces, the paired analyses given below would be appropriate. In this section, we present statistical methods for testing the equality of multiple paired false match rates. We will present two approaches, a bootstrap approach and a randomization approach, for testing

$$H_0 : \nu_1 = \nu_2 = \cdots = \nu_G$$
$$H_1 : \text{ not } H_0.$$

Bootstrap Approach Our approach here is a bootstrap one for testing the equality of G different process FMR's against an alternative that at least one is different from the others. Below we present an algorithm for carrying out the hypothesis test given above based upon the *two-instance bootstrap*.

1. Calculate and store
$$F = \sum_{g=1}^{G} (\hat{\nu}_g - \bar{\nu})^2 \tag{4.62}$$
where $\hat{\nu}_g$ is calculated following (4.2) and
$$\bar{\nu} = \frac{1}{G} \sum_{g=1}^{G} \hat{\nu}_g. \tag{4.63}$$
$\bar{\nu}$ is the estimate of the FMR if we assume that all G groups have the same FMR which is our null hypothesis.
2. Bootstrap the decisions for the first group following the algorithm outlined in Sect. 4.2 to generate $D^{(1)}_{b_i h_{i,k} \ell}$'s. Since each decision from the first group is paired with $G-1$ other decisions, we use the exact same set of bootstrapped indices for all groups that we used for the first group.
3. Calculate and store
$$F_\nu = \sum_{g=1}^{G} (\hat{\nu}^b_g - \hat{\nu}_g)^2 \tag{4.64}$$
where
$$\hat{\nu}^b_g = \frac{\sum_{i=1}^{n_{\mathbb{P}}} \sum_{k=1}^{n_{\mathbb{G}}} \sum_{\ell=1}^{m_{b_i h_{i,k}}} D^{(g)}_{b_i h_{i,k} \ell}}{\sum_{i=1}^{n_{\mathbb{P}}} \sum_{k=1}^{n_{\mathbb{G}}} m_{b_i h_{i,k}}}. \tag{4.65}$$
4. Repeat Steps 2 and 3 above a large number of times, say M times.
5. Then the *p-value* for this test is given by
$$p = \frac{1 + \sum_{\varsigma=1}^{M} I_{\{F_\nu \geq F\}}}{M+1}. \tag{4.66}$$
6. We will reject the null hypothesis if the *p-value* is small.

Example 4.19 To illustrate the methodology above for testing multiple paired FMR's, we compare three, $G = 3$, FMR's calculated on group g1 from the G protocol of the BANCA database. These three classifiers are all voice classifiers. Specifically they are: IDIAP_voice_gmm_auto_scale_25_100_pca (IDIAP-25-100-PCA), IDIAP_voice_gmm_auto_scale_25_100 (IDIAP-25-100), and IDIAP_voice_gmm_auto_scale_25_10_pca (IDIAP-25-10-PCA). These classifiers were tested at the following thresholds 0.1, 0.0 and 0.2, respectively. From each of these thresholds, the FMR's that were calculated are $\hat{\nu}_1 = 0.0545$, $\hat{\nu}_2 = 0.0801$ and $\hat{\nu}_3 = 0.0704$. This sample consists of $N_\nu = 312$ decision for each process. The goal of the hypothesis test we are undertaking is to determine if those estimated false match rates are

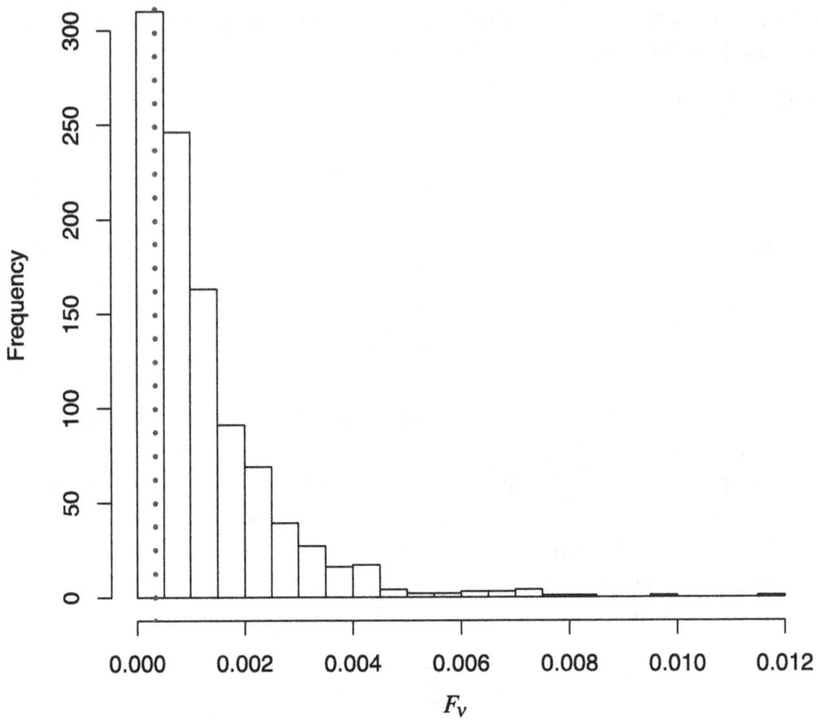

Fig. 4.14 Distribution of F_ν, the values for the bootstrapped test statistic, for comparing IDIAP-25-100-PCA, IDIAP-25-100 and IDIAP-25-10-PCA from the BANCA database

significantly different. Following the bootstrap hypothesis test methodology given above, we created $M = 999$ bootstrap replicates of the false match decision data. Figure 4.14 shows the resulting distribution of the F_ν's. The dashed line represents the value of $F = 0.00034$ which is calculated on the original, i.e. not bootstrapped, data. The *p-value* for this test is 0.817 and, consequently, we fail to reject the null hypothesis that these three process FMR's are equal.

Randomization Approach We are testing the equality of three or more FMR's based upon decisions collected in samples of the processes as we did above. This test is

$$H_0 : \nu_1 = \nu_2 = \cdots = \nu_G$$
$$H_1 : \text{ not } H_0.$$

Due to the nature of the paired comparisons, it is necessary to keep intact the association between the decisions for a given pair of individuals, i and k, across the G groups. With the randomization approach, for each decision in each comparison pair we randomize to which group each decision belongs. We do this by concatenating all Gm_{ik} decisions for a given pair of individuals and permuting them among the

G groups. Having done that all individuals, we recalculate the variability between the FMR's to determine how unusual the original set of observed FMR's is. This algorithm is given below.

1. Calculate F

$$F = \sum_{g=1}^{G} (\hat{\nu}_g - \hat{\nu}_p)^2 \tag{4.67}$$

   where $\hat{\nu}_g$ is calculated following (4.2) and

$$\hat{\nu}_p = \frac{1}{G} \sum_{g=1}^{G} \hat{\nu}_g. \tag{4.68}$$

   where $\hat{\nu}_g$ is calculated following (4.2).
2. For the Gm_{ik} decisions that are 'connected' for the ordered pair of individuals i and k, randomly permute the decisions between the G groups. Take the gth m_{ik} decisions from the list of permuted decisions and assign them to the gth group. Denote these decisions by $D_{ik1}^{(g)*}, \ldots, D_{ikm_{ik}}^{(g)*}$ for $g = 1, \ldots, G$.
3. Calculate and store

$$F_\nu = \sum_{g=1}^{G} (\hat{\nu}_g^b - \hat{\nu}_p^b)^2 \tag{4.69}$$

   where

$$\hat{\nu}_g^b = \frac{\sum_{i=1}^{n_{\mathbb{P}}} \sum_{k=1}^{n_{\mathbb{G}}} \sum_{\ell=1}^{m_{ik}} D_{ik\ell}^{(g)*}}{\sum_{i=1}^{n_{\mathbb{P}}} \sum_{k=1}^{n_{\mathbb{G}}} m_{ik}} \tag{4.70}$$

   and

$$\hat{\nu}_p^b = \frac{1}{G} \sum_{g=1}^{G} \hat{\nu}_g^b. \tag{4.71}$$

4. Repeat Steps 2 and 3 above a large number of times, say M times.
5. Then the *p-value* for this test is given by

$$p = \frac{1 + \sum_{\varsigma=1}^{M} I_{\{F_\nu \geq F\}}}{M+1}. \tag{4.72}$$

6. If the *p-value* is small, we will reject the null hypothesis, H_0. If a significance level α has been specified, we will reject the null hypothesis if $p < \alpha$. Otherwise we fail to reject the null hypothesis.

Example 4.20 The randomization analysis approach that is done for this example is performed on the same matchers that were previously analyzed in Example 4.19. As in that example, we are interested in testing the equality

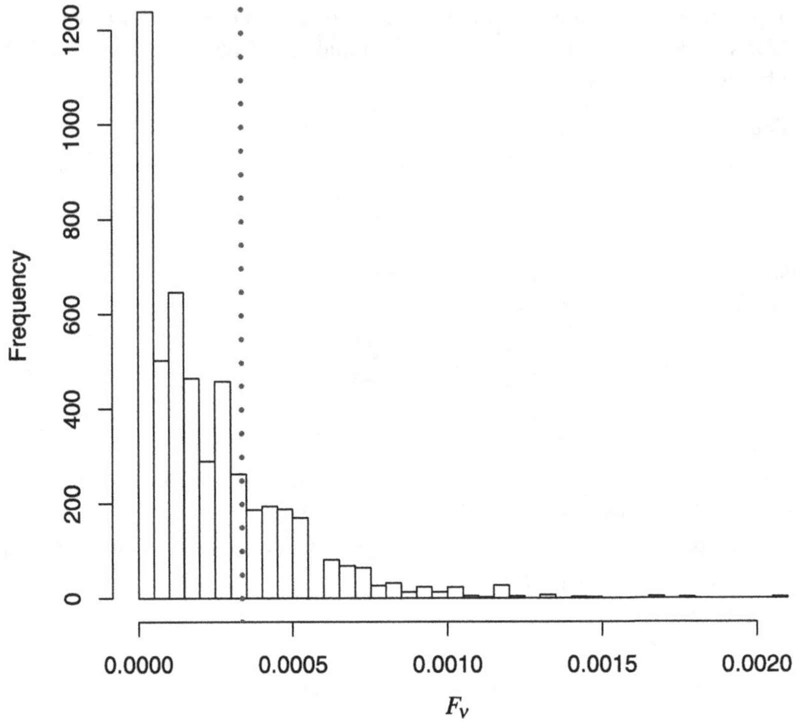

Fig. 4.15 Distribution of F_ν, the randomized values of the test statistic, for comparing IDIAP-25-100-PCA, IDIAP-25-100 and IDIAP-25-10-PCA from the BANCA database

of three voice FMR's. The three voice matchers that comprised that analysis were all collected under the G protocol of the BANCA database. Those matchers are IDIAP_voice_gmm_auto_scale_25_100_pca (IDIAP-25-100-PCA), IDIAP_voice_gmm_auto_scale_25_100 (IDIAP-25-100), and IDIAP_voice_gmm_auto_scale_25_10_pca (IDIAP-25-10-PCA). Tested at the following thresholds 0.1, 0.0 and 0.2, respectively, the FMR's that were calculated are $\hat{\nu}_1 = 0.0545$, $\hat{\nu}_2 = 0.0801$ and $\hat{\nu}_3 = 0.0704$. This sample consists of $N_\nu = 312$ decision for each process. Figure 4.15 gives the distribution of F_ν from $M = 4999$ randomization replicates. The dashed line represents F which is calculated from the original, i.e. unpermuted, match decisions. The *p-value* here is 0.2772. From this we fail to reject the null hypothesis and conclude that there is not a significant different for these three process FMR's based upon the decisions in this sample. We do note that the *p-value* here, 0.2640 is noticeably different from the bootstrapped *p-value* with $p = 0.817$, from Example 4.19. While the difference is striking, it is likely due to the different structures present in the two different testing methodologies, to the discrete nature of the decisions and the small number of decisions. It is also important to note that both approaches yield the same conclusion here.

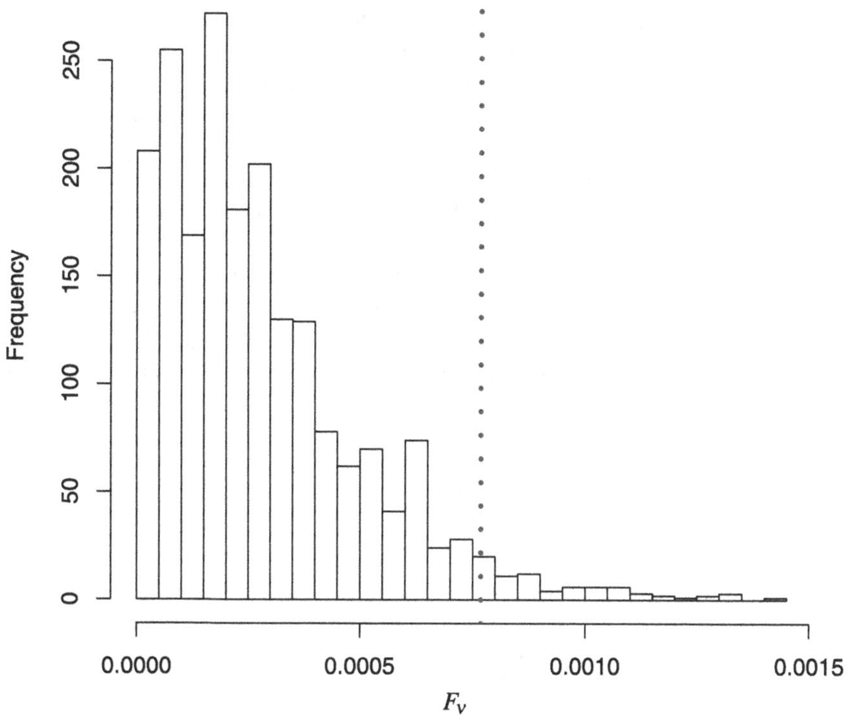

Fig. 4.16 Distribution of F_ν, the randomized values for the test comparing IDIAP-33-100-PCA, IDIAP-33-100, IDIAP-33-10-PCA, IDIAP-33-10 applied to group g2 of the BANCA database

Example 4.21 In this example we are comparing four voice matchers in this example. Those matchers are IDIAP_voice_gmm_auto_scale_33_100_pca (IDIAP-33-100-PCA), IDIAP_voice_gmm_auto_scale_33_100 (IDIAP-33-100), IDIAP_voice_gmm_auto_scale_33_10_pca (IDIAP-33-10-PCA), IDIAP_voice_gmm_auto_scale_33_10 (IDIAP-33-10). We are comparing the false match rates of those matchers applied to group g1 of the BANCA database. Using the thresholds of 0.0, 0.0, 0.2 and 0.07, we get estimated FMR's of $\hat{\nu}_g = 0.0769, 0.0449, 0.0513$, and 0.0737, respectively. We randomized $M = 1999$ times to obtain the distribution of F_ν that is summarized in Fig. 4.16. The test statistic, F, takes the value 0.0007679. That value is represented by the dashed vertical line on that graph. The *p-value* for this test 0.034 which suggests that the differences between these FMR's are significantly different.

4.4 Sample Size and Power Calculations

One important component of planning an evaluation of the performance of a matching system is the amount of data to be collected. There are a number of factors that need to be considered for determining how much data to collect. In this section we present statistical methods for the calculation of the number of individuals that need to be tested to achieve certain criteria relevant to a single FMR. We will assume here that the data collection is done using all possible cross comparisons so that $n_{\mathbb{P}} = n_{\mathbb{G}} = n$. Then the total number of individual pairs from which decisions could be drawn is $n(n-1)$. If this is not the case, then generalizations can be derived following the outline of the process given below. Knowing n is something that is often important to those charged with evaluating a classification or matching algorithm. The methodology here is dependent upon some knowledge or assumption about the matching process under study. Without those assumptions, it is not possible to derive *a priori* calculations of n. Below we distinguish between two type of calculations: those for a confidence interval (sample size calculations) and those for a hypothesis test (power calculations).

4.4.1 Sample Size Calculations

It is possible to derive confidence intervals by assuming a simplified correlation structure. The approach we present here was first proposed by Schuckers [87]. If we assume that $m_{ik} = m$ for all i and that $\omega_1 = \omega_2 = \omega_3 = \omega$, then the variance given in (4.5) becomes

$$V[\hat{\nu}] = [(1 + \hat{\xi}_1) + (\hat{\eta} + \hat{\xi}_2)(m-1) + 4\hat{\omega}(n-2)]. \tag{4.73}$$

Because the ω's represent correlations between decisions where one (and only one) of the individuals involved in those decisions is the same from each pair, it might be reasonable to assume that these correlations would be similar. Making this assumption is effectively averaging ω_1, ω_2 and ω_3 to get a single ω. We can then invert (4.14) and solve for n. As we did above, we will assume that the number of decisions per comparison pair is constant and known (or specified), $m_{jk} = m$, and then determine the value for n based upon this. We then want to solve the following equation:

$$B = z_{\alpha/2}\left[\frac{\nu(1-\nu)}{n(n-1)m}\right][(1 + \hat{\xi}_1) + (\hat{\eta} + \hat{\xi}_2)(m-1) + 4\hat{\omega}(n-2)]. \tag{4.74}$$

Here, as before, B is the margin of error desired and $z_{\alpha/2}$ is the $1-\alpha/2$th percentile of a standard Gaussian distribution. (That is, the desired confidence interval is $\hat{\nu} \pm B$.) Estimates of the parameters ν, ω, ξ_1, ξ_2 and η need to be determined beforehand so that we can determine the necessary n. To solve (4.74), we must solve a quadratic. We use only the 'plus' part of the solution to the quadratic equation. That is, we take only $(-b + \sqrt{b^2 - 4ac})/2a$ as a solution to $ax^2 + bx + c = 0$ since the only way for the 'minus' part of the quadratic equation to yield a positive result is for the sum $\eta + \xi_2$ to be negative. Not believing that correlation parameters will be negative, we discard this possible solution to the quadratic equation. Assuming that

$n \approx n-1 \approx n-2$, we get

$$n = \left\lceil \frac{2z_{1-\alpha/2}^2 \omega \nu(1-\nu)}{B^2 m} + \left(\left(\frac{2z_{\alpha/2}^2 \omega \nu(1-\nu)}{B^2 m} \right)^2 + \frac{z_{\alpha/2}^2 \nu(1-\nu)}{B^2 m} [(1+\xi_1) + (\eta+\xi_2)(m-1)] \right)^{1/2} \right\rceil. \tag{4.75}$$

The calculations above are based upon the general correlation model given in (4.4). It is possible to derive other sample size calculations for other correlation structures. We recognize that there are additional uncertainties that are part of any biometric system evaluation. Therefore, sample size calculations such as (4.75) should be viewed as approximations and viewed in the context of the data collection.

4.4.2 Power Calculations

We now consider power calculations for a hypothesis test of a single FMR. Recall that for a hypothesis of a single FMR we have the following hypothesis test:

$$H_0 : \nu = \nu_0$$
$$H_1 : \nu < \nu_0.$$

Our goal here is to ascertain the number of individuals that are needed to carry out the above test for a significance level α and a power of $1-\beta$. In order to obtain the number of individuals from whom data will be collected, it is necessary to specify a particular value for ν_a in the alternative hypothesis. That is, $\nu_a < \nu_0$. As with all power calculations, only by designating a specific value can we determine the number of individuals to test explicitly. We will make the same assumptions about the data collection here that we made for the sample size calculations in the previous section. Specifically, we will assume that the probe and the gallery are comprised of the same individuals and so $n_{\mathbb{P}} = n_{\mathbb{G}} = n$.

As was the case for the FNMR, the difference between ν_0 and ν_a is known as the 'effect size'. As we will see, the quantity $\nu_0 - \nu_a$ takes the place of the margin of error, B, in the sample size calculations. If the sample size is large, then we can assume that the sampling distribution of $\hat{\nu}$ is approximately Gaussian. To build our power calculation, we must first specify that

$$P\left(\frac{\hat{\nu} - \nu_0}{s_{\hat{\nu}0}} < -z_\alpha | \nu = \nu_a \right) = 1 - \beta \tag{4.76}$$

where $s_{\hat{\nu}0} = \sqrt{\hat{V}[\hat{\nu}] \,|_{\hat{\nu}=\nu_0}}$. This is the formal statement of the probability we wish to attain. We would like to reject the null hypothesis, H_0, with probability $1-\beta$ assuming that the alternative hypothesis, $\nu = \nu_a$ is true. We reject that same null hypothesis for a given significance level α when

$$\frac{\hat{\nu} - \nu_0}{s_{\nu 0}} < -z_\alpha. \tag{4.77}$$

If the alternative hypothesis is true, we can write that

$$P(\hat{\nu} < \nu_0 - z_\alpha s_{\hat{\nu}0} \mid \nu = \nu_a) = 1 - \beta. \tag{4.78}$$

Subtracting ν_a from both sides of the inequality and dividing by $s_{\nu a}$, we get

$$P\left(\frac{\hat{\nu} - \nu_a}{s_{\hat{\nu}a}} < \frac{\nu_0 - z_\alpha s_{\hat{\nu}0} - \nu_a}{s_{\hat{\nu}a}}\right) = 1 - \beta \tag{4.79}$$

where $s_{\hat{\nu}a} = \sqrt{\hat{V}[\hat{\nu}] \mid_{\nu=\nu_a}}$. This becomes

$$P\left(Z < \frac{\nu_0 - z_\alpha s_{\hat{\nu}0} - \nu_a}{s_{\hat{\nu}a}}\right) = 1 - \beta. \tag{4.80}$$

Assuming that we have a sufficiently large sample, then Z is a standard Gaussian random variable and, thus, we can equate the right hand size to the appropriate percentile, z_β of that distribution. We then have

$$\begin{aligned} z_\beta &= \frac{\nu_0 - z_\alpha s_{\hat{\nu}0} - \nu_a}{s_{\hat{\nu}a}} \\ &= \frac{\nu_0 + z_\alpha \sqrt{\frac{\nu_0(1-\nu_0)}{n(n-1)m}[(1+\hat{\xi}_1) + (\hat{\eta}+\hat{\xi}_2)(m-1) + 4\hat{\omega}(n-2)]} - \nu_a}{\sqrt{\frac{\nu_a(1-\nu_a)}{n(n-1)m}[(1+\hat{\xi}_1) + (\hat{\eta}+\hat{\xi}_2)(m-1) + 4\hat{\omega}(n-2)]}} \end{aligned} \tag{4.81}$$

where ω, η, $\hat{\xi}_1$ and $\hat{\xi}_2$ along with the other correlation parameters here represent the estimated values for those parameters. Estimation of these parameters should be done based upon existing data or data from similar studies. See Comment 2.9 for additional information on estimation of parameters for sample size and power calculations. We can then write

$$\begin{aligned} \nu_a - \nu_0 = &\frac{1}{\sqrt{n(n-1)m}} \left(z_\alpha \sqrt{\nu_0(1-\nu_0)[(1+\hat{\xi}_1) + (\hat{\eta}+\hat{\xi}_2)(m-1) + 4\hat{\omega}(n-2)]}\right. \\ &\left. + z_\beta \sqrt{\nu_a(1-\nu_a)[(1+\hat{\xi}_1) + (\hat{\eta}+\hat{\xi}_2)(m-1) + 4\hat{\omega}(n-2)]}\right). \end{aligned} \tag{4.82}$$

Squaring both sides, assuming that $n - 1 \approx n$ and solving (partially) for n we get

$$\begin{aligned} n = &\, m^{-1/2} \mid \nu_0 - \nu_a \mid^{-1} \\ &\times \left(z_\alpha \sqrt{\nu_0(1-\nu_0)[(1+\hat{\xi}_1) + (\hat{\eta}+\hat{\xi}_2)(m-1) + 4\hat{\omega}(n-2)]}\right. \\ &\left. + z_\beta \sqrt{\nu_a(1-\nu_a)[(1+\hat{\xi}_1) + (\hat{\eta}+\hat{\xi}_2)(m-1) + 4\hat{\omega}(n-2)]}\right). \end{aligned} \tag{4.83}$$

We first note that we have not completely solved (4.83) for n, the number of individuals. The right hand side of this equation still contains n. We have chosen to leave (4.83) in this form because it mirrors the general form for many power calculations. However, in order to obtain a specific value for n it is necessary to follow an iterative process. An algorithm for doing this is given below:

1. Choose a reasonable starting value for n_0.
2. Calculate

$$n_{t+1} = m^{-1/2} \mid \nu_0 - \nu_a \mid^{-1} \times \Big(z_\alpha \sqrt{\nu_0(1-\nu_0)[(1+\hat{\xi}_1) + (\hat{\eta}+\hat{\xi}_2)(m-1) + 4\hat{\omega}(n_t - 2)]} + z_\beta \sqrt{\nu_a(1-\nu_a)[(1+\hat{\xi}_1) + (\hat{\eta}+\hat{\xi}_2)(m-1) + 4\hat{\omega}(n_t - 2)]} \Big). \tag{4.84}$$

3. Repeat the previous step if $\mid n_{t+1} - n_t \mid > 1$.

Next we observe that the power calculation given here is significantly more complex than other power calculations in this text. This is a direct consequence of the correlation structure for an FMR. The final remark of this section is a reminder that the power calculation specified in (4.83) is for the testing and evaluation of a single FMR. Power calculations for testing the difference between multiple FMR's are not presented here but can be derived following the same approach that we have used for the single FMR case.

4.5 Prediction Intervals

Prediction intervals provide a way to create confidence intervals for FMR's of future observations based upon already observed false match data. Recall that confidence intervals, like those given in (4.26), are for estimation of the mean of a process but are not meant to capture values of a sample FMR based upon future observations. Prediction intervals on the other hand are constructed to reflect the variability inherent both in estimation of the process FMR and in future observations. To those ends, we add the following notation. Let $\hat{\nu}^\diamond$ be the future unobserved FMR based upon $n^\diamond$ individuals yielding $m^\diamond_{ik}$ decisions from individual i and individual k. We assume here that the probe and gallery are composed of the same individuals. Generalizations of this assumption are readily accommodated. Thus we create below a prediction interval for $\hat{\nu}^\diamond$ where

$$\hat{\nu}^\diamond = \frac{\sum_{i=1}^{n^\diamond} \sum_{k=1}^{n^\diamond} \sum_{\ell=1}^{m^\diamond_{ik}} D^\diamond_{ik\ell}}{\sum_{i=1}^{n^\diamond} \sum_{k=1}^{n^\diamond} m^\diamond_{ik}} \tag{4.85}$$

and $D^\diamond_{ik\ell}$ represents an unobserved false match decision. Let

$$\bar{m}^\diamond = \frac{\sum_{i=1}^{n^\diamond} \sum_{k \neq i}^{n^\diamond} m^\diamond_{ik}}{n^\diamond(n^\diamond - 1)} \tag{4.86}$$

which will represent the average number of future decisions per pair of individuals to be compared. For simplicity, we will assume that the number of decisions per pair is constant. That is, for all i and all k, $m^{\diamond}_{ik} = m^{\diamond}$. Then the estimated variance for an FMR based upon $n^{\diamond}(n^{\diamond} - 1)m^{\diamond}$ decisions is

$$\begin{aligned}
\hat{V}[\hat{\nu}^{\diamond}] &= \frac{\hat{\nu}(1-\hat{\nu})}{n(n-1)\bar{m}}[(1+\hat{\xi}_1) + (\hat{\eta}+\hat{\xi}_2)(m-1) + 4\hat{\omega}(n-2)] \\
&\quad + \frac{\hat{\nu}(1-\hat{\nu})}{n^{\diamond}(1-n^{\diamond})m^{\diamond}}[(1+\hat{\xi}_1) + (\hat{\eta}+\hat{\xi}_2)(m^{\diamond}-1) + 4\hat{\omega}(n^{\diamond}-2)] \\
&= \hat{\nu}(1-\hat{\nu})\Bigg[(1+\hat{\xi}_1)\left(\frac{1}{n(n-1)\bar{m}} + \frac{1}{n^{\diamond}(n^{\diamond}-1)m^{\diamond}}\right) \\
&\quad + (\hat{\eta}+\hat{\xi}_2)\left(\frac{m-1}{n(n-1)\bar{m}} + \frac{m^{\diamond}-1}{n^{\diamond}(n^{\diamond}-1)m^{\diamond}}\right) \\
&\quad + 4\hat{\omega}\left(\frac{n-2}{n(n-1)\bar{m}} + \frac{n^{\diamond}-2}{n^{\diamond}(n^{\diamond}-1)m^{\diamond}}\right)\Bigg].
\end{aligned} \tag{4.87}$$

Example 4.22 To illustrate the prediction interval methodology above, we apply it to the voice matcher IDIAP_voice_gmm_auto_scale_25_100_pca (IDIAP-25-100-PCA) from the BANCA database. At a threshold of 0.1, the FMR is 0.0513. A 95% confidence interval for the FMR here is $0.0531 \pm 1.96(0.0148) = (0.0223, 0.0803)$. This is derived from correlation parameter estimates of $\hat{\omega}_1 = -0.0181$, $\hat{\omega}_2 = 0.0298$, $\hat{\omega}_3 = -0.0121$ and $\hat{\xi}_1 = 0.0777$. These data will not support estimation of η and ξ_2 since the combinations decisions that lead to these parameters are not present in these data. Suppose that we have planned a data collection using this matcher that will include 20 new individuals and each pair of individuals will be compared 3 times. Then, following the methodology above we get a 95% prediction interval for $\hat{\nu}^{\diamond}$ that is $0.0531 \pm 1.96(0.0167) = (0.0185, 0.0840)$. We can be 95% confident that the estimated FMR for this new collection assuming the process is stationary is between 1.85% and 8.40%.

Note that the prediction interval here is slightly wider than the confidence interval that was calculated—0.0167 as compared to 0.0148. We have assumed that all pairs of the 20 new individuals will be compared 3 times. This gives $20(19)(3) = 1140$ future decisions which is a good deal more than the 312 decisions that went into the confidence interval.

4.6 Discussion

This chapter has presented an array of methods for estimation and testing of verification false match rates(vFRM). These methods both parametric and non-parametric are build upon the correlation structure found in (4.4) for two instance matching, that is, for the matching of one capture from each of two different individuals. From that correlation structure, we have devised the methodology in this chapter including confidence intervals, hypothesis tests, sample size and power calculations and prediction intervals. To accomplish these methods, we have followed the correlation structure proposed by Schuckers [87]. The bootstrap methodology that we proposed and described in Sect. 4.2, the *two-instance bootstrap*, is meant to provide variance estimates similar to those given by the correlation model in (4.4). Building upon these methods, we have developed techniques and approaches that allows for estimation of a single FMR as well as the comparison of two or more. The mechanisms that we have given in this chapter for carrying out those methodologies are *new* to the biometric authentication literature.

A related topic of interest to testers of biometric authentication devices is the estimation of an identification false match rate (iFMR) . Identification false match rates occur when taking a particular biometric signal capture and comparing across a database or gallery of signals rather than doing a one-to-one comparison as we do in the vFMR. We assume that the decision resulting from such a search is either a match or a non-match where the tester knows whether the individual from which the probe signal was obtained is in the gallery. As a consequence, we really to not have a two instance correlation structure but rather we have a single instance structure. Because we have only information on the single individual from the probe, we would use a repeated measures correlation structure similar to the one used for estimation of both the FNMR and FTA. Thus, we can let $D_{i\ell}$ be the decision from submitting the ℓth capture from individual i against a gallery. The correlation structure would then be

$$Corr(D_{i\ell}, D_{i'\ell'}) = \begin{cases} 1 & \text{if } i = i', \ell = \ell' \\ \phi & \text{if } i = i', \ell \neq \ell' \\ 0 & \text{otherwise.} \end{cases} \tag{4.88}$$

Using this correlation structure, it is possible to develop methods for estimation and comparison of multiple iFMR's. Methods based upon this approach would follow those from Chap. 3 where the correlation structure is similar.

The methodology that we have provided for sample size and power calculations are based upon the general correlation structure for an asymmetric matcher. A different correlation structure will result in different formulae for determining the relevant sample size. The basic procedure for deriving these formulae would be the same as those used in this chapter.

4.7 Appendix: Estimation of Simplified Correlation Structures

In this appendix, we give additional details on two simplified models for the general correlation structure we gave earlier in this chapter.

4.7.1 Symmetric Matcher Correlation Structure

Here, we derive the correlation structure for decisions from a symmetric matching algorithm for false match decisions. This structure is a special case of the general false match correlation model in (4.4). Let $D_{ik\ell}$ represent the ℓth decision from the comparison pair (i,k), $\ell = 1,\dots,m_{ik}$, $i = 1,\dots,n_{\mathbb{P}}$, $k = 1,\dots,n_{\mathbb{G}}$ where i and $k \neq i$ are individuals. Here we use essentially the same notation as for the general case; however, we truncate our data to avoid replications due to symmetry.

$$Corr(D_{ik\ell}, D_{i'k'\ell'}) = \begin{cases} 1 & \text{if } i=i', k=k', \ell=\ell', \\ 1 & \text{if } i=k', i'=k, \ell=\ell', \\ \eta & \text{if } i=i', k=k', \ell\neq\ell', \\ \eta & \text{if } i=k', i'=k, \ell\neq\ell', \\ \omega_1 & \text{if } i=i', k\neq k', i\neq k', \\ \omega_2 & \text{if } k=k', i\neq i', k\neq k', \\ \omega_3 & \text{if } i'=k, i\neq k', i\neq i', k\neq k' \\ \omega_3 & \text{if } i=k', i'\neq k, i\neq i', k\neq k' \\ 0 & \text{otherwise.} \end{cases} \tag{4.89}$$

The correlation in (4.89) is a simplification of the more general case. This simplification is the direct result of the symmetry of the decisions. Effectively, this means that $\xi_1 = 1$ and $\xi_2 = \eta$. See the discussion in Sect. 4.1. Under this model,

$$\begin{aligned} V[\hat{\nu}] &= N^{-2}\nu(1-\nu)\mathbf{1}^T\boldsymbol{\Phi}_\nu\mathbf{1} \\ &= N_\nu\nu(1-\nu)\Bigg[\left(\sum_{i=1}^{n_{\mathbb{P}}}\sum_{k\neq i}^{n_{\mathbb{G}}} m_{ik}\right) + \eta\left(\sum_{i=1}^{n_{\mathbb{P}}}\sum_{k\neq i}^{n_{\mathbb{G}}} m_{ik}(m_{ik}-1)\right) \\ &\quad + \omega_1\sum_{i=1}^{n}\sum_{k\neq i}^{n_{\mathbb{P}}} m_{ik}\sum_{k'=k+1}^{n_{\mathbb{G}}} m_{ik'} + \omega_2\sum_{i=1}^{n_{\mathbb{P}}}\sum_{k\neq i,i'}^{n_{\mathbb{G}}} m_{ik}\sum_{\substack{i'=1\\ i'\neq i}}^{\mathbb{P}} m_{i'k} \\ &\quad + \omega_3\left(\sum_{i=1}^{n_{\mathbb{P}}}\sum_{k\neq i}^{n_{\mathbb{G}}} m_{ik}\left\{\sum_{\substack{k'\neq k\\ k'\neq i}}^{n_{\mathbb{G}}} m_{kk'} + \sum_{\substack{i'\neq k\\ i'\neq i}}^{n_{\mathbb{P}}} m_{i'i}\right\}\right)\Bigg]. \end{aligned} \tag{4.90}$$

To obtain estimators for these parameters, we set $(\mathbf{D}_\nu - \hat{\nu}\mathbf{1})(\mathbf{D}_\nu - \hat{\nu}\mathbf{1})^T$, the estimated second central moment for $\mathbf{D}$, equal to Σ, the model variance defined by (4.89) and (4.90), and solve for the correlation parameters. The resulting estimators are

$$\hat{\eta} = \left(\hat{\nu}(1-\hat{\nu})\sum_{i=1}^{n_{\mathbb{P}}}\sum_{k\neq i}^{n_{\mathbb{G}}} m_{ik}(m_{ik}-1)\right)^{-1}$$

$$\times \sum_{i=1}^{n_{\mathbb{P}}} \sum_{k \neq i}^{n_{\mathbb{G}}} \sum_{\ell=1}^{m_{ik}} \sum_{\substack{\ell'=1 \\ \ell' \neq \ell}}^{m_{ik}} (D_{ik\ell} - \hat{\nu})(D_{ik\ell'} - \hat{\nu}), \tag{4.91}$$

$$\hat{\omega}_1 = \left(\hat{\nu}(1-\hat{\nu}) \sum_{i=1}^{n_{\mathbb{P}}} \sum_{k \neq i}^{n_{\mathbb{G}}} m_{ik} \sum_{\substack{k'=i+1 \\ k' \neq k}}^{n_{\mathbb{G}}} m_{ik'} \right)^{-1}$$

$$\times \sum_{i=1}^{n_{\mathbb{P}}} \sum_{k \neq i}^{n_{\mathbb{G}}} \sum_{\substack{k' \neq i \\ k' \neq k}}^{n_{\mathbb{G}}} \sum_{\ell=1}^{m_{ik}} \sum_{\ell'=1}^{m_{i'k}} (D_{ik\ell} - \hat{\nu})(D_{ik'\ell'} - \hat{\nu}), \tag{4.92}$$

$$\hat{\omega}_2 = \left(\hat{\nu}(1-\hat{\nu}) \sum_{i=1}^{n_{\mathbb{P}}} \sum_{k \neq i}^{n_{\mathbb{G}}} m_{ik} \sum_{\substack{i' \neq k \\ i' \neq i}}^{n_{\mathbb{P}}} m_{i'k} \right)^{-1}$$

$$\times \sum_{i=1}^{n_{\mathbb{P}}} \sum_{k \neq i}^{n_{\mathbb{G}}} \sum_{\substack{i' \neq k \\ i \neq i}}^{n_{\mathbb{P}}} \sum_{\ell=1}^{m_{ik}} \sum_{\ell'=1}^{m_{ik'}} (D_{ik\ell} - \hat{\nu})(D_{i'k\ell'} - \hat{\nu}), \tag{4.93}$$

and

$$\hat{\omega}_3 = \left[\hat{\nu}(1-\hat{\nu}) \left(\sum_{i=1}^{n_{\mathbb{P}}} \sum_{k \neq i}^{n_{\mathbb{G}}} m_{ik} \left\{ \sum_{\substack{k' \neq k \\ k' \neq i}}^{n_{\mathbb{G}}} m_{kk'} + \sum_{\substack{i' \neq i \\ i' \neq k}}^{n_{\mathbb{P}}} m_{i'i} \right\} \right) \right]^{-1}$$

$$\times \left(\sum_{i=1}^{n_{\mathbb{P}}} \sum_{k \neq i}^{n_{\mathbb{G}}} \sum_{\substack{k' \neq k \\ k' \neq i}}^{n_{\mathbb{G}}} \sum_{\ell=1}^{m_{ik}} \sum_{\ell'=1}^{m_{kk'}} (D_{ik\ell} - \hat{\nu})(D_{kk'\ell'} - \hat{\nu}) \right.$$

$$\left. + \sum_{i=1}^{n_{\mathbb{P}}} \sum_{k \neq i}^{n_{\mathbb{G}}} \sum_{\substack{i' \neq i \\ i' \neq k}}^{n_{\mathbb{P}}} \sum_{\ell=1}^{m_{ik}} \sum_{\ell'=1}^{m_{i'i}} (D_{ik\ell} - \hat{\nu})(D_{i'i\ell'} - \hat{\nu}) \right). \tag{4.94}$$

4.7.2 Simplified Asymmetric Correlation Structure

In this section, we propose a simplified correlation structure for false match decisions from an asymmetric matching algorithm. This is a simplified version of the correlation structure in (4.4). Here we assume that $\omega = \omega_1 = \omega_2 = \omega_3$. This is formally equivalent to supposing that the correlation between decisions is the same when one and only one individual is the same in both of the decisions being compared. This correlation is

$$
Corr(D_{ik\ell}, D_{i'k'\ell'}) = \begin{cases} 1 & \text{if } i = i', k = k', \ell = \ell' \\ \eta & \text{if } i = i', k = k', \ell \neq \ell' \\ \omega & \text{if } i = i', k \neq k', i \neq k, i \neq k' \\ \omega & \text{if } i \neq i', k = k', k \neq i, k \neq i' \\ \omega & \text{if } i = k', i' \neq k, i \neq i', i \neq k \\ \omega & \text{if } i' = k, i \neq k', i' \neq i, i' \neq k' \\ \xi_1 & \text{if } i = k', k = i', i \neq i', k \neq k', \ell = \ell' \\ \xi_2 & \text{if } i = k', k = i', i \neq i', k \neq k', \ell \neq \ell' \\ 0 & \text{otherwise.} \end{cases} \tag{4.95}
$$

Under the correlation structure in (4.95), the variance becomes

$$
\begin{aligned}
V[\hat{\nu}] &= N_\nu^{-2} V[\mathbf{1}^T \mathbf{D}_I] = N_\nu^{-2} \mathbf{1}^T \boldsymbol{\Sigma}_I \mathbf{1} = N_\nu^{-2} \nu(1-\nu) \mathbf{1}^T \boldsymbol{\Phi}_I \mathbf{1} \\
&= N_\nu^{-2} \nu(1-\nu) \Bigg[N_\nu + \eta \sum_{i=1}^{n_{\mathbb{P}}} \sum_{\substack{k=1 \\ k \neq i}}^{n_{\mathbb{G}}} m_{ik}(m_{ik} - 1) \\
&\quad + \omega \Bigg\{ \sum_{i=1}^{n_{\mathbb{P}}} \sum_{\substack{k=1 \\ k \neq i}}^{n_{\mathbb{G}}} m_{ik} \Bigg(\sum_{\substack{k'=1 \\ k' \neq i, k' \neq k}}^{n_{\mathbb{G}}} m_{ik'} \Bigg) + \sum_{i=1}^{n_{\mathbb{P}}} \sum_{\substack{k=1 \\ k \neq i}}^{n_{\mathbb{G}}} m_{ik} \Bigg(\sum_{\substack{i'=1 \\ i' \neq i, i' \neq k}}^{n_{\mathbb{P}}} m_{i'k} \Bigg) \\
&\quad + \sum_{i=1}^{n_{\mathbb{P}}} \sum_{\substack{k=1 \\ k \neq i}}^{n_{\mathbb{G}}} m_{ik} \Bigg(\sum_{\substack{k'=1 \\ k' \neq i, k' \neq k}}^{n_{\mathbb{G}}} m_{k'i} \Bigg) + \sum_{i=1}^{n_{\mathbb{P}}} \sum_{\substack{k=1 \\ k \neq i}}^{n_{\mathbb{G}}} m_{ik} \Bigg(\sum_{\substack{i'=1 \\ i' \neq i, i' \neq k}}^{n_{\mathbb{P}}} m_{ki'} \Bigg) \Bigg\} \\
&\quad + \xi_1 \sum_{i=1}^{n_{\mathbb{P}}} \sum_{\substack{k=1 \\ k \neq i}}^{n_{\mathbb{G}}} m_{ki} + \xi_2 \sum_{i=1}^{n_{\mathbb{P}}} \sum_{\substack{k=1 \\ k \neq i}}^{n_{\mathbb{G}}} m_{ki}(m_{ki} - 1) \Bigg].
\end{aligned} \tag{4.96}
$$

We again derive estimators for ω while the estimates for η, ξ_1 and ξ_2 remain the same as those found in (4.7), (4.11), and (4.12). Thus,

$$
\begin{aligned}
\hat{\omega} &= (\hat{\nu}(1-\hat{\nu}))^{-1} \\
&\quad \times \Bigg(\sum_{i=1}^{n_{\mathbb{P}}} \sum_{\substack{k=1 \\ k \neq i}}^{n_{\mathbb{G}}} m_{ik} \Bigg(\sum_{\substack{k'=1 \\ k' \neq i, k' \neq k}}^{n_{\mathbb{P}}} m_{ik'} + \sum_{\substack{i'=1 \\ i' \neq i, i' \neq k}}^{n_{\mathbb{G}}} m_{i'k} \\
&\quad + \sum_{\substack{i'=1 \\ i' \neq i, i' \neq k}}^{n_{\mathbb{P}}} m_{i'i} + \sum_{\substack{k'=1 \\ k' \neq i, k' \neq k}}^{n_{\mathbb{G}}} m_{kk'} \Bigg) \Bigg)^{-1}
\end{aligned}
$$

$$\times \left(\sum_{i=1}^{n_{\mathbb{P}}} \sum_{\substack{k=1 \\ k \neq i}}^{n_{\mathbb{G}}} \sum_{\substack{k'=1 \\ k' \neq i, k' \neq k}}^{n_{\mathbb{G}}} \sum_{\ell=1}^{m_{ik}} \sum_{\ell'=1}^{m_{ik'}} (D_{ik\ell} - \hat{\nu})(D_{ik'\ell'} - \hat{\nu}) \right.$$

$$+ \sum_{i=1}^{n_{\mathbb{P}}} \sum_{\substack{k=1 \\ k \neq i}}^{n_{\mathbb{G}}} \sum_{\substack{i'=1 \\ i' \neq i, i' \neq k}}^{n_{\mathbb{P}}} \sum_{\ell=1}^{m_{ik}} \sum_{\ell'=1}^{m_{i'k}} (D_{ik\ell} - \hat{\nu})(D_{i'k\ell'} - \hat{\nu})$$

$$+ \sum_{i=1}^{n_{\mathbb{P}}} \sum_{\substack{k=1 \\ k \neq i}}^{n_{\mathbb{G}}} \sum_{\substack{i'=1 \\ i' \neq i, i' \neq k}}^{n_{\mathbb{P}}} \sum_{\ell=1}^{m_{ik}} \sum_{\ell'=1}^{m_{i'i}} (D_{ik\ell} - \hat{\nu})(D_{i'i\ell'} - \hat{\nu})$$

$$\left. + \sum_{i=1}^{n_{\mathbb{P}}} \sum_{\substack{k=1 \\ k \neq i}}^{n_{\mathbb{G}}} \sum_{\substack{k'=1 \\ k' \neq k, k' \neq k}}^{n_{\mathbb{G}}} \sum_{\ell=1}^{m_{ik}} \sum_{\ell'=1}^{m_{kk'}} (D_{ik\ell} - \hat{\nu})(D_{kk'\ell'} - \hat{\nu}) \right). \tag{4.97}$$

The models we have presented here are all simplifications that occur by setting one or more correlation parameters equal to each other. Another manner in which to reduce the model would be to set certain parameters of the general model to zero which suggests that any observed correlation is simply noise.

Chapter 5
Receiver Operating Characteristic Curve and Equal Error Rate

The focus of this chapter is statistical estimation and comparison of receiver operating characteristic (ROC's) curves and equal error rates (EER's) which can be derived from ROC's. We will begin this chapter with an introduction to ROC curves. An ROC curve is a plot of the false match rate (FMR) on the x-axis against the true match rate or 1-false non-match rate (1-FNMR) along the y-axis. This plot is made by varying the threshold, τ, at which a decision is made. As such the ROC is a way to represent the performance of a matching or classification system. (Note that we can also plot false accept rates versus one minus false reject rates to obtain an ROC.) Figure 5.1 has the estimated ROC for the face matcher (DCTb, GMM) from the XM2VTS database. A variant of the ROC is the Detection Error Tradeoff (DET) Curve. In the previous chapters on FNMR (Chap. 3) and FMR (Chap. 4) we have suppressed the different thresholds both from our notation and from our analyses, though we have mentioned the thresholds in the examples found in those chapters. The equal error rate is the value for the FMR when the FMR = FNMR.

The rest of this chapter is organized in the following manner. We begin with an introduction to the ROC with special focus on a polar coordinates representation of the ROC. We then propose a *new* bootstrap methodology for estimation of the variability in an estimated ROC. Next we discuss our methodology for making curvewise confidence regions for the ROC. This methodology forms the basis for our approach to inference for the rest of the chapter. Having presented our methodology for a single ROC, we move to methods for comparing two ROC's. We do this both for the case when the ROC's are collected independently as well as when the matching scores are collected in a paired manner. Comparisons of three or more ROC's whether paired or independent are the last topic on ROC's. We then move to a discussion of statistical methods for equal error rates (EER's). Our organization for the part of this chapter on EER's is similar to our structure for ROC inference. We start with estimation for a single EER. This is followed by methodology for comparing two EER's and then comparing three or more EER's. Both sections have descriptions of comparisons for independent and paired data collection. This chapter ends with a discussion of these topics.

M.E. Schuckers, *Computational Methods in Biometric Authentication*, Information Science and Statistics,
DOI 10.1007/978-1-84996-202-5_5, © Springer-Verlag London Limited 2010

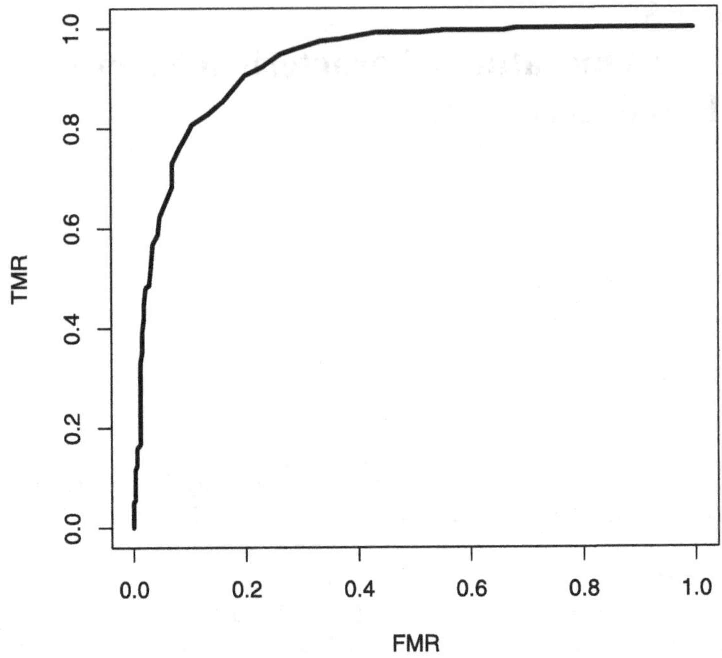

Fig. 5.1 A receiver operating characteristic (ROC) curve for the voice matcher IDIAP-25-10-PCA from the BANCA database collected under the G protocol and applied to group g1

5.1 Notation

In the previous two chapters we focused on error rates based upon dichotomizing two different distributions. For the FNMR, we used decisions based upon match scores, Y_{iij}'s, when the same individual i was involved. We refer to these match scores as a *genuine* match scores. For the FMR, we used decisions based upon match scores $Y_{ik\ell}$ when two different individuals i and k are involved. We refer to these match scores as *imposter* match scores. For ROC and EER estimation we will use both distributions. For this discussion we will assume that genuine match scores are *stochastically* larger than imposter match scores meaning that, in general, genuine scores tend to be larger than imposter scores. Table 5.1 has an example of the structure of match score data. This is typically the case for a correlation-type matcher.

To make a decision about whether a given comparison is declared a match we compare the match score to the threshold, τ. If the match score is at least the threshold, $Y_{ik\ell} \geq \tau$, we will declare a match. If the match score is below the threshold, $Y_{ik\ell} < \tau$ we will declare a non-match. We formally outline this methodology below.

The distributions of match scores, $Y_{ik\ell}$'s, are denoted by $g()$ and $f()$ for the *genuine* and *imposter* distributions, respectively. From these distributions, an ROC

Table 5.1 Example of match scores for ROC

i	k	ℓ	Y_{ikl}	i	k	ℓ	Y_{ikl}
1	1	1	4.24	3	3	1	4.35
1	1	2	3.38	3	3	2	3.41
1	1	3	3.01	3	3	3	3.51
1	2	1	6.78	3	3	4	4.05
1	2	2	7.56	3	5	1	7.16
1	2	3	6.41	3	5	2	5.93
1	3	1	8.71	4	1	1	6.05
1	3	2	8.94	4	1	1	8.82
1	3	3	8.88	4	1	1	6.53
1	4	1	5.93	4	1	2	6.72
1	4	2	7.18	4	2	1	7.90
1	5	1	8.87	4	2	2	5.34
1	5	2	4.95	4	2	3	8.71
1	5	3	5.76	4	2	4	5.15
1	5	4	6.51	4	5	1	6.87
1	5	5	5.12	4	5	2	7.14
2	1	1	6.56	5	1	1	5.17
2	1	2	6.53	5	1	2	8.31
2	1	3	6.53	5	1	3	7.46
2	1	4	5.95	5	2	1	5.25
2	1	5	6.16	5	2	2	5.73
2	1	6	7.15	5	3	1	4.58
2	2	1	4.52	5	3	2	5.09
2	2	2	4.39	5	4	1	8.01
2	2	3	3.37	5	4	2	3.44
2	2	4	3.64	5	4	3	5.55
2	2	5	4.37	5	4	4	7.82
2	2	6	4.09	5	4	5	7.52
2	3	1	4.64	5	5	1	4.31
2	3	2	5.28	5	5	2	3.85
2	3	3	8.92	5	5	3	3.40
2	4	1	6.63	5	5	4	3.42
2	4	2	7.73	5	5	5	4.37
2	5	1	4.71	5	4	6	4.69
3	1	1	6.76				
3	1	2	6.62				
3	2	1	5.64				
3	2	2	6.37				
3	2	3	4.86				
3	2	4	5.25				

is plotted as the false match rate (FMR) on the *x-axis* against the true match rate (TMR = 1-FNMR) on the *y-axis*, by varying a threshold τ, and calculating

$$\text{FMR}(\tau) = \int_{\tau}^{\infty} f(t)dt \quad \text{and} \quad \text{TMR}(\tau) = 1 - \text{FNMR} = \int_{\tau}^{\infty} g(t)dt. \quad (5.1)$$

If $f(t)$ and $g(t)$ are known then our ROC can be calculated based upon (5.1) by varying τ, the threshold. In general, this is rarely the case and these distributions must be estimated. In that case we can calculate, $\hat{R}$ based upon an approximation of $f(t)$ and $g(t)$.

$$\widehat{\text{FMR}}(\tau) = \int_{\tau}^{\infty} \hat{f}(x)dx = \frac{\sum_{i=1}^{n_{\mathbb{P}}} \sum_{k \neq i}^{n_{\mathbb{G}}} \sum_{\ell=1}^{m_{ik}} I_{\{Y_{ik\ell} \geq \tau\}}}{\sum_{i=1}^{n_{\mathbb{P}}} \sum_{k \neq i}^{n_{\mathbb{G}}} m_{ik}}, \quad (5.2)$$

and

$$\widehat{\text{TMR}}(\tau) = 1 - \widehat{\text{FNMR}}(\tau) = 1 - \int_{-\infty}^{\tau} \hat{g}(y)dy = \frac{\sum_{i=1}^{n} \sum_{j=1}^{m_i} I_{\{Y_{iij} < \tau\}}}{\sum_{i=1}^{n} m_i} \quad (5.3)$$

where $I_{\{\cdot\}}$ represents an indicator function, cf. Definition 2.10. We use here notation for estimation of an FMR and an FNMR that was originally used in Chaps. 3 and 4. x and y are observed scores from the imposter and genuine distributions, respectively. As with R, the estimated ROC, $\hat{R}$ is computed by calculating $\widehat{\text{FMR}}$ and $\widehat{\text{TMR}}$ for various values of the threshold τ. The equal error rate (EER) is then the value at which $\widehat{\text{FMR}} = \widehat{\text{FNMR}}$ or to use the notation from the previous chapters when $\hat{\pi} = \hat{\nu}$. In this way, the sample ROC curve summarizes the performance of the biometric matching process based upon the sample from which it is calculated. The FMR and FNMR estimation calculations above are based on a criterion of accept if a match score is greater than τ, the threshold. An alternative formulation can easily be made for other decision making structures. Thus, we will treat this ROC, $\hat{R}$, as an estimate of the population ROC, R.

A good deal has been written about the ROC in a variety of contexts. See, for example, Zhou et al. [108] or Hernández-Orallo et al. [45]. Fawcett [32] provides a well-written introduction to ROC's and their usage. Macskassy et al. [61] compared three methods that have been proposed for ROC confidence bands/regions. These include: a simultaneous joint confidence region and a fixed width approach both due to Campbell [13], and a Working-Hotelling band technique due to Ma and Hall [60]. The simultaneous joint confidence approach converts confidence rectangles at various points along the ROC to create a joint region. Schuckers et al. [88] created a curvewise confidence band based upon the use of polar coordinate. It is this last approach that we will follow and extend here.

Additionally, there are quite a few variants of the ROC which are the result of transforming the false match rate or the true match rate or both. Among these are the Detection Error Tradeoff (DET) Curve introduced by Martin et al. [65]. The DET is a plot of the $\Phi^{-1}(\widehat{\text{FMR}})$ versus $\Phi^{-1}(\widehat{\text{FNMR}})$ where $\Phi^{-1}(\cdot)$ is the inverse Gaussian function. An example of a DET curve is given in Fig. 5.2. Bengio and Mariéthoz [6] have proposed the use of an Expected Performance Curve (EPC) which is another alternative similar to the DET but is based upon plotting a weight value between zero

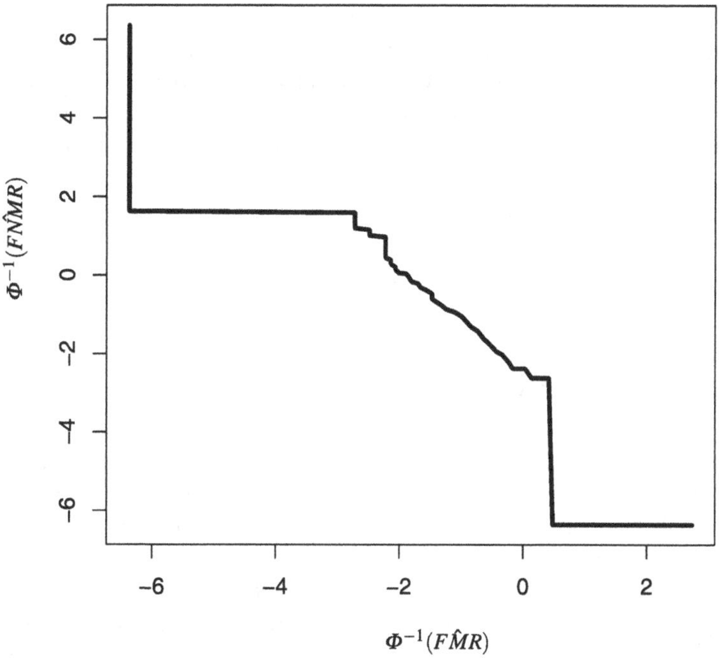

Fig. 5.2 A detection error tradeoff (DET) curve for the voice matcher IDIAP-25-10-PCA from the BANCA database collected under the G protocol and applied to group g1

and one against the corresponding weighted average of the FMR and FNMR. A plot which supplements the ROC with explicit costs for a false match and a false non-match was proposed by Drummond and Holte [26]. For this chapter we will focus on the ROC as a plot of FMR against 1-FNMR. In general, we will think of any plot that is based upon changing thresholds for matching error rates as an ROC. We trust the reader to appropriately interpret a well-labeled graph. Since the alternatives and extensions listed above and others are based upon linear combinations of the FMR and the FNMR or upon one-to-one transformations of these same rates, the procedures that we present in this chapter should be readily extendable to those methods.

The approach here is a curvewise one for the ROC as a function of both the FMR and FNMR based on a polar coordinates approach. By curvewise, we mean that our statistical inference will be based upon the entire curve rather than on any individual point. The approach, described below, transforms each curve from the (FMR, 1-FNMR) space to polar coordinates, (radius (r_θ), angle (θ)). This polar representation creates an appropriate confidence region for all the FMR's and FNMR's with a single confidence region. Adler and Schuckers [1] originally used this approach to average across several DET's to create a composite DET or mean DET. Those same authors, along with Minev, then extended that work to create a confidence region for a single ROC in, Schuckers et al. [88]. Since our focus is inference for the entire curve, we present curvewise methodology. Pointwise methodology is appropriate if we are interested in inference about a single point. For instance, if our interest is the FMR, ν, at a FNMR of $\pi = 0.01$, then it is reasonable and appropriate to create a pointwise confidence interval. See Chaps. 3 and 4 for pointwise methodology for a given FNMR or FMR, respectively. We present our curvewise approach for a single ROC below and extend that basic methodology to the comparison of multiple ROC's.

One commonly used derivative of the ROC is the equal error rate. The second half of this chapter is devoted to the presentation of statistical methods for the estimation and comparison of one or more EER's. We will use χ to denote an EER.

5.2 Bootstrap for Two-Instance ROC

In order to do appropriate inference for an ROC curve, it is necessary to estimate the variability in the entire ROC curve. That variability is bivariate, having a component due to the variability in estimation of the FNMR and another component due to the variability in the estimation of the FMR. We note that these variabilities are potentially correlated and so it is important to resample them appropriately. This section outlines an approach for resampling ROC data that comes from a two-instance matcher in order to estimate this variability. The approach here is aimed at bioauthentication match score data but there are simplified versions that are appropriate for independent and identically distributed *iid* data as well as repeated measures data. The change to the methods given below that would result from these different correlation structures is a change in the bootstrap methodology for resampling the match scores.

In this section, we introduce a new method for bootstrapping ROC match score data. The methods given here maintain the correlation structure between the distribution of genuine match scores and the distribution of imposter match scores as well as the correlations within each of those distributions. Our fundamental approach is to sample *with replacement* individuals from the probe. For each selected individual in the resampled probe we will sample *with replacement* the gallery to get a completely resampled set of match scores. We then repeat this process a large number of times. We will refer to this resampling approach as a *two-instance bootstrap*. We used a similar bootstrap mechanism for the variability of the false match rate. For the ROC *two-instance bootstrap*, we allow both genuine and imposter match scores to be part of our bootstrap approach. For the FMR *two-instance bootstrap*, we were only resampling from the match decisions resulting from the imposter match score distribution. This is not to be confused with the 'two sample bootstrap' proposed by Wu [105] which neither accounts for the correlations within match scores distributions, nor accounts for potential correlations between those distributions.

Before we enumerate this approach explicitly we need to introduce some notation. The number of individuals in the probe is $n_{\mathbb{P}}$ and the number of individuals in gallery is $n_{\mathbb{G}}$. Also let m_{ik} be the number of match scores for the pair of individuals i and k. More formally this algorithm is:

1. Sample *with replacement* from the $n_{\mathbb{P}}$ individual in the probe. Call the ith selected individual b_i where $i = 1, \ldots, n_{\mathbb{P}}$.
2. For each of the selected individuals, the b_i's, in the previous step, resample *with replacement* $n_{\mathbb{G}}$ individuals from the gallery and call those individuals $h_{i,k}$'s.
3. For each selected pair of individuals b_i and $h_{i,k}$, take all $m_{b_i,h_{i,k}}$ scores for that pair of selected individuals.
4. Concatenate all of the selected match scores, the $Y_{b_i,h_{i,k}}$'s for all pairs of resampled individuals.
5. The bootstrapped genuine distribution is the collection of match scores $Y_{b_i,h_{i,k}}$ such that $b_i = h_{i,k}$ and the replicated imposter distribution is all of the $Y_{b_i,h_{i,k}}$ such that $b_i \neq h_{i,k}$.

The end result of the above algorithm is a collection of match scores of approximately the same number as the original. We then calculate a bootstrapped ROC from this combination of the replicated genuine and imposter distributions. The methodology given above is a *new* one. It is designed to replicate the variability in match scores from both the genuine and the imposter distributions as well as maintain the correlations between and within those two distributions. We will use this methodology extensively in the rest of this chapter.

5.3 Statistical Methods for ROC

In this section, we start with a confidence region for a single ROC. After presenting our approach for one ROC, we discuss statistical methods for curvewise comparison of two ROC's. The basis for our comparison of two ROC's is a radial difference between the two ROC's. Paired data occurs when each match score from one process has a 'paired' match score from another process. This might occur when you run two different algorithms on the same database of bioauthentication signals. Independent data occurs when the data collections are distinct and do not involve overlap in the individuals. We then investigate and propose methods for comparing three or more ROC's. When we are considering three or more ROC's, it is important to avoid comparing these ROC's two at a time, since this results in increased probability of making a Type I error. So when our focus is three or more ROC's, we will use a radial sum of squares between the ROC's. In this way, we are able to test simultaneously the equality of several ROC's. All of the methods below are based upon the two-instance bootstrap methods presented in the previous section. However, the methods below can be generalized based upon other resampling approaches that may be appropriate for a given data collection and a given classifier. For example, many non-biometric authentication ROC applications have data for which an *iid* bootstrap would be appropriate. In that case, the methods below could be applied using the *iid* bootstrap. Using the *iid* bootstrap would only be appropriate if we have reason to believe that there is not a correlation between the match scores that comprise the ROC.

We begin by outlining the methodology for estimating a single ROC curve. Having calculated an ROC, either sample or population, we next convert the coordinates of the curve from rectangular to polar coordinates. Then our ROC curve over a range of values Θ is $R = \{(r_\theta, \theta) : \theta \in \Theta = (\theta_L, \theta_U)\}$ and θ_L and θ_U represent lower and upper values for the range of the radial angles. In practice we sweep across a range of T θ's using a discrete list of values $\Theta^* = (\theta_L, \theta_L + \delta, \theta_L + 2\delta, \ldots, \theta_L + (T-2)\delta, \theta_U)$ where $\Theta = (\theta_L, \theta_U)$ is given. Then we can write $\hat{R} = \{(\hat{r}_\theta, \theta) : \theta \in \Theta^*\}$. We refer to this approach as radial sweeping. One important issue in radial sweeping is the location of the radial center, (c_x, c_y). Because we would like the confidence region to not depend on which error rate is on which axis and because we also want to derive a confidence interval for the EER, we limited possible center points to the collection of points $\{(c_x, c_y) : c_y = 1 - c_x\}$.

For the examples in this text, we chose to use $(1, 0)$ as the center of our polar coordinates. This selection seems natural as the typical ROC has a curvature that (roughly) follows an arc of radius 1 from the center $(1, 0)$. As a consequence of our choice of center, we use $\theta_L = \pi/2$ and $\theta_U = \pi$. The choice of the center is a particularly important one as it governs the shape of the confidence region. Further, if the EER is of interest, then we will use an odd number of angles, T, to ensure that we have $\theta = 3\pi/4 \in \Theta^*$. It is important to note that other choices for centers will yield some of the other methods that have been proposed in the literature. For example, it is possible to use this methodology to create regions strictly for the FMR or FNMR like those proposed by Dass et al. [18] by allowing the coordinate of the polar center for the error rate of interest to become very small i.e. $(0, -c_y)$ for some large number c_y, say $c_y = 1000$. This choice would produce radii which are nearly parallel vertical lines. Similarly, the approach for a fixed width confidence region due to Campbell [13] can be accommodated by taking the center to be $(c_x, c_y) = (c_x, 1 + bc_x)$ for a value of c_x that is negative and large in magnitude where $b = -\sqrt{N_\nu / N_\pi}$. The values for N_ν and N_π are the number of scores in the *imposter* and *genuine* distributions, respectively. By choosing such a value for the center then we could create a confidence region that is fixed width and follows Campbell [13]; however, it would be slightly different than the variable width method we give below. It might also be reasonable to adjust the sample sizes to account for correlations in the match scores when the match scores are not *iid* within the respective distributions. This adjustment might be done by replacing N_ν and N_π with equivalent effective sample sizes, cf. Definition 2.34.

The choice of Θ^* is also an important one. We have chosen to use a uniformly spaced set of T angles θ's but there might be other specifications that are appropriate to particular applications. It might be plausible to use a set of angles Θ^* that is not uniformly spaced but is more concentrated on angles in a designated region of interest for a particular application. For example, a particular forensic application might lead to much more interest in the region of the ROC where the FMR is below 0.20, the lower left quadrant of the unit square $(0, 1) \times (0, 1)$. Thus, in that case it might be reasonable to have a $\theta_L = 3\pi/4$ and $\theta_U = \pi$. Having chosen the center of the polar coordinate system, (c_x, c_y), and the collection of angles, Θ^*, we can convert the coordinates of every point on the estimated ROC curve, $\hat{R}$, to polar coordinates.

Finally, one practical issue in dealing with the conversion of match scores to a polar coordinates based ROC is interpolation of points along the ROC. In our examples below, we have used a linear interpolation between the points along the ROC. It is certainly possible to other forms of interpolation, e.g. a quadratic or cubic interpolation, to generate our ROC and the resulting confidence region. We next turn our attention to deriving a confidence region for the ROC, R.

5.3.1 Confidence Region for Single ROC

In this section we present curvewise methodology for creating a confidence region for a single ROC. By curvewise, we mean that we want to be $100\%(1-\alpha)$ confident that the entire process ROC curve, R, is within the limits of our confidence bands. We will use the terms confidence band and confidence region interchangeably throughout this chapter. The approach taken for estimation of the sampling variability for a given ROC is the two-instance bootstrap given in Sect. 5.2. In this section, we will assume that the process that generated the match scores for the ROC is stationary in the sense given by Definition 2.23. We will base our confidence region on a polar coordinates representation of the ROC of interest. Schuckers et al. [88] first introduced this approach and we review their approach below. The general outline of this confidence region technique is as follows:

1. Calculate an estimated ROC, $\hat{R}$.
2. Bootstrap both genuine and imposter distributions.
3. Calculate and store a bootstrapped ROC, $\hat{R}^b$.
4. Repeat previous two steps M times.
5. Determine curvewise bounds for confidence region, $\mathfrak{R}$.

Let $R^b = \{(r^b_{m\theta}, \theta) : \theta \in \Theta^*\}$ be the mth bootstrapped ROC where $m = 1, \ldots, M$ for some large number M. $r^b_{m\theta}$ represents the radial length for the mth bootstrapped ROC at the angle θ. Next, we find the adjusted standard deviation of $r_{m\theta}$'s at each θ in Θ^* by taking the square root of the variance plus a slight adjustment, ι_θ. We will denote this by

$$s_{\hat{r}_\theta} = \sqrt{\frac{1}{M-1}\sum_{m=1}^{M}(r^b_{m\theta} - \hat{r}_\theta)^2 + \iota_\theta}. \tag{5.4}$$

This adjustment is made to ensure that $s_{\hat{r}_\theta}$ is positive even when no radial variability is present in the bootstrapped replications. Our specific concern is that it is possible for the unadjusted variance to be 0 at some angles and therefore we use ι_θ to ensure a positive value here. This possibility often occurs at the ends of a given ROC curve near the points $(0,0)$ and $(1,1)$. Here we use

$$\iota_\theta = \frac{|(\theta - \theta_{min})(\theta_{max} - \theta)|}{(\theta_{max} - \theta_{min})^2 10^2 N} \tag{5.5}$$

where $\theta_{max} = \max_{\theta \in \Theta^*}(\theta)$, $\theta_{min} = \min_{\theta \in \Theta^*}(\theta)$ and N is the total number of matching scores in both the genuine and imposter distributions.

We define the ROC confidence region, $\Re$, as the region bounded by the linear interpolation of the $r_{L,\theta}$'s, the lower limits for the region at angle θ, and a linear interpolation of $r_{U,\theta}$'s, the upper limits for the region at angle θ for each $\theta \in \Theta^*$. We want that region to have the property that

$$P(R \in \Re) = 1 - \alpha \tag{5.6}$$

where $1 - \alpha$ is the confidence level. We say the population ROC, R, is captured by the $(1 - \alpha)100\%$ ROC confidence region and denote it by $R \in \Re$ if

$$r_{L,\theta} \leq r_\theta \leq r_{U,\theta} \quad \text{for all } \theta \in \Theta^*. \tag{5.7}$$

In order to determine the ROC confidence region, $\Re$, we need to create a region that captures the entire curve R with probability $(1 - \alpha)$. As Schuckers et al. [88] noted, the approach of calculating pointwise intervals at each angle θ and combining them does not produce a confidence region with the desired confidence level. This approach ignores the correlation between radii at different values of θ, particularly neighboring values of θ in Θ^*. Consequently, it yields confidence regions with $P(R \in \Re) < 1 - \alpha$. The observed confidence levels are often considerably less than $1 - \alpha$.

Several options exist for finding an ROC confidence region, $\Re$, that meets our criteria in (5.6). We prefer a variable width approach that uses a different width at each angle, θ, to match the variability in the ROC. Other options such as the aforementioned fixed width confidence bands could be derived from the radial sweep approach. We have chosen to create a confidence region with variable width at each angle θ that depends only upon two parameters, δ_U and δ_L which govern the number of standard deviations in $\hat{r}_\theta$ to expand the confidence region.

We build our variable width confidence region, $\Re$, by connecting the endpoints $(r_{L,\theta} = \hat{r}_\theta - \delta_U s_{r_\theta}, r_{U,\theta} = \hat{r}_\theta - \delta_L s_{r_\theta})$ at each $\theta \in \Theta^*$ where δ_L and δ_U represent lower and upper constants, respectively, that yield the desired $(1 - \alpha)100\%$ confidence level. Next, we describe the process for determining δ_L and δ_U. First, we calculate a standardized residual, $e_{m\theta}$, at each bootstrapped radii $r^b_{m\theta}$,

$$e_{m\theta} = \frac{r^b_{m\theta} - \hat{r}_\theta}{s_{\hat{r}_\theta}}. \tag{5.8}$$

This residual tells us how many standard deviations, s_{r_θ}'s, each bootstrapped radii, $r^b_{m\theta}$, is from the estimated ROC, $\hat{r}_\theta$, at each θ. Next, the maximum absolute standardized residual, w_m, for the mth ROC curve is calculated. This value, w_m, tells us the furthest amount (in number of standard deviations) that each bootstrapped curve is from the estimated ROC, $\hat{R}$. We calculate this quantity to determine the largest standardized value needed to obtain $(1 - \alpha)100\%$ of the bootstrapped curves within the ROC region, $\Re$. We then define w_m to be:

$$w_m = \begin{cases} \max_\theta(e_{m\theta}) & \text{if } |\min_\theta(e_{m\theta})| < |\max_\theta(e_{m\theta})| \\ \min_\theta(e_{m\theta}) & \text{if } |\min_\theta(e_{m\theta})| \geq |\max_\theta(e_{m\theta})|. \end{cases} \tag{5.9}$$

Once we have all M w_m's we find the $\alpha/2$th and the $1 - \alpha/2$th percentiles of the w_m's and they become δ_L and δ_U respectively. Note that the w_m's can take both positive and negative values. Having found δ_L and δ_U, we then build our confidence region $\mathfrak{R}$, as the region bounded by the line segments created by connecting the upper endpoints, $\hat{r}_\theta - \delta_L s_{\hat{r}_\theta}$, at each angle θ, and the line segments created by connecting the lower endpoints, $\hat{r}_\theta - \delta_U s_{\hat{r}_\theta}$, at each angle θ. For the plots presented here we truncate the spline to ensure it is contained in the unit square defined by $(0, 1) \times (0, 1)$.

The choice of ι_θ is a delicate balance. The primary reason for using an adjustment here is to ensure that we have a non-zero $s_{\hat{r}_\theta}$ and hence, a non-zero denominator in (5.8). Additionally, the choice of ι_θ can alter the distribution of the w_m's, particularly the tails of the distribution. Large ι_θ's relative to $s_{\hat{r}_\theta}$ will result in a significantly less variable distribution for the w_m's and narrower confidence regions. For large sets of matching scores this effect will be quite small, but for smaller sets the effect is noticeable. This discrepancy is our primary motivation for choosing ι_θ to depend upon N the number of match scores in our sample. In effect, we can think of ι_θ as a smoothing parameter and we smooth less as N increases. For a given N, the value of ι_θ is dependent on the angle θ. This choice is intentional to match the variability in ROC curves which tends to be largest at angles near $\theta = 3\pi/4$ and tends to be near zero at $\theta = \pi/2$ and $\theta = \pi$. The specific choice here is analogous to the variability in a binary random variable whose variance $p(1-p)$ is a function of its mean p.

The approach described above is one that results in a curvewise ROC confidence region that has variable width to match the variability in the estimated ROC. The two-instance bootstrap methodology that gives us this variability is the advance here. One of the drawbacks of this methodology is its substantial dependence upon the selection of the angles, Θ^*. This is particularly the case when a significant portion of the sample ROC from the matching process is near the upper limits of performance formed by the line segments $(0, 0)$ to $(0, 1)$ and $(0, 1)$ to $(1, 1)$. In those cases, the choice of δ_L and δ_U is governed by the heavily skewed shape of the sampling distribution of $\hat{r}_\theta$ at those θ's. As noted above, tail behavior drives the choice of δ_U and δ_L. In particular, this can lead to intervals that are overly generous. This is another reason that we use the adjustment ι_θ. Consequently, we recommend that if users are only interested in a specific section of the ROC curve that they specify a particular range of angles that are of interest *a priori* of the analysis and use that range to create their confidence regions.

Example 5.1 For this example, we consider match scores from the BANCA database. We focus on the SURREY_face_nc_man_scale_200 (SURREY-NC-200) facial matcher applied to group g1 under the G protocol. The ROC curve and the resulting confidence region, $\mathfrak{R}$, can be found in Fig. 5.3. The center curve is the estimated ROC curve $\hat{R}$. The upper and lower bounds of our 95% confidence region are also denoted by bold dark lines on either side of $\hat{R}$ and are the result of $M = 2000$ bootstrapped ROC curves. The lighter lines that are present in this graph are the bootstrapped curves, the $\hat{R}^b$'s. Here we used $T = 99$ angles distributed evenly between $\theta_L = \pi/2$ and $\theta_U = \pi$.

Example 5.2 The goal of this example is the creation of a 95% confidence region for the face matcher (FH, MLP) from the XM2VTS database. This confidence region

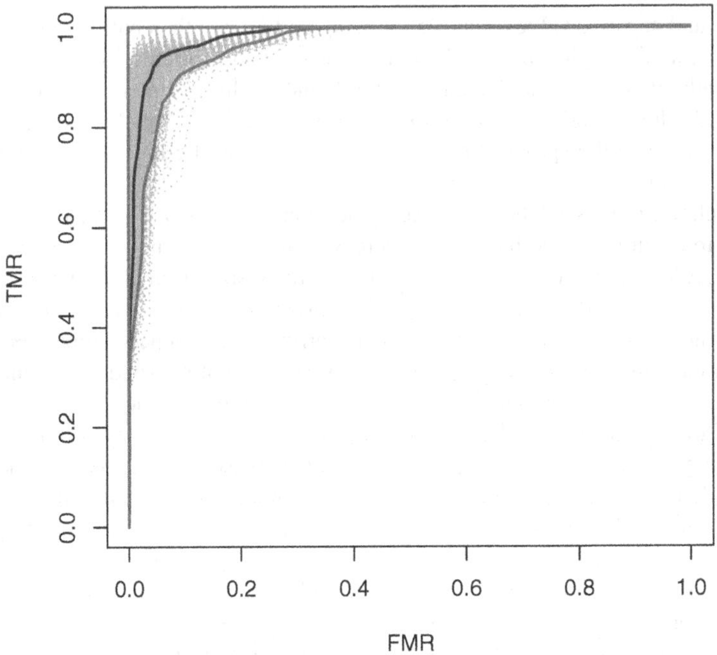

Fig. 5.3 ROC confidence region for the face matcher SURREY-NC-200 from the BANCA database

is given by the outer two lines of Fig. 5.4. To obtain this region, we generated $M = 1000$ resampled ROC curves, $\hat{R}^b$'s. The center solid line in Fig. 5.4 is the observed ROC based upon the sampled data. The solid lines on either side are the bounds for that confidence region. The lighter lines are the resampled ROC's which are included to illustrate the variability in the original ROC curve, $\hat{R}$. Here we calculated our confidence region across the angles from $\pi/2$ to π at $T = 49$ different angles.

Example 5.3 This example is to illustrate the variability that can be present in an ROC. The matcher we are considering here is the voice matcher IDIAP_voice_gmm_auto_scale_25_10_pca (IDIAP-25-10-PCA) from the BANCA database. We note that IDIAP-25-10-PCA does not generally perform as well as the other two matchers that we considered in Examples 5.1 and 5.2. This assessment is made because the ROC for IDIAP-25-10-PCA is further from the optimal performance point at (FMR, TMR) = (0, 1) than the other two matchers. As before, the 95% confidence region bounds are lines that surround the estimated ROC, $\hat{R}$. These bounds are much wider for this ROC than for the ROC's in the previous examples. The primary reason for this width is the variability in the sampling distribution of $\hat{R}$ the sample ROC.

Although we do not do so here, it is possible to build a hypothesis test for a single ROC using the approach above. The basic structure of such a test would be

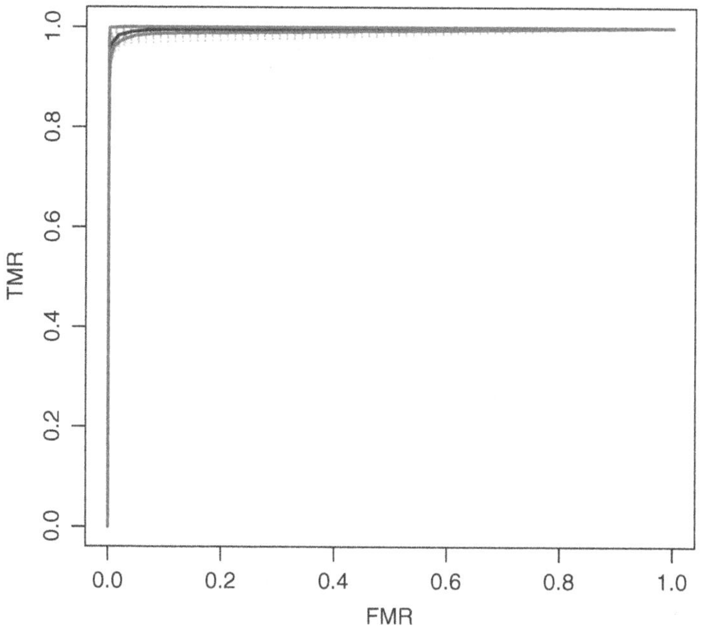

Fig. 5.4 ROC confidence region for the face matcher (FH, MLP) from the XM2VTS database

to specify the hypothesized ROC as a sequence of radii. We could then build the variability in the observed radii centered at the hypothesized radii and determine if the actual ROC is a reasonable realization of the hypothesized ROC. Since we would be evaluating the ROC at multiple angles, it will be necessary to adjust the significance level for the multiple comparisons.

5.3.2 Two Sample Methods for ROC's

In this section we present methodology for comparing two process ROC's: R_1 and R_2. Specifically, we are interested in making inference about the difference between these two ROC's. We will do so by considering the difference in the estimated ROC's at each angle θ. We will denote the difference in the radii at each angle θ by $r_\theta^{(1)}$ and $r_\theta^{(2)}$ respectively. For two ROC's, as with two means or two rates, it is necessary to recognize that there are two sources of sampling variability, one from each ROC. The focus here for inference is the difference between the two ROC's at each angle θ, $r_\theta^{(1)} - r_\theta^{(2)}$. The question of interest is then whether those differences are curvewise significantly different from zero at each θ. We decide that the two ROC curves are different if the confidence interval for the difference of the two curves does not include 0 for all angles. To highlight that the difference is our focus, we will introduce a new plot for illustrating the difference between the two ROC's. This graph will display the differences between the two curves at each angle

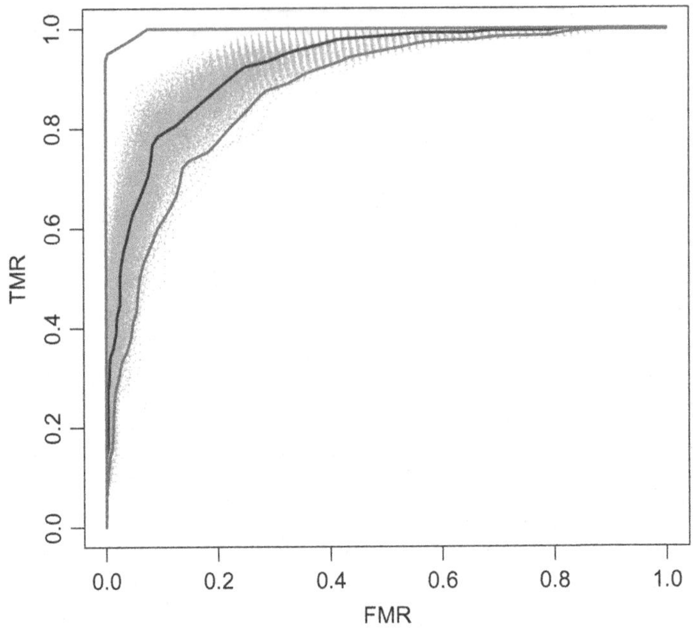

Fig. 5.5 ROC confidence region for the voice matcher IDIAP-25-10-PCA from the BANCA database

θ in Θ^*. We note that it is important to use the same set of angles Θ^* for each of the two ROC curves in order to develop a methodology that works across the groups. For comparing two ROC using the methodology here, it is important to ensure that the angles under consideration are the angles of interest. This selectivity is necessary because the radial sweep approach makes a confidence region dependent upon tail behavior of the sampling distribution of the differences, $r_\theta^{(1)} - r_\theta^{(2)}$. At certain angles, particularly those close to $\pi/2$ and π the sampling distribution of the difference in radii can be quite skewed which results in wide confidence regions. Broader confidence bands make it more difficult to detect real differences between two ROC curves. As above, the methodology we present for statistical inference is based upon the two-instance bootstrap. All of the methods below for comparing two ROC are new. They are extensions of the work on inference for a single ROC. They are based upon the radial sweep methodology in Schuckers et al. [88] that was given above.

5.3.2.1 Confidence Region for Difference of Two Independent ROC's

The approach that we take for generating a confidence region for the difference of two ROC's, R_1 and R_2, is to create a curvewise confidence region for the difference at each angle, θ. For the case of two independent ROC's, we bootstrap each of those curves separately and calculate the difference between the two curves. We might use this methodology below when we are comparing two ROC's from implementations of the same device at two different locations. For example, we might want to compare the ROC's for a facial recognition system that has been implemented at an internal office entryway in Pittsburgh, PA, USA and to one that has been implemented at an external loading dock in Melbourne, VIC, Australia. Since the individuals involved in both collections will be distinct, we have an independent data collection. The approach here is a bootstrap one. Having generated the two replicated ROC's, we create a difference curve by subtracting the radii for one ROC from the radii for the other and create a curvewise confidence region based upon these difference curves. Because we are dealing with independent data collections, we bootstrap each curve individually following the approach given in Sect. 5.2.

Bootstrap Approach To create a $100 \times (1-\alpha)\%$ confidence region for the difference of two ROC's, $R_1 - R_2$, we use the following:

1. Calculate $\hat{r}_\theta^{(1)} - \hat{r}_\theta^{(2)}$ for each $\theta \in \Theta$.
2. Resample from the first process to generate a replicate ROC following the procedure given in Sect. 5.2. We will call this ROC, $\hat{R}_1^b$ which is the collection of radii $r_\theta^{(1)b}$ for all $\theta \in \Theta^*$.
3. Resample from the second process to generate a replicate ROC following the procedure given in Sect. 5.2. Denote this ROC by $\hat{R}_2^b$ which is the collection of radii $r_\theta^{(2)b}$ for all $\theta \in \Theta^*$.
4. Calculate and store the replicated normalized difference by

$$e_{m\theta} = \frac{(r_\theta^{(1)b} - \hat{r}_\theta^{(1)}) - (r_\theta^{(2)b} - \hat{r}_\theta^{(2)})}{s_{\hat{r}_\theta^{(1)} - \hat{r}_\theta^{(2)}}} \tag{5.10}$$

   where

$$s_{\hat{r}_\theta^{(1)} - \hat{r}_\theta^{(2)}} = \sqrt{\frac{1}{M-1}\sum_{m=1}^{M}\left(r_{m\theta}^{(1)b} - r_{m\theta}^{(2)b} - (\overline{r_\theta^{(1)b} - r_\theta^{(2)b}})\right)^2 + \iota_\theta} \tag{5.11}$$

   for each $\theta \in \Theta^*$ and

$$\iota_\theta = \frac{|(\theta - \theta_{min})(\theta_{max} - \theta)|}{(\theta_{max} - \theta_{min})^2 10^2 (N_1 + N_2)} \tag{5.12}$$

   where $\theta_{max} = \max_{\theta \in \Theta^*}(\theta)$, $\theta_{min} = \min_{\theta \in \Theta^*}(\theta)$ and N_g is the total number of match scores in both the genuine and imposter distributions from the gth group. Note that $\hat{r}_\theta^{(g)}$ is the radius at angle θ of the gth ROC based upon the original data and $r_\theta^{(g)b}$ is the equivalent for the mth bootstrapped ROC of group g.
5. Next for each $m = 1, \ldots, M$ find the maximal normalized difference of each replicate ROC from the sample difference $\hat{R}_1 - \hat{R}_2$ in the following manner

$$w_m = \begin{cases} \max_\theta(e_{m\theta}) & \text{if } |\min_\theta(e_{m\theta})| < |\max_\theta(e_{m\theta})| \\ \min_\theta(e_{m\theta}) & \text{if } |\min_\theta(e_{m\theta})| \geq |\max_\theta(e_{m\theta})|. \end{cases} \tag{5.13}$$

6. Find the $\alpha/2$th and $1-\alpha/2$th percentiles for w_m and call them δ_L and δ_U, respectively.

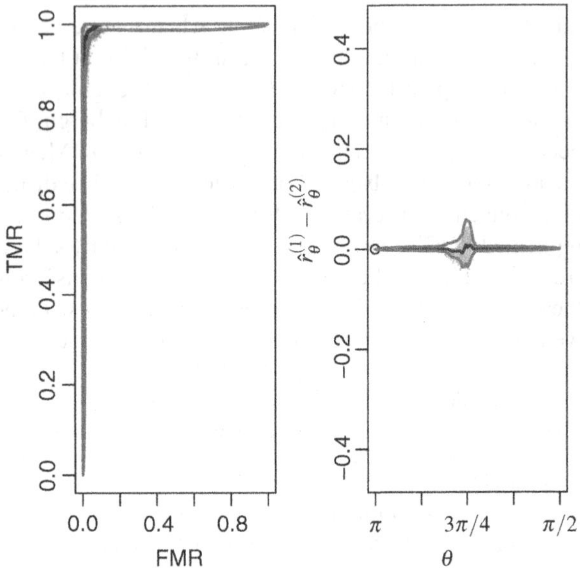

Fig. 5.6 95% confidence region for difference in ROC's for the voice matcher IDIAP-33-300 applied to two groups, g1 and g2, from the BANCA database under protocol G

7. Our $100(1-\alpha)\%$ confidence region for the difference between the two ROC curves $R_1 - R_2$ is given by connecting the upper limits at each θ

$$\hat{r}_\theta^{(1)} - \hat{r}_\theta^{(2)} - \delta_L s_{\hat{r}_\theta^{(1)} - \hat{r}_\theta^{(2)}} \tag{5.14}$$

and the lower limits at each θ

$$\hat{r}_\theta^{(1)} - \hat{r}_\theta^{(2)} - \delta_U s_{\hat{r}_\theta^{(1)} - \hat{r}_\theta^{(2)}}. \tag{5.15}$$

Below we introduce a new graph that plots the differences between the two ROC's on the vertical axis against the angle θ on the horizontal axis. We will center the vertical axis at 0 so that we assess how different the two curves are and whether those differences are significant. We can then evaluate the differences in the two ROC's by plotting our confidence region for the differences in the ROC's on that graph. We will plot the ROC difference graph alongside a graph that depicts the two ROC's that are being compared.

Example 5.4 For this example, we compare the voice matcher IDIAP_voice_gmm_auto_scale_33_300 (IDIAP-33-300) from the BANCA database to two groups, group g1 and group g2, following the G protocol. Figure 5.6 has two graphs of the output of the methodology given above including a 90% confidence region for the difference of the two ROC's. From the graph on the right of that figure, we can see that the line at a radial difference of zero is completely bounded by the 95% confidence region for the difference. Consequently, we can conclude that there is not a significant difference between these two ROC's. To obtain this region, we created $M = 1000$ replicated ROC curves for each group and took the difference of those curves. We used $T = 50$ angles spread uniformly between $\pi/2$ and π.

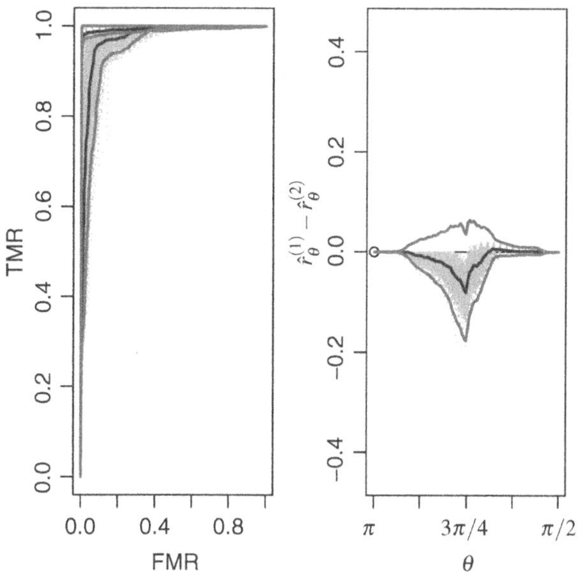

Fig. 5.7 95% confidence region for the difference of ROC's from two face matchers: SURREY-NC-200 from the BANCA database and (DCTb, GMM) from the XM2VTS database

Example 5.5 A 95% confidence region for the difference of two facial matchers is the goal of this example. The matcher SURREY_face_nc_man_scale_200 (SURREY-NC-200) from the BANCA database is the first of these two matchers and the second is the matcher (DCTb, GMM) from the XM2VTS database. For the former, the matching scores that we use here are from group g1 following the G protocol. Figure 5.7 has a summary of our application of the bootstrap methodology given above to these data for $T = 100$ angles and $M = 1000$ repetitions. In that figure, we can see the individual ROC curves for these two classifiers in the graph on the left along with their individual 95% confidence regions. In the graph on the right, we observe the confidence region for the difference of the two curves plotted against the angles, the θ's. We note that since the line at $r_\theta^{(1)} - r_\theta^{(2)} = 0$ is included in this confidence region for all θ's, we can conclude that these two ROC's are not significantly different.

Example 5.6 This example illustrates the differences in the confidence regions that we get for the difference of two ROC's when we use different ranges for Θ^*. Both confidence regions are built upon comparing the ROC from the voice matcher IDIAP_voice_gmm_auto_scale_25_10_pca (IDIAP-25-10-PCA) applied to group g1 from the BANCA database to the ROC from the voice matcher IDIAP_voice_gmm_auto_scale_25_200_pca (IDIAP-25-200-PCA) applied to group g2 from the BANCA database. Both of sets of data were collected under the G protocol. Figure 5.8 shows the confidence region for the difference of these two ROC's if we restrict the range of angles to $\theta_L = 5\pi/8$ to $\theta_U = 7\pi/8$, while Fig. 5.9 shows the confidence region for the difference of these two ROC's using the full range of the angles $\theta_L = \pi/2$ to $\theta_U = \pi$. In the former figure, we denote the limits of the angles by light vertical lines. Visually, we can see that there are large differences in these two confidence regions.

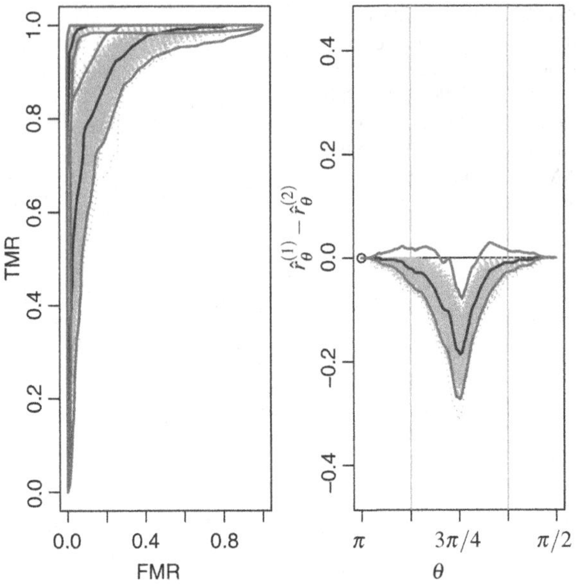

Fig. 5.8 95% confidence region for the difference in ROC's between IDIAP-25-10-PCA applied to group g1 of the BANCA database and IDIAP-25-200-PCA applied to group g2 of the BANCA database with $\theta_L = 5\pi/8$ to $\theta_U = 7\pi/8$

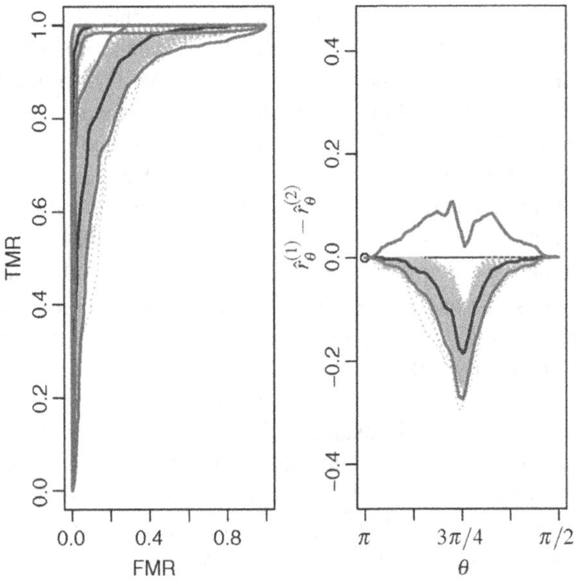

Fig. 5.9 95% confidence region for the difference in the independent ROC's between IDIAP-25-10-PCA applied to group g1 of the BANCA database and IDIAP-25-200-PCA applied to group g2 of the BANCA database with $\theta_L = \pi/2$ to $\theta_U = \pi$

5.3.2.2 Confidence Region for Difference of Two Paired ROC's

In this section, as above, we take a bootstrap approach to the estimation of the curve-wise difference between two ROC's, $R_1 - R_2$. That is, we would like to create a curvewise confidence region for the differences between the curves at each angle in Θ^* based upon our sample ROC's $\hat{R}_1$ and $\hat{R}_2$. The notable difference here is that with paired data collection when we resample individuals and match scores following the *two-instance bootstrap* methodology given in Sect. 5.2, we take the scores that are paired or connected between groups. That is, for the ℓth matching score from the resampled pair of individuals b_i and $h_{b_i,k}$, we will take both $Y^{(1)}_{b_i h_{b_i,k}\ell}$ and $Y^{(2)}_{b_i h_{b_i,k}\ell}$. Thus, we will only have a single bootstrapping process. One possible application of the methodology we give below is a comparison of two different voice matching algorithms applied to the same database of signals. Since each algorithm will give a match score for each pair of signals that are compared, then we have a 'paired' data collection.

Bootstrap Approach The bootstrap approach to making a confidence interval for the difference of two paired ROC's is similar in concept to the other bootstrap approaches to creating curvewise ROC confidence regions. We first calculate the original curve of differences, then we bootstrap the match scores to create bootstrapped ROC curves. Based upon the bootstrap replicated curves, we obtain a confidence region for the difference in the process ROC's. This algorithm for deriving the confidence region for the difference of two paired ROC is given below.

1. Calculate $\hat{R}_1 - \hat{R}_2$ by which we mean the collection of differences $\{(\hat{r}^{(1)}_\theta - \hat{r}^{(2)}_\theta, \theta) : \theta \in \Theta^*\}$.
2. Resample from the first process to generate a replicate ROC following the procedure given in Sect. 5.2. For each pair of individuals selected as part of this resampling process, take all of the match scores that go with first process and all of the match scores that go with the second process. That is for each selected pair of individuals, say i and k, take all of the $Y^{(1)}_{ik\ell}$'s from the first process and all of the $Y^{(2)}_{ik\ell}$'s from the second process. We will call replicate ROC's from the first process, $\hat{R}^b_1$ which will be the collection of radii $r^{(1)b}_\theta$ for all $\theta \in \Theta$. Call replicate ROC's from the second process $\hat{R}^b_2$. $\hat{R}^b_2$ will represent the collection of radii $r^{(2)b}_\theta$ for all $\theta \in \Theta$.
3. Repeat the two previous steps some large number of times M storing the mth ROC replicate $\hat{R}^b_g$ where $m = 1, \dots, M$.
4. Calculate and store the replicated normalized difference by

$$e_{m\theta} = \frac{(r^{(1)b}_{m\theta} - \hat{r}^{(1)}_\theta) - (r^{(2)b}_{m\theta} - \hat{r}^{(2)}_\theta)}{s_{\hat{r}^{(1)}_\theta - \hat{r}^{(2)}_\theta}} \tag{5.16}$$

where

$$s_{\hat{r}^{(1)}_\theta - \hat{r}^{(2)}_\theta} = \sqrt{\frac{1}{M-1}\sum_{m=1}^{M}\left(r^{(1)b}_{\theta m} - r^{(2)b}_{\theta m} - (\overline{r^{(1)b}_\theta - r^{(2)b}_\theta})\right)^2 + \iota_\theta} \tag{5.17}$$

for each $\theta \in \Theta^*$

$$\overline{r_\theta^{(1)b} - r_\theta^{(2)b}} = \frac{1}{M}\sum_{m=1}^{M} r_{\theta m}^{(1)b} - r_{\theta m}^{(2)b}$$

and

$$\iota_\theta = \frac{|(\theta - \theta_{min})(\theta_{max} - \theta)|}{(\theta_{max} - \theta_{min})^2 10^2 N} \tag{5.18}$$

where $\theta_{max} = \max_{\theta \in \Theta^*}(\theta)$, $\theta_{min} = \min_{\theta \in \Theta^*}(\theta)$ and N is the total number of match scores in both the genuine and imposter distributions.

5. Next for each $m = 1, \ldots, M$ find the maximal normalized difference of each replicate ROC from the sample difference $\hat{R}_1 - \hat{R}_2$ in the following manner

$$w_m = \begin{cases} \max_\theta(e_{m\theta}) & \text{if } |\min_\theta(e_{m\theta})| < |\max_\theta(e_{m\theta})|, \\ \min_\theta(e_{m\theta}) & \text{if } |\min_\theta(e_{m\theta})| \geq |\max_\theta(e_{m\theta})|. \end{cases} \tag{5.19}$$

6. Find the $\alpha/2$th and $1 - \alpha/2$th percentiles for w_m and call them w_L and w_U, respectively.
7. Our $100(1-\alpha)\%$ confidence region for the difference between the two paired ROC curves $R_1 - R_2$ is given by connecting the upper limits

$$\hat{r}_\theta^{(1)} - \hat{r}_\theta^{(2)} - \delta_L s_{\hat{r}_\theta^{(1)} - \hat{r}_\theta^{(2)}} \tag{5.20}$$

at each θ using linear interpolation and connecting the lower limits

$$\hat{r}_\theta^{(1)} - \hat{r}_\theta^{(2)} - \delta_U s_{\hat{r}_\theta^{(1)} - \hat{r}_\theta^{(2)}} \tag{5.21}$$

at each θ in the same manner.

Example 5.7 In this example, we compare two voice matchers from the BANCA database. Those matchers are IDIAP_voice_gmm_auto_scale_25_50_pca (IDIAP-25-50-PCA) and IDIAP_voice_gmm_auto_scale_25_50 (IDIAP-25-50). The match scores that are analyzed here, both the genuine and imposter distributions, are from captures taken from group g1 of the BANCA database under the G protocol. Two graphs of the relationship between the two estimated ROC's are found in Fig. 5.10. On the right is a 95% confidence region for the difference of those two ROC's plotted against the angle θ. This confidence region was made for the entire region from $\theta_U = \pi/2$ to $\theta_L = \pi$ with $T = 100$ using $M = 1000$ bootstrapped ROC's. This region contains the difference $r_\theta^{(1)} - r_\theta^{(2)} = 0$ at each angle θ. Consequently, we conclude that these two ROCs are not significantly different.

Example 5.8 The goal of this example is to create a 90% confidence region for the difference of two paired ROC's from the BANCA database. The two classifiers that we will use are IDIAP_voice_gmm_auto_scale_25_10_pca (IDIAP-25-10-PCA) and IDIAP_voice_gmm_auto_scale_25_200_pca (IDIAP-25-200-PCA). The match scores that we will use are from the G protocol of the BANCA database applied to group g2. To create this confidence region, we repeated the bootstrap

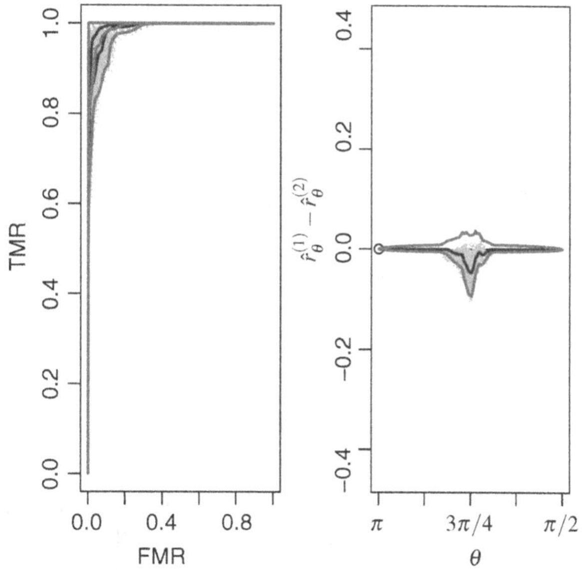

Fig. 5.10 95% confidence region for the difference of two ROC's from the voice matchers IDIAP-25-50-PCA and IDIAP-25-50 from the BANCA database applied to group g1 under protocol G

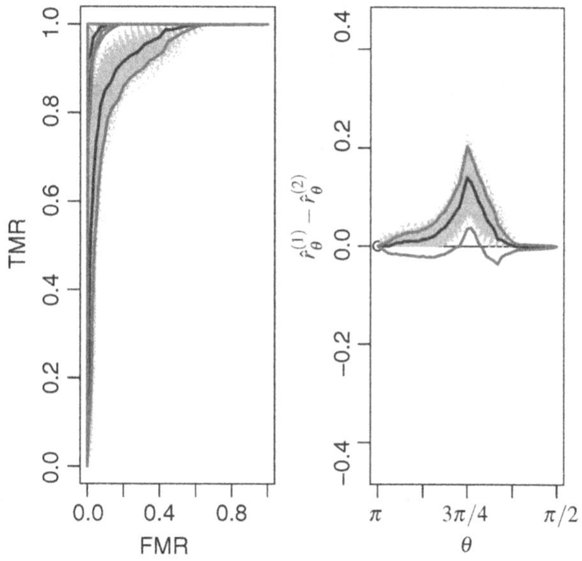

Fig. 5.11 99% confidence region for the difference of the IDIAP-25-10-PCA and the IDIAP-25-200-PCA paired matchers from the BANCA database applied to group g2 following the G protocol

approach given above $M = 2000$ times with $T = 49$ angles which were uniformly spaced between $\pi/2$ and π. Our 90% confidence region is represented in the graph on the right of Fig. 5.11 by the thickest lines. The left graph in that figure displays the two ROC's individually along with their individual 90% confidence regions denoted by the thicker lines. The light lines in each of these graphs represents the bootstrapped ROC's, the R_g^b's, on the left, and the bootstrapped differences in the ROC's, the $R_1^b - R_2^b$'s, on the right. From the graph on the right, we can see that this confidence region does not include a difference of $R_1 - R_2 = 0$ at several angles. Consequently, we can conclude that there is a significant difference between these two ROC curves.

5.3.3 Multiple Sample Methods for ROC's

We now turn our attention to comparing three or more ROC curves and testing whether there are significance differences among these curves. We will denote the gth process curve here by R_g where $g = 1, \ldots, G$ and we will denote the estimated process curve by $\hat{R}_g$. The hypotheses that we will be testing are

$$H_0 : R_1 = R_2 = \cdots = R_G$$
$$H_1 : \text{ not } H_0.$$

The alternative hypothesis here is equivalent to saying that at least one of these ROC's is significantly different from the others. For these tests, we assume that each of the G matching processes is stationary. The tests given in this section are analogous to an analysis of variance-type (ANOVA) type test. Since we are comparing multiple curves, it is no longer sufficient to use the difference between the curves. We will use a measure, F_θ, similar to a traditional F-statistic to compare the G ROC curves. Our F_θ is the variability between the ROC curves at the angle θ. Since we are considering this statistic at each angle θ, we'll have to address the multiple comparison problem, cf. Sect. 8.2.2 for further discussion. The multiple comparison problem or *multiplicity problem* results from the fact the probability that we reject at least one of the hypotheses being tested increasing as the number of tests increases. That is, the chance that we make a type I error increases the more hypotheses tests that we conduct. Here we are planning T tests, one at each angle $\theta \in \Theta^*$. This leads us to Holm's method for comparing a large number of *p-values* Holm [48]. The basic ideas is to sort the *p-values* from each angle and then adjust the significance level to account for the fact that we are assessing a large number of *p-values*.

5.3.3.1 Hypothesis Test for Multiple Independent ROC's

Below we present bootstrap methodology for comparing multiple independent ROC curves. The statistical focus here is the following.

$$H_0 : R_1 = R_2 = \cdots = R_G$$
$$H_1 : \text{ not } H_0.$$

These hypotheses are equivalent to

$$H_0 : r_\theta^{(1)} = r_\theta^{(2)} = \cdots r_\theta^{(G)} \quad \text{for all } \theta \in \Theta^*$$
$$H_1 : \text{ not } H_0.$$

The outline of the methodology for this test is that we first take and calculate the variability between the original ROC curves, the $\hat{r}_\theta^{(g)}$'s, at each angle θ. We next repeatedly generate replicate versions of the match score distributions, imposter and genuine using the bootstrap methodology given in Sect. 5.2. At each iteration, we similarly calculate the variability between the bootstrapped ROC's, the $r_\theta^{(g)b}$'s, assuming that the null hypothesis of equality between the ROC, H_0, is true. Having done this numerous times, we consider what percentage of times the variability between the original ROC's is exceeded by the variability in the bootstrapped ROC's. If this percentage is small, then we reject the null hypothesis of equality between the ROC's.

One application of this approach is the comparison of different groups of individuals from a single data collection. For example, we might want to compare the ROC's for five different occupational groups. The approach given below is appropriate for that scenario as long as we only analyze match scores from signals within each group.

Bootstrap Approach To test the hypotheses given above, we present a methodology based upon bootstrapping the match scores from which each ROC is derived for each of the G groups. This is given below.

1. Derive the estimated ROC's for each of the G groups and denote them by $\hat{R}_g$.
2. Calculate the test statistic, F_θ, at each angle $\theta \in \Theta^*$ using

$$F_\theta = \sum_{g=1}^{G} N_g(\hat{r}_\theta^{(g)} - \bar{r}_\theta)^2 \tag{5.22}$$

   for each $\theta \in \Theta^*$ where

$$\bar{r}_\theta = \frac{\sum_{g=1}^{G} N_g r_\theta^{(g)}}{\sum_{g=1}^{G} N_g} \tag{5.23}$$

   and N_g is the total number of match scores in the gth group.
3. Bootstrap the indices that comprise the match score distributions for each of the G groups separately and derive ROC curves from the resampled genuine and imposter distributions that result from following the methods in Sect. 5.2. Call these resampled ROC's, R_g^b's.
4. Calculate F_θ^b for each $\theta \in \Theta^*$ following

$$F_\theta^b = \sum_{g=1}^{G} N_g(r_\theta^{(g)b} - \hat{r}_\theta^{(g)})^2. \tag{5.24}$$

5. Repeat the previous two steps some large number of times, M, each time storing F_θ^b.
6. Then the *p-value* at each θ is defined to be

$$p_\theta = \frac{1 + \sum_{\varsigma=1}^{M} I_{\{F^b \geq F\}}}{M+1} \tag{5.25}$$

where $I_{\{\ \}}$ is an indicator function. See Definition 2.10.

7. Sort the *p-values* and label the sorted *p-values* such that

$$p_{(1)} \leq p_{(2)} \leq \cdots \leq p_{(T)}. \tag{5.26}$$

Further let $\theta_{(t)}$ be the angle associated with $p_{(t)}$ for $t = 1, 2, \ldots, T$.

8. For a given significance value, α, we follow these steps below to determine the angles at which there are significant differences between the curves.
   a. We will reject the null hypothesis of equality for each of the G ROC's at the angle $\theta = \theta_{(1)}$ if $p_{(1)} < \alpha/T$.
   b. If $p_{(1)} \geq \alpha/T$, then we conclude there are no significant differences between the G curves and we proceed no further.
   c. If $p_{(2)} \leq \alpha/(T-1)$, we reject the null hypothesis that $r^{(1)}_{\theta_{(2)}} = r^{(2)}_{\theta_{(2)}} = \cdots = r^{(G)}_{\theta_{(2)}}$. Otherwise, proceed no further and conclude that the only significant differences between the ROC's occur at $\theta_{(1)}$.
   d. Continue testing the equality of ROC's until $p_{(t)} > \alpha/(T - t + 1)$. Conclude that there are significant differences at angles $\theta_{(1)}, \theta_{(2)}, \ldots, \theta_{(t-1)}$.
9. We will reject the overall hypothesis $H_0 : R_1 = R_2 = \cdots = R_G$ if we reject the hypothesis of equality for at least one angle θ.

In this methodology, we use Holm's method to deal with multiple comparisons. There are other methods for comparing multiple hypotheses as we do here. Our choice of Holm's method is because it does not depend on the independence of the tests that we are considering. It seems reasonable to suspect that because of the similarity of the curves at different radii that there is dependence between the *p-values* at adjacent as well as nearby angles. One consideration from the use of Holm's methods is that we should ensure that it is possible to reject the null hypothesis, $H_0 : R_1 = R_2 = \cdots = R_G$. This will only be possible if $1/(M + 1) < \alpha/T$. Therefore, we need at least $M > T/\alpha - 1$ bootstrap replicates to have any chance of rejecting the null hypothesis following Holm's method.

To summarize the results from this section on comparing multiple ROC's, we have developed a new graphical tool. The output of this tool consists of two graphs juxtaposed. Examples of this tool can be seen in Fig. 5.12 and 5.14. The subfigure on the left is the individual ROC curves plotted on a single axis as is typical. The subfigure on the right is a representation of the variability in our statistic F_θ. The dark line represents the F_θ calculated on the original data. The gray lines there are the bootstrapped versions of F_θ. Above we have called these the F^b_θ's. The x-axis on this subfigure is the angle at which we are comparing the ROC's. Note that we have reversed the order of the angles—π on the left and $\pi/2$ on the right—which seems to agree more naturally with the order of those angles in the subfigure on the left. The utility of this graphical tool is that we can visually assess how distinctive each F_θ is from the distribution of the F^b_θ's at the same angle θ. If, at least, some of the F_θ's are at the upper tail of their respectively distributions, then we can conclude that there is enough evidence to suggest that the ROC's are not all the same. On the other hand, if all of the F_θ's are in or near the center of the distribution of the F^b_θ at each angle θ, then there is not enough evidence to reject a null hypothesis of equality of the ROC's.

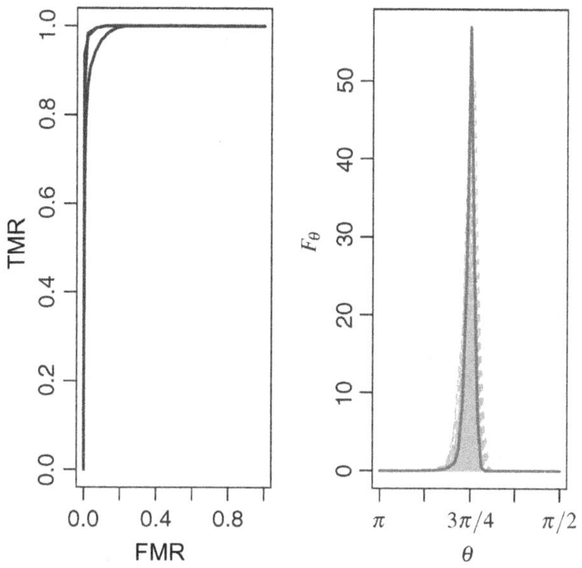

Fig. 5.12 Summary of test of equality of three independent ROC's: IDIAP-33-300-PCA from the BANCA database, UC3M-34-300 from the BANCA database and (PAC, GMM) from the XM2VTS database

Example 5.9 For this example, we consider three voice/speaker matching algorithms. The first of these matchers is the IDIAP_voice_gmm_auto_scale_33_300_pca (IDIAP-33-300-PCA) matcher from the BANCA database following the G protocol applied to the individuals in group g1. The second matcher also from the BANCA database following the G protocol but using individuals from group g2 is UC3M_voice_gmm_ auto_scale_34_300 (UC3M-34-300). The third matcher is the speaker matcher (PAC, GMM) from the XM2VTS database. We bootstrapped the matching scores from these matchers following the two-instance *two-instance bootstrap*. A summary of the ROC's from these matchers as well as the distribution of bootstrapped F_θ's is given in Fig. 5.12. The smallest *p-value*, $p_{(1)} = 1/1001 = 0.000999$, from that process is less than $\alpha/T = 0.05/49 = 0.00102$. Following Holm's method, we are able to reject the null hypothesis here and conclude that there is a significant difference between these ROC's.

Example 5.10 For this example, we took the NIST database and broke the 517 individual in that database into $G = 5$ groups. We did this based solely upon their individual ID. The ID's numbered 1 to 100 were assigned to the first group, ID's 101 to 200 to the second, ID's 201 to 300 to the third, ID's 301 to 400 to the fourth, ID's 401 to 517 to the five. (If we had demographic information for these individuals, we could group them in a similar way using that information.) To obtain five different sets of matching scores, we used all possible combinations of match scores within each group. With a significance level of $\alpha = 0.05$, we are formally testing

$$H_0 : R_1 = R_2 = R_3 = R_4 = R_5$$
$$H_1 : \text{not } H_0.$$

We implemented the bootstrap methodology given above to test this hypothesis. We bootstrapped these five different groups $M = 1000$ times using $T = 49$ different

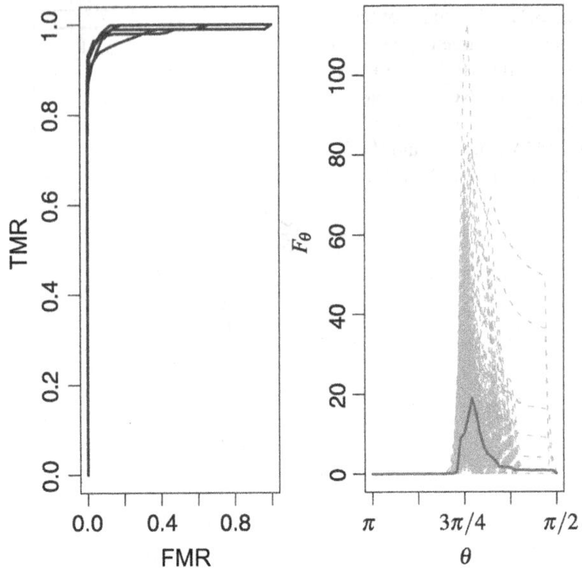

Fig. 5.13 Summary of test of equality of five ROC's from independent groups applied to the Face_C classifier from the NIST database

angles. The results of this process are given in Fig. 5.13. In that figure the graph on the left shows the five individual ROC's, while the graph on the right is a plot of F_θ, the dark line, against the angle θ. The lighter lines in the graph on the right represent the value for the bootstrapped versions of our test statistic, the F_θ^b's. From this graph we can see that the values for F_θ are not atypical given the variability in the five ROC's. The smallest *p-value* in this case is $p = 0.1758$ which is larger than $\alpha/T = 0.05/49 = 0.0010$. As a consequence, at a significance level of 0.05 we do not have enough evidence to reject the null hypothesis, H_0, that these five ROC's are the same. Another interpretation of this result would be that the matching process here is stationary across all five groups.

5.3.3.2 Hypothesis Test for Multiple Paired ROC's

Our interest in this section will be the testing of the equality of G ROC's when those ROC's are paired. As above this test is given by

$$\begin{aligned} &H_0 : R_1 = R_2 = \cdots = R_G \\ &H_1 : \text{ not } H_0. \end{aligned}$$

This is equivalent to the test.

$$\begin{aligned} &H_0 : r_\theta^{(1)} = r_\theta^{(2)} = \cdots = r_\theta^{(G)} \quad \text{for all } \theta \in \Theta^* \\ &H_1 : \text{ not } H_0. \end{aligned}$$

Our technique for carrying out this test, which is enumerated below, is based upon the *two-instance bootstrap* for the match scores from which an estimated ROC curve is calculated.

Bootstrap Approach The bootstrap approach here is similar to other approaches for analyzing paired data. We will bootstrap the individuals following the two-instance bootstrap described in Sect. 5.2. Then for a selected pair of individuals i and k, we will take all of the match scores, $Y_{ik1}^{(g)}, \ldots, Y_{ikm_{ik}}^{(g)}$ for all $g = 1, \ldots, G$. We then combine all of the scores from the gth group and we then calculate a bootstrapped ROC for that group.

1. Derive the estimated ROC's for each of the G groups and denote them by $\hat{R}_g$.
2. Calculate the test statistic, F_θ, at each angle $\theta \in \Theta^*$ using

$$F_\theta = \sum_{g=1}^{G} N_g(\hat{r}_\theta^{(g)} - \bar{r}_\theta)^2 \tag{5.27}$$

   where

$$\bar{r}_\theta = \frac{1}{G}\sum_{g=1}^{G} r_\theta^{(g)} \tag{5.28}$$

   for each $\theta \in \Theta^*$.
3. Bootstrap the match scores that comprise the G ROC curves following the *two-instance bootstrap* methods in Sect. 5.2, so that the scores associated with a pair of selected individuals i and k are taken from all G groups. Keep the scores from each group separate and calculate a replicated ROC from each group. Call the ROC's that result from the bootstrapping process R_g^b's.
4. Calculate F_θ^b for each $\theta \in \Theta^*$ following

$$F_\theta^b = \sum_{g=1}^{G} N_g(r_\theta^{(g)b} - \hat{r}_\theta^{(g)})^2. \tag{5.29}$$

5. Repeat the previous two steps some large number of times, M, each time storing F_θ^b.
6. Then the *p-value* at each θ is defined to be

$$p_\theta = \frac{1 + \sum_{\varsigma=1}^{M} I_{\{F^b \geq F\}}}{M+1}. \tag{5.30}$$

7. Sort the *p-values* and label the resorted *p-values* such that

$$p_{(1)} \leq p_{(2)} \leq \cdots \leq p_{(T)}. \tag{5.31}$$

   Further let $\theta_{(t)}$ be the angle associated with $p_{(t)}$ for $t = 1, 2, \ldots, T$.
8. For a given significance value, α, we follow these steps to determine if there is a significance difference between the ROC's:
   a. We will reject the null hypothesis of equality for each of the G ROC's at the angle $\theta = \theta_{(1)}$ if $p_{(1)} < \alpha/T$.
   b. If $p_{(1)} \geq \alpha/T$, then we conclude there are no significant differences between the G curves and we proceed no further.

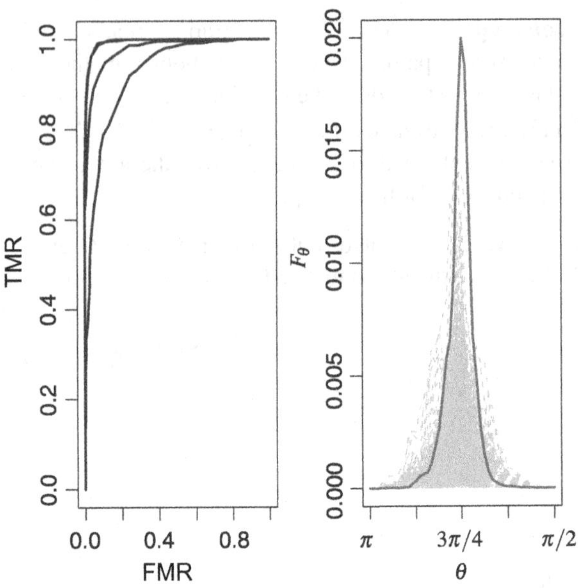

Fig. 5.14 Summary of results from testing the equality of ROC's from four matchers: IDIAP-25-100-PCA IDIAP-25-10-PCA, IDIAP-25-20-PCA, and IDIAP-25-25-PCA, from the BANCA database applied to group g1 following protocol G

c. If $p_{(2)} \le \alpha/(T-1)$, we reject the null hypothesis that $r^{(1)}_{\theta_{(2)}} = r^{(2)}_{\theta_{(2)}} = \cdots = r^{(G)}_{\theta_{(2)}}$. Otherwise proceed no further and conclude that the only significant differences between the ROC's occur at $\theta_{(1)}$.
d. Continue testing the equality of ROC's until $p_{(t)} > \alpha/(T-t+1)$. Conclude that there are significant differences at angles $\theta_{(1)}, \theta_{(2)}, \ldots, \theta_{(t-1)}$.

9. We will reject the overall hypothesis $H_0 : R_1 = R_2 = \cdots = R_G$ if we reject the hypothesis of equality for at least one angle θ.

Example 5.11 Here we illustrate the bootstrap approach to testing the equality of multiple paired ROC's at a significance level of $\alpha = 0.05$ by applying this methodology to four different voice matchers applied to group g1 of the BANCA database following protocol G. These four matchers are IDIAP_voice_gmm_auto_scale_25_100 _pca (IDIAP-25-100-PCA), IDIAP_voice_gmm_auto_scale_25_10_pca (IDIAP-25-10-PCA), IDIAP_voice_gmm_auto_scale_25_200_pca, (IDIAP-25-200-PCA), IDIAP_voice_gmm_auto_scale_25_25_pca (IDIAP-25-25-PCA). The results of $M = 1000$ repetitions of the bootstrap algorithm given above are represented in Fig. 5.10. For this application we used $T = 49$ different angles that were uniformly spaced between $\pi/2$ and π. We reject the null hypothesis that these four voice matchers have the same ROC since we have sufficient evidence to reject this null hypothesis. In particular, we have four angles for which the *p-value* is $1/1001 = 0.000999$. Since this value is less than $\alpha/49 = 0.00102$, we can conclude that there is at least one angle at which these ROC's are significantly different. Since they are different at one or more angles, they we can say that they are significantly different.

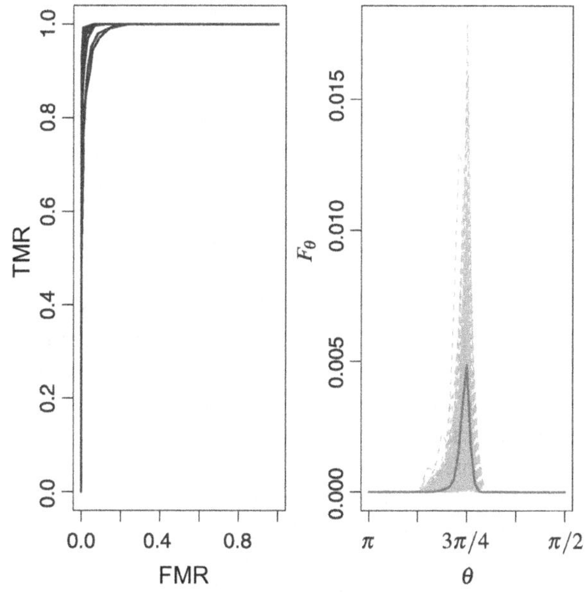

Fig. 5.15 Distribution of the bootstrapped test statistic, F^b, from simultaneously comparing ten voice matchers from the BANCA database

Example 5.12 In this example, we take the bootstrap methodology given above and apply it to ten voice matchers from the BANCA database. Those ten matchers are IDIAP_voice_gmm_auto_scale_25_100 (IDIAP-25-100), IDIAP_voice_gmm_auto_scale_25_10 (IDIAP-25-10), IDIAP_voice_gmm_auto_scale_25_200 (IDIAP-25-200), IDIAP_voice_gmm_auto_scale_25_25 (IDIAP-25-25), IDIAP_voice_gmm_auto_scale_25_300 (IDIAP-25-300), IDIAP_voice_gmm_auto_scale_25_50 (IDIAP-25-50), IDIAP_voice_gmm_auto_scale_25_75 (IDIAP-25-75), IDIAP_voice_gmm_auto_scale_33_100 (IDIAP-33-100), IDIAP_voice_gmm_auto_scale_33_10 (IDIAP-33-10), IDIAP_voice_gmm_auto_scale_33_200 (IDIAP-33-200). Bootstrapping the paired matching scores from these distributions $M = 5000$ times, we are testing null hypothesis that these 10 ROC curves are the same. For this test, we used a significance level of $\alpha = 0.05$ and we tested at $T = 49$ different angles. Figure 5.15 contains a summary of the distribution of the F_θ^b's as well as a summary of the ROC's themselves. The smallest *p-value* across all $T = 49$ of the angles is $p_{(1)} = 124/5001 = 0.0248$ which is not less than $\alpha/T = 0.05/49 = 0.0010$. Therefore, we are do not have enough evidence to reject the null hypothesis, H_0, that these 10 ROC's are equal.

5.4 Statistical Methods for EER

In this section we present methodology for estimating and testing equal error rates, EER's. The EER, as mentioned previously, is the value of the FMR and the FNMR for a given matching process when the FMR = FNMR. Following the notation of previous chapters, we can rewrite this inequality as $\nu = \pi$. The EER is often used as a single overall measure of the quality of the matching or classification mechanism. It is a single metric that summarizes the entire ROC curve. We denote a single process EER by χ. The statistical mechanisms for the EER in this section are similar to the statistical ROC mechanisms above but with interest limited to the value of the radius when $\theta = \frac{3}{4}\pi$. It is at this radius that $\chi = \nu = \pi$. For that radius, we can find the estimated EER by

$$\hat{\chi} = 1 - \frac{\hat{r}_{\theta=3\pi/4}}{\sqrt{2}} \tag{5.32}$$

if the center point is $(1, 0)$. We have chosen this specific value for the center of our polar coordinates. Values for the center point along the line defined by $(c, 1-c)$ also seem reasonable, since they would also give a confidence interval for an EER. The values along that line will give estimates where the false match rate and the false non-match rate are equal. It is straightforward to determine the approximate EER for other center points, (c_x, c_y) using basic trigonometry; however, the methodology for creating a confidence interval for the EER given below would not to appropriate for those values. The methodology give below for a confidence interval for a single EER was first introduced by Schuckers et al. [88]. We extend this methodology to other inferential cases in the remainder of this chapter. Since the EER is a scalar quantity, we do not need to use the curvewise adjustments made above for the entire ROC curve. We, however, need to ensure that $3\pi/4 \in \Theta^*$ and we can make a pointwise interval since we're interested in an inferential interval for a scalar, χ. One advantage of the ROC methodology given above is that we can use the output from those methods for inference about the EER.

5.4.1 One Sample Methods for EER

In this section we present methods for estimation of a single process EER, χ. The approach taken in this section is a bootstrapping one. We rely upon this methodology to calculate the sampling variability in our estimates of χ. From this sampling variability, we can approximate the sampling distributions required for the confidence interval and hypothesis given below. We will assume here that we have a single process, so that there is stationarity of the EER. In later sections, we present methods that are capable of distinguishing multiple groups that are part of a single process and multiple processes.

5.4.1.1 Confidence Interval for Single EER

The aim of this section is the creation of a confidence interval for the EER, χ, of a bioauthentication matching process. To construct this interval, we bootstrap the genuine and imposter match score distributions following the *two-instance bootstrap* methodology above in Sect. 5.2.

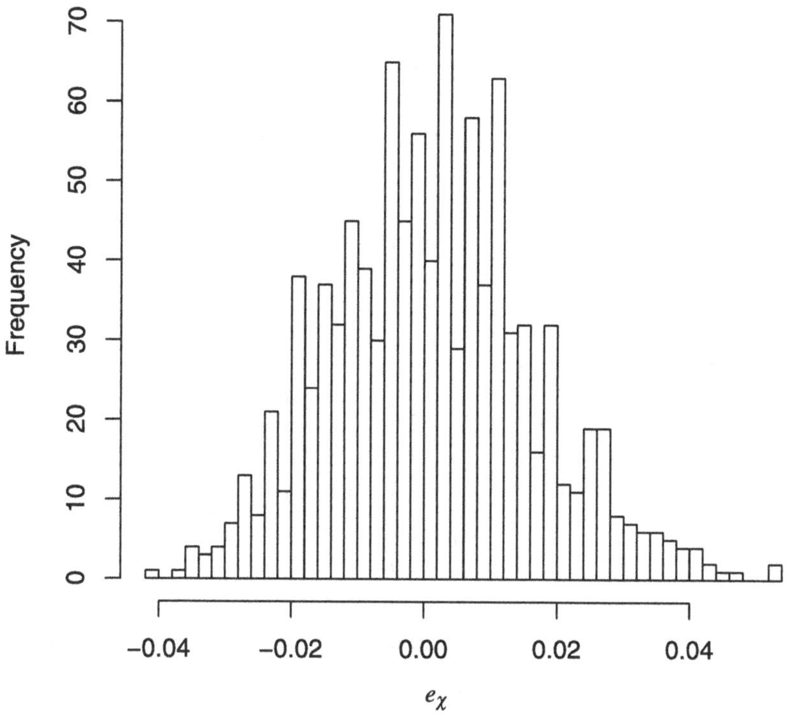

Fig. 5.16 Distribution of the bootstrapped differences, the e_χ's, from the estimated EER for the face matcher SURREY-NC-200 from the BANCA database

Bootstrap Approach Following the *two-instance bootstrap* approach to estimation of an ROC, we can create a $100(1-\alpha)\%$ confidence interval for a process EER, χ.

1. Calculate $\hat{\chi} = 1 - \hat{r}_{\theta=3\pi/4}/\sqrt{2}$ from the observed ROC, $\hat{R}$.
2. Bootstrap the match scores to obtain an ROC to yield $\hat{R}^b$ and then the resampled EER is $\hat{\chi}^b = 1 - \hat{r}^b_{\theta=3\pi/4}/\sqrt{2}$.
3. Determine and store $e_\chi = \hat{\chi}^b - \hat{\chi}$.
4. Repeat Steps 2 and 3 above a large number of times, M.
5. A $100(1-\alpha)\%$ confidence interval for χ is then formed by taking the interval from $\hat{\chi} - e_U$ to $\hat{\chi} - e_L$ where e_L and e_U represent the $\alpha/2$th and the $1-\alpha/2$th percentiles of the distribution of e_χ.

Example 5.13 In this example, we consider a face matcher, SURREY-NC-200, from the BANCA database applied to group g1 following the G protocol. Our goal is to make a 90% confidence interval for the process EER, χ. Following the algorithm above, we bootstrapped the distribution of the match scores for this matcher $M = 1000$ times. The distribution of the e_χ is given in Fig. 5.16. The estimated EER from these data is $\hat{\chi} = 0.0595$. To make our 90% confidence interval, we find the 5th

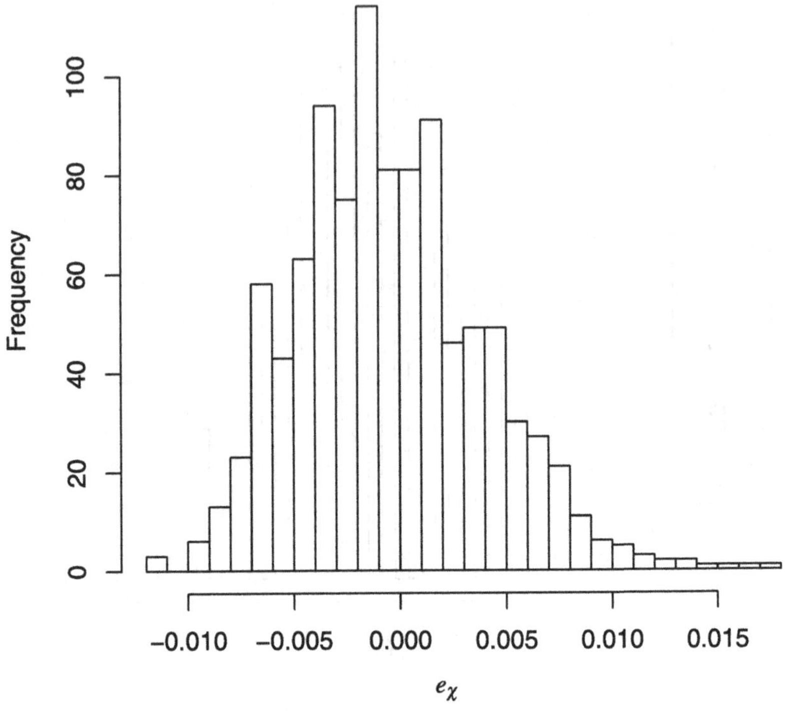

Fig. 5.17 Distribution of the bootstrapped differences from the estimated EER, the e_χ's, for the face matcher (FH, MLP) from the XM2VTS database

and 95th percentiles of e_χ which are -0.0253 and 0.0264, respectively. Subtracting both from $\hat{\chi}$, we get a 90% confidence interval that is from 0.0331 to 0.0848. Thus, we can be 90% confident that the EER for this process is between 3.31% and 8.48%.

Example 5.14 This example creates a 98% confidence interval for the EER of the (FH, MLP) matcher from the XM2VTS database. Following the algorithm above, we resampled the 40600 match scores obtained from this facial matcher $M = 1000$ times and the distribution of the e_χ's is give in Fig. 5.17. Our estimated EER for the original data is $\hat{\chi} = 0.0166$. Finding the 99th and 1st percentiles of e_χ which are 0.0111 and -0.0086, respectively, we then create our confidence interval to be $(0.0055, 0.0252)$. Thus, we are 98% confident that the process EER for this matcher is between 0.55% and 2.52%.

5.4.1.2 Hypothesis Test for Single EER

One consideration that is commonly of interest for the performance of a biometric identification system is to determine whether a system has an EER, χ, that is below some specified value. Formally we are testing:

$$H_0: \chi = \chi_0$$
$$H_1: \chi < \chi_0.$$

Relying upon the *two-instance bootstrap* approach outlined above for estimation of variability in an ROC, we can test this hypothesis in the following manner.

Bootstrap Approach This method uses a bootstrap approach to test whether a process EER, χ, is significantly less than a hypothesized EER, χ_0.

1. Calculate $\hat{\chi} = 1 - \hat{r}_{3\pi/4}/\sqrt{2}$ from the observed sample ROC, R.
2. Bootstrap the match scores to generate a replicated ROC, $\hat{R}^b$, and then the resampled EER is
$$\hat{\chi}^b = 1 - \hat{r}^b_{3\pi/4}/\sqrt{2}. \quad (5.33)$$
3. Calculate and store
$$e_\chi = \hat{\chi}^b - \hat{\chi} + \chi_0. \quad (5.34)$$
4. Repeat Steps 2 and 3 above a large number of times M.
5. Then the *p-value* for this test is
$$p = \frac{1 + \sum_{\varsigma=1}^{M} I_{\{e_\chi \le \hat{\chi}\}}}{M+1}. \quad (5.35)$$
6. If *a priori* we have specified a significance level then we will reject the null hypothesis, H_0, if $p < \alpha$. Otherwise, we reject if the *p-value* is small.

Example 5.15 To illustrate this approach, we consider the facial matcher SURREY_face_svm_man_scale_2.00 (SURREY-SVM-MAN-2.00) from the BANCA database. The particular match scores that are used here are from group g1 of that database using the G protocol. Testing the hypothesis that the EER is less than 0.05 at the significance level $\alpha = 0.05$ is the goal here. Formally we want to test

$$H_0: \chi = 0.05$$
$$H_1: \chi < 0.05.$$

We resampled the 545 match scores $M = 1000$ times. The distribution of e_χ is summarized in Fig. 5.18. The sample EER is $\hat{\chi} = 0.0300$ and is represented by the dashed vertical line in Fig. 5.18. The *p-value* is 0.0300 which is less than $\alpha = 0.05$. Therefore, we reject the null hypothesis, H_0, and conclude that the EER for the matcher SURREY-SVM-MAN-2.00 is significantly less than 0.05.

5.4.2 Two Sample Methods for EER's

We now turn to the comparison of two equal error rates. We will differentiate between those error rates by using χ_1 and χ_2. Following the methodology above, cf.

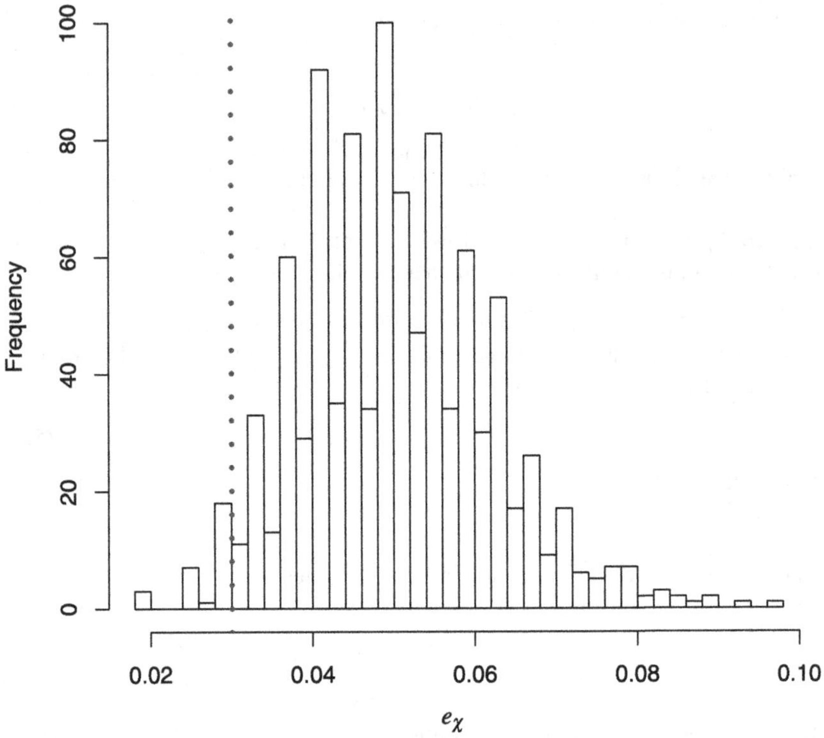

Fig. 5.18 Distribution of the bootstrapped differences from the estimated EER, the e_χ's, for the face matcher SURREY-SVM-MAN-2.00 from the BANCA database

Sect. 5.2, that relies upon the bootstrap to estimate the variability in a given sample EER, we will derive methodology for both independent samples as well as paired samples. We provide both confidence interval and hypothesis testing methods.

5.4.2.1 Confidence Interval for Difference of Two Independent EER's

In this section, we provide a method for estimating the differences between two independent EER's. We might have independent EER's if we are comparing the EER's of two different age groups for the same device. For example, we might want to compare the EER's for individuals who are over 30 years of age and those 30 years of age and under. The bootstrap approach below gives the variability in the estimated difference between the two EER's, $\hat{\chi}_1 - \hat{\chi}_2$. This variability is obtained by bootstrapping the ROC's *separately* using the methodology given in Sect. 5.2. It is conceivable that the sampling distribution for the difference of two estimated EER's would follow a Gaussian distribution. However, an explicit formulation of the variance of the EER has not been derived.

Bootstrap Approach Below we present a bootstrap approach for two independent EER's. The basics of this approach are that we bootstrap the two sets of matching scores from the two groups separately and then use those bootstrapped error rates to create an approximate sampling distribution for the difference of the two sample EER's, $\hat{\chi}_1 - \hat{\chi}_2$. We do this in the following way:

1. Calculate
$$\hat{\chi}_1 - \hat{\chi}_2 = (1 - \hat{r}^{(1)}_{3\pi/4}/\sqrt{2}) - (1 - \hat{r}^{(2)}_{3\pi/4}/\sqrt{2}) \tag{5.36}$$
from the estimated ROC's, $\hat{R}_1$ and $\hat{R}_2$.
2. Bootstrap the match scores that form the ROC from the first group to yield a replicated ROC, $\hat{R}^b_1$, and then the resampled EER is
$$\hat{\chi}^b_1 = 1 - \hat{r}^{(1)b}_{3\pi/4}/\sqrt{2}. \tag{5.37}$$
3. Bootstrap the matching scores from the second group and derive from those resampled scores a bootstrap denoted by $\hat{R}^b_2$. Then the resampled EER is
$$\hat{\chi}^b_2 = 1 - \hat{r}^{(2)b}_{3\pi/4}/\sqrt{2}. \tag{5.38}$$
4. Determine and store
$$e_\chi = \hat{\chi}^b_1 - \hat{\chi}^b_2 - (\hat{\chi}_1 - \hat{\chi}_2). \tag{5.39}$$
5. Repeat steps 2, 3, and 4 above a large number of times M.
6. A $100(1-\alpha)\%$ confidence interval for $\chi_1 - \chi_2$ is then formed by taking the interval from $\hat{\chi}_1 - \hat{\chi}_2 - e_U$ to $\hat{\chi}_1 - \hat{\chi}_2 - e_L$ where e_L and e_U represent the $\alpha/2$th and the $1-\alpha/2$th percentiles of the distribution of e_χ.

Example 5.16 For this example, we are using two voice matchers: IDIAP_voice_gmm_auto_scale_25_10_pca (IDIAP-25-10-PCA) and IDIAP_voice_gmm_auto_scale_25_200_pca (IDIAP-25-200-PCA). Figure 5.19 has a histogram of the distribution of e_χ for these two matchers. The estimated difference between these two EER's is $\hat{\chi}_1 - \hat{\chi}_2 = -0.1293$. To make a 90% confidence interval for the difference of the two process EER's, $\chi_1 - \chi_2$, we take the 5th and 95th percentiles of the distribution of e_χ and subtract them from our estimated difference to give an interval of -0.1695 and -0.0928. Then, with 90% confidence we can be sure that the difference in the process EER's for these two matchers is between -16.95% and -9.28%.

Example 5.17 In this example, we will create a 95% confidence interval for the difference of two independent face matcher's EER's, $\chi_1 - \chi_2$. Specifically, the estimand for this interval is the difference in the process EER's for the facial matchers SURREY_face_nc_man_scale_100 (SURREY-NC-100) from the BANCA database and (DCTb, GMM) matcher from the XM2VTS database. We bootstrapped the match scores from each of these matchers independently $M = 1000$ times and calculated the resulting distribution of the e_χ's. Figure 5.20 has a summary of this distribution. Our sample difference in the EER's is $\hat{\chi}_1 - \hat{\chi}_2 =$

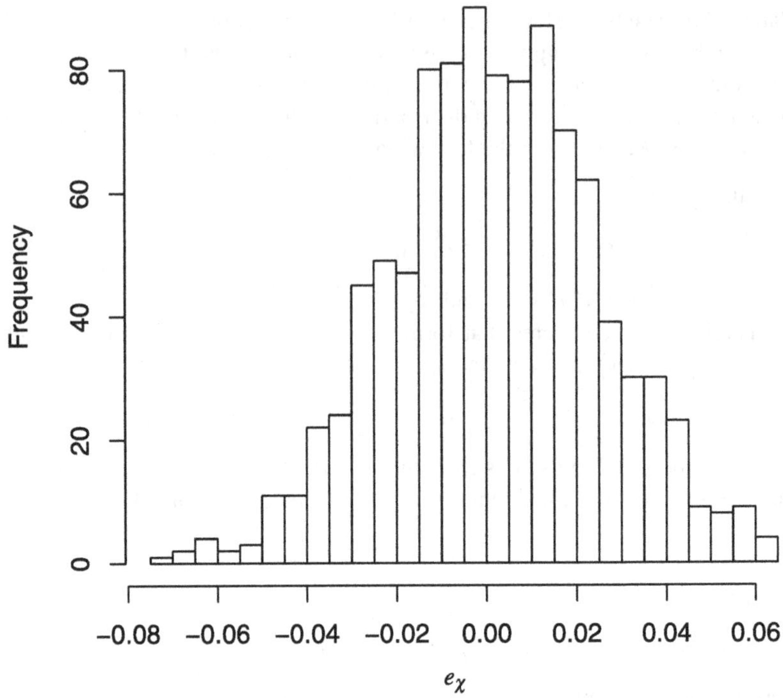

Fig. 5.19 Distribution of e_χ for the difference of two EER's from the voice matchers IDI-AP-25-10-PCA and IDIAP-25-200-PCA applied to groups g1 and g2, respectively, from the BANCA database under the G protocol

-0.0511. Taking the 2.5th and 97.5th percentiles of the distribution of e_χ gives us an interval of $(-0.0796, -0.0149)$. Then, we can be 95% confident that the difference in the process EER's for these two matchers is between -7.96% and -1.49%.

5.4.2.2 Hypothesis Test for Difference of Two Independent EER's

The next topic is a hypothesis test for comparing two EER's that come from distinct data collections. Independence here assumes that the two data collections are distinct. For example, this might occur if we are comparing the performance of a single fingerprint sensor at two distinct locations such that individuals involved in each data collection do not overlap. For the hypothesis test here, we will order the EER's so that $\hat{\chi}_1$ is less than $\hat{\chi}_2$. Our goal is to determine if the difference in the sample EER's is significant. Accordingly, we are formally testing the following hypotheses:

$$H_0 : \chi_1 = \chi_2$$
$$H_1 : \chi_1 < \chi_2.$$

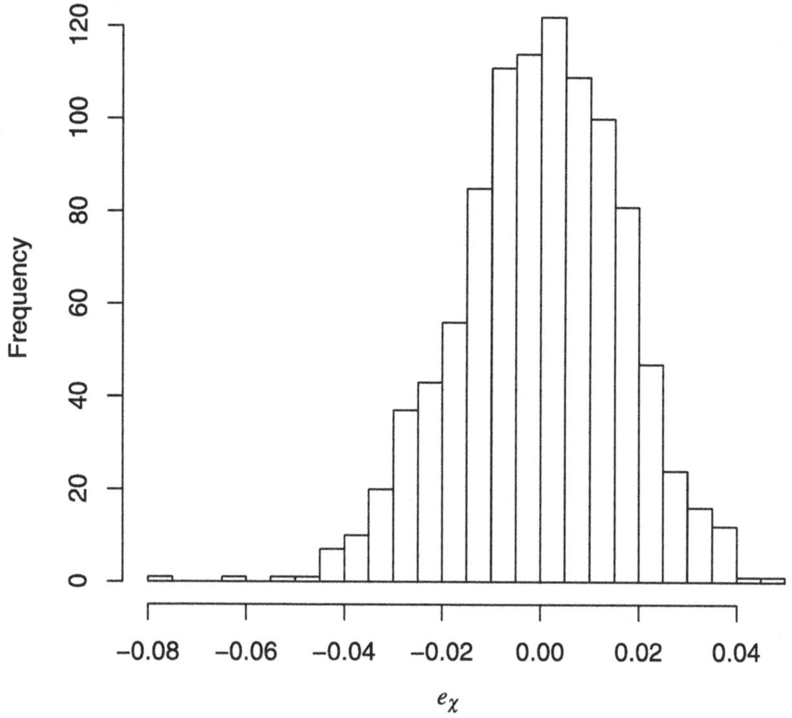

Fig. 5.20 Distribution of e_χ for the difference of the EER's for comparing the voice matchers (IDIAP-25-10-PCA) and (IDIAP-25-200-PCA) applied to groups $g1$ and $g2$ respectively from the BANCA database under the G protocol

Below we present a bootstrap methodology for this type of test. One application of this methodology would be to test whether two demographic groups have the same EER for a given matcher.

Bootstrap Approach Our bootstrap methodology using the two-instance bootstrap is the following.

1. Calculate

$$\hat{\chi}_1 - \hat{\chi}_2 = (1 - \hat{r}^{(1)}_{3\pi/4}/\sqrt{2}) - (1 - \hat{r}^{(2)}_{3\pi/4}/\sqrt{2}) \tag{5.40}$$

from the observed sample ROC's, $\hat{R}_1$ and $\hat{R}_2$.

2. We next follow the two-instance bootstrap for bootstrapping the matching scores to produce a replicated ROC, $\hat{R}^b_1$, for the first group. then the resampled EER is $\hat{\chi}^b_1 = 1 - \hat{r}^{(1)b}_{3\pi/4}/\sqrt{2}$.

3. For the second group, we, likewise, create a bootstrapped ROC, $\hat{R}_2^b$, by resampling the match scores following the methods given in Sect. 5.2. From the bootstrapped ROC, we can calculate the resampled EER to be

$$\hat{\chi}_2^b = 1 - \hat{r}_{3\pi/4}^{(2)b}/\sqrt{2}. \tag{5.41}$$

4. Determine $e_\chi = \hat{\chi}_1^b - \hat{\chi}_2^b - (\hat{\chi}_1 - \hat{\chi}_2)$.
5. Repeat steps 2, 3, and 4 above a large number of times M.
6. Then the *p-value* for this test is

$$p = \frac{1 + \sum_{\varsigma=1}^{M} I_{\{e_\chi \leq \hat{\chi}_1 - \hat{\chi}_2\}}}{M+1}. \tag{5.42}$$

7. We will then reject if the *p-value* is small, say $p < \alpha$ for a given significance level α.

Example 5.18 In this example, we are testing the equality of a matcher applied to two independent groups. The matcher is the face matcher SURREY_face_nc_man_scale_100 (SURREY-NC-100) from the BANCA database. We are comparing this matcher applied to groups g1 and g2 from that database following the G protocol. Our goal here is to test

$$H_0 : \chi_1 = \chi_2$$
$$H_1 : \chi_1 < \chi_2.$$

Following the algorithm above, we bootstrapped the match scores from these two matchers $M = 999$ times and obtained a distribution for e_χ. This distribution is summarized in Fig. 5.21. The sample difference is represented by the dashed vertical line at $\hat{\chi}_1 - \hat{\chi}_2 = -0.0494$. The resulting *p-value* is 0.007 which is small enough for us to reject the null hypothesis, H_0, and conclude that the EER's here are significantly different.

5.4.2.3 Confidence Interval for Difference of Two Paired EER's

The approach here to a confidence interval of the difference between two paired EER's, χ_1 and χ_2, is similar to the approach taken for independent EER's above. The difference is in how we resample the match scores and, consequently, the ROC's from which we obtain the EER. The paired nature of the match scores is retained through bootstrapping that takes all of the scores that are connected for a given sampled pair of individuals. One circumstance where we might want to use the methodology described below is when we have two matchers applied to the same database of biometric signals. For example, a test of two different fingerprint systems would be paired, if every time that an individual contributed a signal to one system, they contributed a signal to the other. Then, the signals would be paired and the resulting match scores from the two systems would likewise be paired.

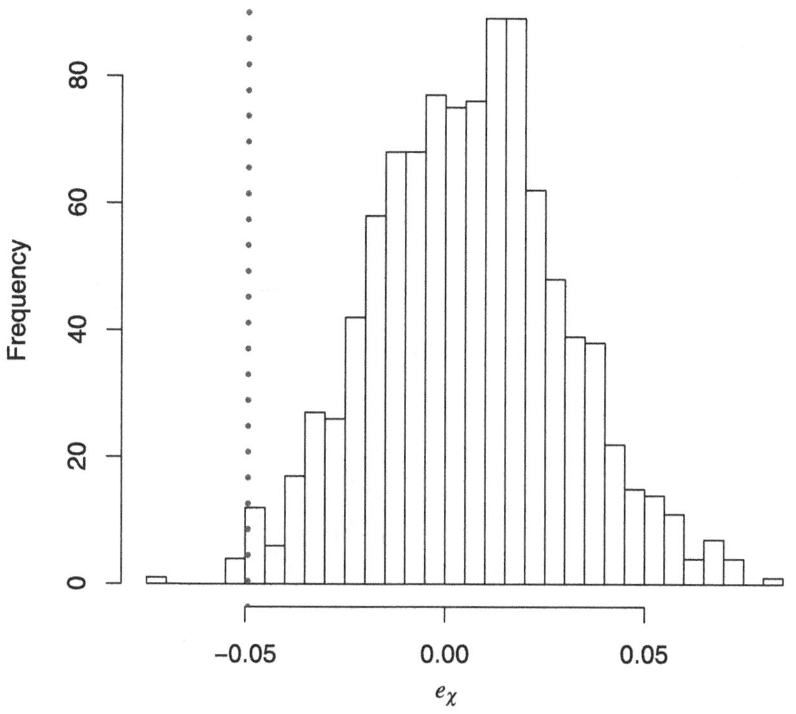

Fig. 5.21 Distribution of e_χ, the bootstrapped differences in the EER's from the observed differences in the EER's, for comparing the matcher SURREY-NC-100 applied to two groups, g1 and g2, from the BANCA database

Bootstrap Approach The goal of the bootstrap methodology here is to create an approximation to the sampling distribution for the estimated difference of the two process EER's, $\hat{\chi}_1 - \hat{\chi}_2$. We then use this distribution to create our confidence interval for $\chi_1 - \chi_2$. Our approach is given below.

1. Calculate
$$\hat{\chi}_1 - \hat{\chi}_2 = (1 - \hat{r}^{(1)}_{3\pi/4}/\sqrt{2}) - (1 - \hat{r}^{(2)}_{3\pi/4}/\sqrt{2}) \tag{5.43}$$
from the observed ROC's, $\hat{R}_1$ and $\hat{R}_2$.
2. Bootstrap the individuals for each group following the two-instance bootstrap described in Sect. 5.2. For a selected pair (i, k), we take both the matching scores from the first group, $g = 1$, and the matching scores from the second group, $g = 2$, associated with that specific pair of individuals. Those scores would be $Y^{(1)}_{ik1}, \dots, Y^{(1)}_{ikm_{ik}}$ and $Y^{(2)}_{ik1}, \dots, Y^{(2)}_{ikm_{ik}}$, respectively. We then use these match scores for all of the selected pairs of individuals to get bootstrapped versions, $\hat{R}^b_1$ and $\hat{R}^b_2$, of the original ROC's.
3. Calculate and store
$$e_\chi = \hat{\chi}^b_1 - \hat{\chi}^b_2 - (\hat{\chi}_1 - \hat{\chi}_2) \tag{5.44}$$

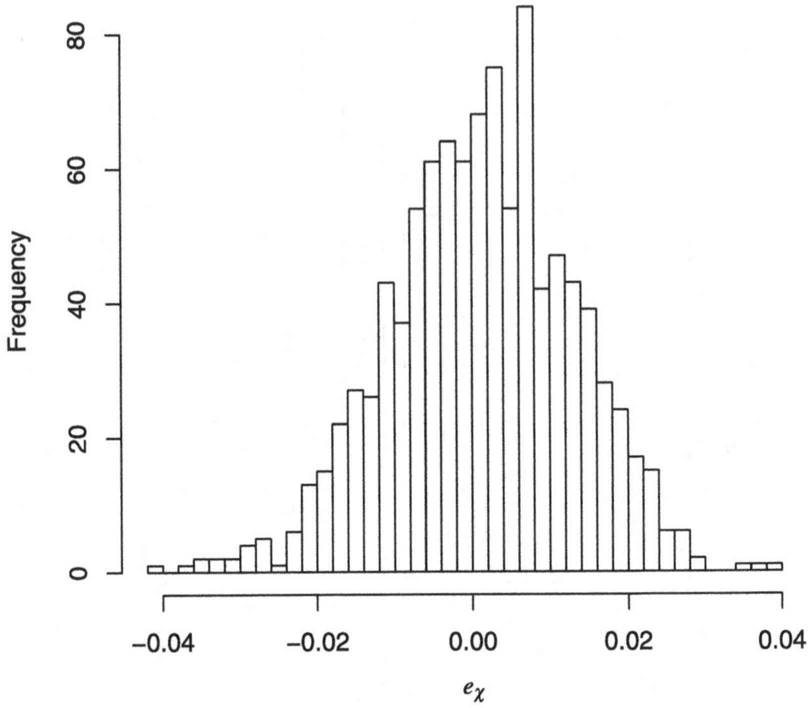

Fig. 5.22 Distribution of e_χ, the bootstrapped differences from the estimated difference between the EER's, for the two voice matchers IDIAP-25-50-PCA and IDIAP-25-50 from the BANCA database applied to group g1

which is the distribution of the bootstrapped difference of the EER's from the observed difference of the EER's.

4. Repeat Steps 2 and 3 above a large number of times M.
5. A $100(1-\alpha)\%$ confidence interval for $\chi_1 - \chi_2$ is then formed by taking the interval from $\hat{\chi}_1 - \hat{\chi}_2 - e_U$ to $\hat{\chi}_1 - \hat{\chi}_2 - e_L$ where e_L and e_U represent the $\alpha/2$th and the $1-\alpha/2$th percentiles of the distribution of e_χ.

Example 5.19 Here we are comparing two voice matchers, IDIAP-25-50-PCA and IDIAP-25-50, from the BANCA database. In particular, our goal is a 95% confidence interval for the difference of the EER's from paired data collection applied to group g1 of this database under the G protocol. The data here are paired since we have an associated match score from IDIAP-25-50-PCA for every match score from IDIAP-25-50. The estimated EER's for the two matchers is $\hat{\chi}_1 = 0.0275$ for IDIAP-25-50-PCA and $\hat{\chi}_2 = 0.0687$ for the IDIAP-25-50. Thus, the estimated difference is $\hat{\chi}_1 - \hat{\chi}_2 = 0.0275 - 0.0687 = -0.0412$. Having estimated the difference, we resampled the EER's to obtain a sampling distribution for the difference. Figure 5.22 has a summary of this distribution. Based upon this distribution, we find that the 95% confidence interval for this difference is $(-0.0718, -0.0250)$. Based

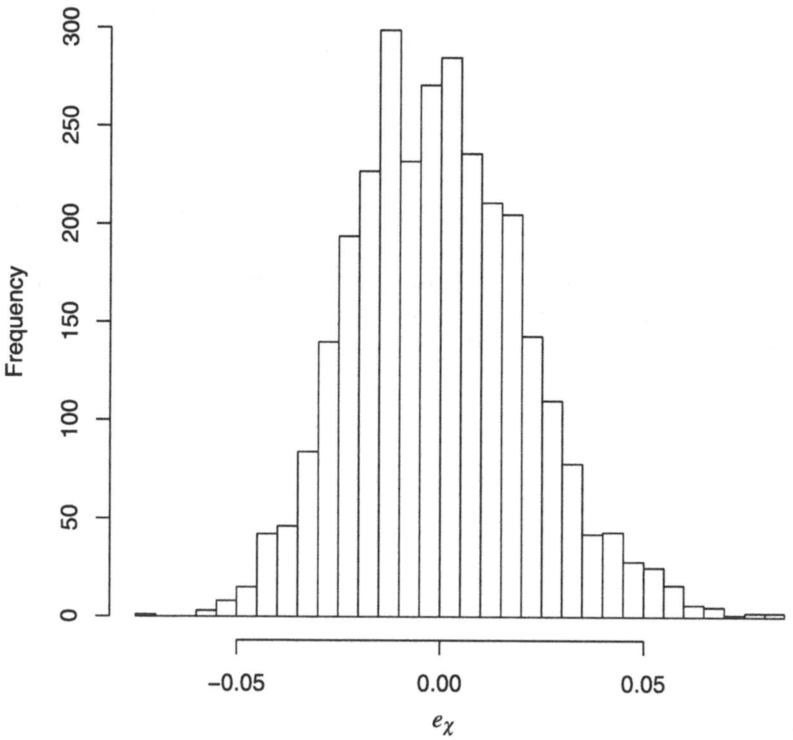

Fig. 5.23 Distribution of the bootstrapped differences from the estimated differences in the EER's, the e_χ's, for two face matchers SURREY-AUTO and SURREY-SVM-MAN-2.00 from the BANCA database applied to group g1

upon this interval, we can be 95% confidence that the difference between these two EER's is between -7.18% and -2.50% . Further, since 0 is not inside this interval, we can conclude that there is a significant difference between these two EER's at the $\alpha = 0.05$.

Example 5.20 For this example, we apply the bootstrap methodology for comparing two paired EER's to two face matchers from the BANCA database. These matchers are SURREY_face_svm_auto (SURREY-AUTO) and SURREY_face_svm_man_scale_2.00 (SURREY-SVM-MAN-2.00). The match scores from these two matchers come from their application to group g2 using the G protocol. The estimated EER for SURREY-SVM-AUTO is $\hat{\chi}_1 = 0.1169$ and for SURREY-SVM-MAN-2.00 is $\hat{\chi}_2 = 0.0321$, then the difference of these is 0.0848. Bootstrapping the match scores $M = 3000$ times, we approximate the distribution of the difference in the EER's, $\hat{\chi}_1 - \hat{\chi}_2 = 0.0848$. A histogram of the e_χ's from this process is given in Fig. 5.23. The resulting 90% confidence interval is 0.0474 and 0.1178. Since the value 0 which represents $\chi_1 - \chi_2 = 0$ is not inside this interval, we can conclude at a significance level of $\alpha = 0.10$ that the matcher SURREY-SVM-AUTO has a significantly higher EER than the EER of SURREY-SVM-MAN-2.00.

5.4.2.4 Hypothesis Test for Difference of Two Paired EER's

In this section, we derive a hypotheses test of the equality of two paired EER's. Paired EER's might be the result of comparing two matching algorithms on the same database of biometric captures. We assume here that we are interested in testing whether the first process EER, χ_1 is significantly less than the second, χ_2. Implicitly this assumes that $\hat{\chi}_1 < \hat{\chi}_2$. This test is

$$H_0 : \chi_1 = \chi_2$$
$$H_1 : \chi_1 < \chi_2.$$

We employ the *two-instance bootstrap* methodology to evaluate these hypotheses. This method is shown below.

Bootstrap Approach Our bootstrap approach is the following:

1. Calculate
$$\hat{\chi}_1 - \hat{\chi}_2 = (1 - \hat{r}^{(1)}_{3\pi/4}/\sqrt{2}) - (1 - \hat{r}^{(2)}_{3\pi/4}/\sqrt{2}) \tag{5.45}$$
from the observed ROC's, $\hat{R}_1$ and $\hat{R}_2$.
2. Bootstrap the individuals from the probe and the gallery following the two-instance bootstrap given in Sect. 5.2. For a selected pair of individuals (i, k) take all the scores from both processes for that pair. There will be an equal number from each process. Having done this we can concatenate all of the scores for each process and derive *separately* a bootstrapped ROC for each of those processes. These bootstrapped ROC's we will call $\hat{R}^b_1$ and $\hat{R}^b_2$.
3. From the bootstrapped ROC's, $\hat{R}^b_1$ and $\hat{R}^b_2$, calculate
$$e_\chi = \hat{\chi}^b_1 - \hat{\chi}^b_2 - (\hat{\chi}_1 - \hat{\chi}_2). \tag{5.46}$$
4. Repeat Steps 2 and 3 above a large number of times M.
5. Then the *p-value* for this test is
$$p = \frac{1 + \sum_{\varsigma=1}^{M} I_{\{e_\chi \le \chi^b_1 - \chi^b_2\}}}{M + 1}. \tag{5.47}$$

Example 5.21 In this example we are testing whether two voice matchers applied to the same database have significantly different EER's. The two matchers we are testing are IDIAP_voice_gmm_auto_scale_25_ 10_pca (IDIAP-25-10-PCA) IDIAP_voice_gmm_auto_scale_25_200_pca (IDIAP-25-200-PCA). Example 5.8 previously considered these same matchers as part of a comparison of the full ROC's. Here we our focus is on testing the difference of the two EER's. The

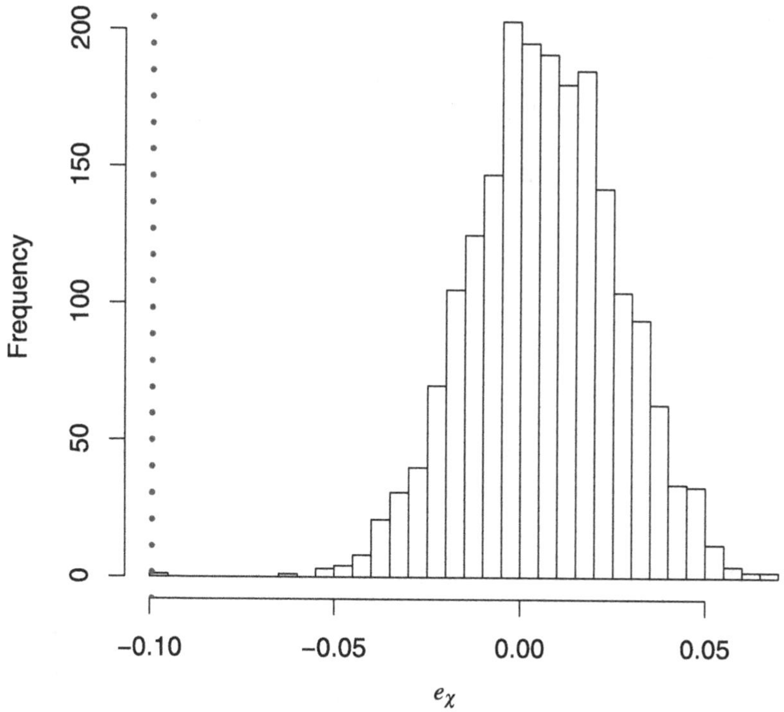

Fig. 5.24 Bootstrapped distribution of e_χ, the differences from the estimated difference in the EER's, for two voice matchers (IDIAP-25-10-PCA) and (IDIAP-25-200) applied to group g1 of the BANCA database

observations that we will use are from the G protocol of the BANCA database applied to group g2. To test our hypotheses, we repeated the bootstrap approach given above $M = 1999$ times with $T = 49$ angles which were uniformly spaced between $\pi/2$ and π. For each of the bootstrap iterations, we calculated e_χ. Our *p-value* for this analysis is $p = 0.0005$. Figure 5.24 has a summary of the distribution of the e_χ's. The dashed vertical line on that graph represents the observed difference in the EER's $\hat{\chi}_1 - \hat{\chi}_2 = -0.0993$. Given that the *p-value* here is small, we can conclude that there is a significant difference between the two EER's of these matchers.

5.4.3 Multiple Sample Methods for EER's

Multiple EER's is the focus of this section and we will present two methods for testing the hypotheses below:

$$H_0 : \chi_1 = \chi_2 = \cdots = \chi_G$$
$$H_1 : \text{not } H_0.$$

The *two-instance bootstrap* will be used both for the case of independent processes and for paired processes. The template here will follow the basic outline above. We will bootstrap the match scores and find the EER from each. Repeating this process, we obtain a distribution of EER's and we can use that distribution to assess the null hypothesis given above. The test statistic that we will use for the methods here is the variation between the EER's. That measure is

$$F = \sum_{g=1}^{G} N_g (\hat{\chi}_g - \bar{\chi})^2 \tag{5.48}$$

where

$$\bar{\chi} = \frac{\sum_{g=1}^{G} N_g \hat{\chi}_g}{\sum_{g=1}^{G} N_g} \tag{5.49}$$

and N_g is the total number of match scores in the gth group. This is a scalar version of the test statistic used for comparing multiple ROC's above.

5.4.3.1 Hypothesis Test for Multiple Independent EER's

Here we describe a procedure for testing the equality of G independent EER's. Our approach is to generate bootstrapped EER's, χ_g^b's, for each process multiple times and from that build a sampling distribution for our test statistic, F. One application of the procedure given below would be to compare the EER's of G distinct demographic groups, for example, G age groups.

Bootstrap Approach We can test the equality of multiple EER's using the bootstrap approach given below.

1. Calculate the EER's, $\hat{\chi}_g$, for each of the g EER's, $g = 1, 2, \ldots, G$ and derive the test statistic F following

$$F = \sum_{g=1}^{G} N_g (\hat{\chi}_g - \bar{\chi})^2 \tag{5.50}$$

   where

$$\bar{\chi} = \frac{\sum_{g=1}^{G} N_g \hat{\chi}_g}{\sum_{g=1}^{G} N_g}. \tag{5.51}$$

   The total number of individuals in the gth group is represented by N_g.
2. Bootstrap the match score distributions for all G processes separately and calculate the EER based upon these replicated distributions. Call these EER's, $\hat{\chi}_g^b$.
3. Calculate and store the test statistic F_χ using

$$F_\chi = \sum_{g=1}^{G} N_g (\hat{\chi}_g^b - \hat{\chi}_g)^2. \tag{5.52}$$

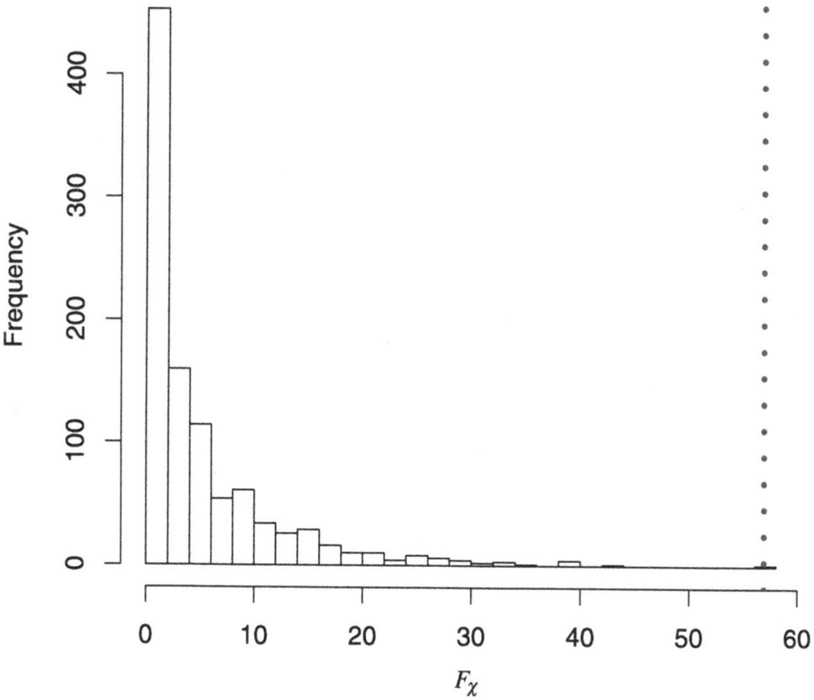

Fig. 5.25 Distribution of the bootstrapped test statistic, F_χ, for comparing the EER's for two voice matchers from the BANCA database, IDIAP-33-300-PCA and UC3M-34-300, and a speaker matcher, (PAC, GMM), from the XM2VTS database

4. Repeat the previous two steps some large number of times, M.
5. Define the *p-value* to be

$$p = \frac{1 + \sum_{\varsigma=1}^{M} I_{\{F_\chi \geq F\}}}{M+1}. \tag{5.53}$$

6. We reject the null hypothesis if the *p-value* is small or less than a specified significance level α.

Example 5.22 We will compare the equal error rates, EER's, of three independent data collections in this example. The matchers upon which data was collected are the following: IDIAP_voice_gmm_auto_scale_33_300_pca (IDIAP-33-300-PCA), UC3M_voice_gmm_auto_scale_34_300 (UC3M-34-300) and (PAC, GMM). The first of these matchers is from the BANCA database and the match scores we use here are from applying that matcher to group g1 following the G protocol. The second matcher is also from the BANCA database following the G protocol but the application is to group g2. Both of these are voice matchers. The third matcher, (PAC, GMM), is a speaker matcher from the XM2VTS database. The algorithm above outlines the methodology that we are using to test whether there is a significant difference in the EER's for these three matchers. Using $M = 1000$ replications, we derived a distribution for F_χ following this algorithm. That distribution is summarized in Fig. 5.25. The test statistic here is $F = 56.86967$, the dashed vertical line in Fig. 5.25, and the resulting *p-value* is $p = 0.0010$. From this *p-value*, we can reject the null hypothesis, H_0, and conclude that one or more of the EER's are significantly different.

5.4.3.2 Hypothesis Test for Multiple Paired EER's

The goal of this section is to test the equality of G paired EER's. A comparison of three or more paired EER's might arise as the result of applying multiple matching algorithms to a database of biometric captures or as the result of having the same individuals test the same set of G devices. For example, we might want to compare the EER's for six different facial recognition algorithms applied to the same database of signals. The hypotheses we will test are the following:

$$H_0: \chi_1 = \chi_2 = \cdots = \chi_G$$
$$H_1: \text{ not } H_0.$$

We will present a bootstrap approach for evaluating these hypotheses. The basics of the algorithm for this test are similar to those above for the case of independent EER's; however, the bootstrap methodology is different to account for the paired nature of the match scores.

Bootstrap Approach Our bootstrap methodology here is to resample the match scores for each of the G groups by following the *two-instance bootstrap* algorithm presented in Sect. 5.2. For a selected pair of individuals i and k, we will take all of the match scores, $Y_{ik\ell}^{(g)}, \ell = 1, \ldots, m_{ik}$ for all of the matchers $g = 1, \ldots, G$. Repeating this process for all resampled individuals, we derive the distribution of F^b which is a measure of variability between the different EER's. This distribution is then compared to the test statistic F from the original—non-bootstrapped—data. This procedure is given next.

1. Calculate the G EER's, $\hat{\chi}_g$, $g = 1, 2, \ldots, G$ and derive the test statistic, F, following

$$F = \sum_{g=1}^{G} (\hat{\chi}_g - \bar{\chi})^2 \tag{5.54}$$

where

$$\bar{\chi} = \frac{1}{G} \sum_{g=1}^{G} \hat{\chi}_g. \tag{5.55}$$

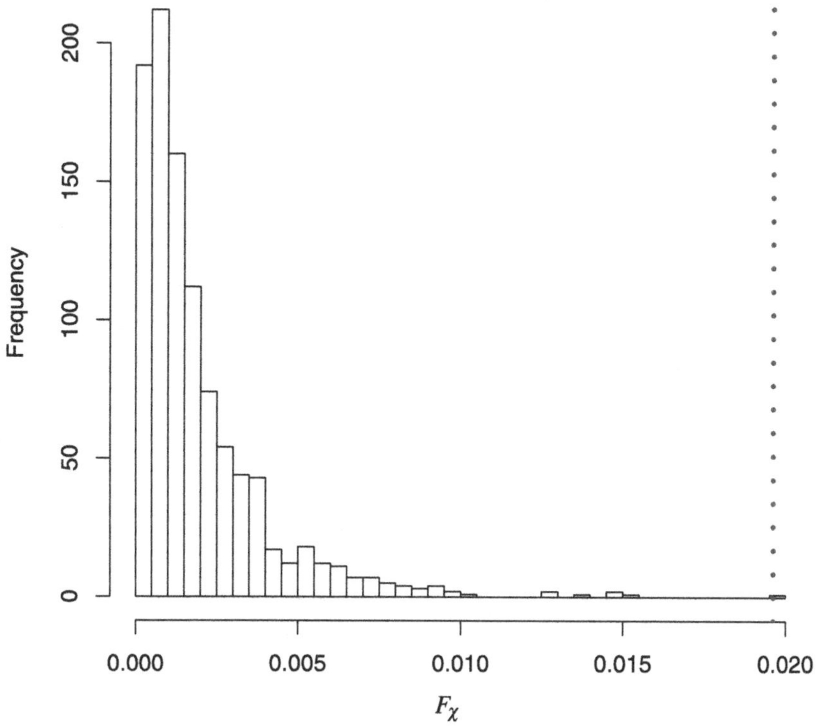

Fig. 5.26 Distribution of the bootstrapped test statistics, F_χ, for comparing the EER's of the four voice matchers: IDIAP-25-100-PCA IDIAP-25-10-PCA IDIAP-25-200-PCA and IDIAP-25-25-PCA applied to group g1 of the BANCA database

2. Bootstrap the individuals following the *two-instance bootstrap* and take all of the match scores for a selected pair of individuals from all G groups. Keeping the match scores from each group separate, calculate the EER from the complete set of bootstrapped match scores. Call these EER's, $\hat{\chi}_g^b$ for $g = 1, \ldots, G$.
3. Calculate the test statistic F_χ using

$$F_\chi = \sum_{g=1}^{G} (\hat{\chi}_g^b - \hat{\chi}_g)^2. \tag{5.56}$$

4. Repeat the previous two steps some large number of times, M.
5. Finally our *p-value* is defined to be

$$\bar{\chi} = \frac{1}{G} \sum_{g=1}^{G} \hat{\chi}_g. \tag{5.57}$$

6. Reject the null hypothesis if the *p-value* is small or less than a specified significance level α.

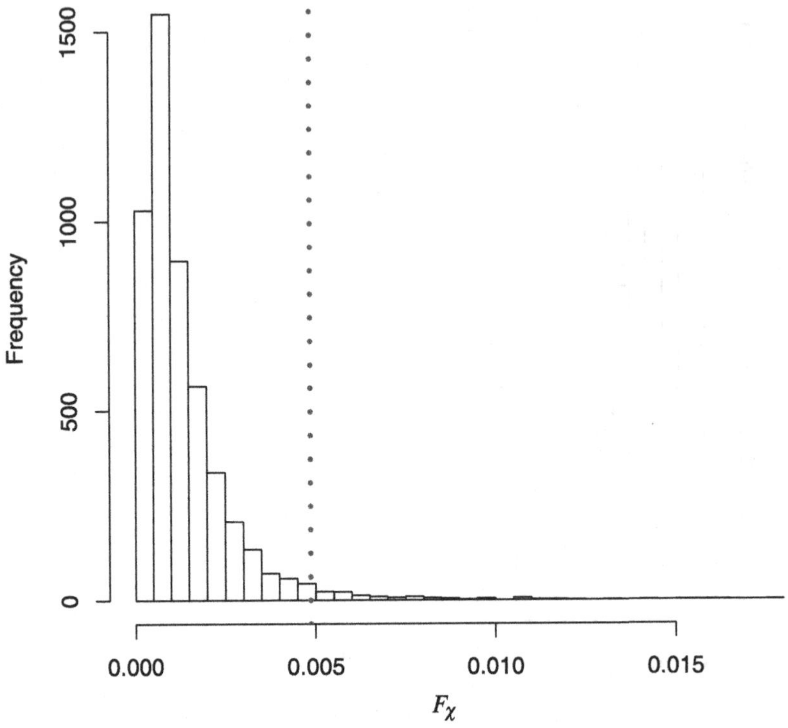

Fig. 5.27 Distribution of F_χ for comparing the EER's of IDIAP-25-100, IDIAP-25-10, IDIAP-25-200, IDIAP-25-25, IDIAP-25-300, IDIAP-25-50, IDIAP-25-75, IDIAP-33-100, IDIAP-33-10, and IDIAP-33-200 applied to group g2 from the BANCA database

Example 5.23 For this example we are comparing the EER for four voice classifiers. These are: IDIAP_voice_gmm_auto_scale_25_100_pca (IDIAP-25-100-PCA) IDIAP_voice_gmm_auto_scale_25_10_pca (IDIAP-25-10-PCA) IDIAP_voice_gmm_auto_scale_25_200_pca (IDIAP-25-200-PCA) and IDIAP_voice_gmm_auto_scale_25_25_pca (IDIAP-25-25-PCA). The EER's from applying these four matchers to group g1 of the BANCA database were 0.0362, 0.1595, 0.0405 and 0.0807, respectively. We simulated $M = 1000$ bootstrapped replicate sets of scores for each matcher. Figure 5.26 gives a summary of the distribution of F_χ for these data. The dashed vertical line represents $F = 0.0196$. The *p-value* for this test is then 0.0010. Since this *p-value* is small we reject the null hypothesis and conclude that there are significant differences among these EER's. Although there is not a very large number of match scores—$N = 545$—in this example, the differences here are large enough for those differences to be significant. In particular, $\hat{\chi}_2 = 0.1595$ seems to stand out.

Example 5.24 In this example we are testing the equality of $G = 10$ different voice matchers from the BANCA database. Data from group g2 of that database following the G protocol were analyzed. Those matchers are: IDIAP_voice_gmm_auto_scale

_25_100 (IDIAP-25-100), IDIAP_voice_gmm_auto_scale_25_10 (IDIAP-25-10), IDIAP_voice_gmm_auto_scale_25_200 (IDIAP-25-200), IDIAP_voice_gmm_auto _scale_25_25 (IDIAP-25-25), IDIAP_voice_gmm_auto_scale_25_300 (IDIAP-25-300), IDIAP_voice_gmm_auto_scale_25_50 (IDIAP-25-50), IDIAP_voice_ gmm_auto_scale_25_75 (IDIAP-25-75), IDIAP_voice_gmm_auto_scale_33_100 (IDIAP-33-100), IDIAP_voice_gmm_auto_scale_33_10 (IDIAP-33-10), and IDIAP_voice_gmm_auto_scale_33_200 (IDIAP-33-200). The EER's, respectively, are 0.0179, 0.0557, 0.0115, 0.0304, 0.0121, 0.0306, 0.0221, 0.0121, 0.0486, and 0.0077. There are $N = 546$ matching scores for each of the 10 voice matchers here. Figure 5.27 shows the distribution of F^b for $M = 5000$ bootstrapped versions of the original data. F from the original data is 0.0049 and is represented in Fig. 5.27 by the dashed vertical line. Our *p-value* here is 0.0250. This value is small and suggests that there are significant differences between the process EER's, the χ_g's. Consequently, we reject the null hypothesis and conclude that there is at least one EER that is different from the others among these 10 EER's.

Note that the data that was analyzed in Example 5.24 was also analyzed for comparing the entire ROC curve in Example 5.12. In the ROC case, we failed to reject differences between the ROC curves, while the *p-value* here suggests that we should reject the null hypothesis of equality of the EER's. The reason for the difference is that for comparing ROC curves we are simultaneously looking at differences at multiple angles and, therefore, we must account for that multiplicity when we consider our *p-values*.

5.5 Discussion

In this chapter, we have presented statistical methods for ROC's and EER's. All of these methods depend upon an understanding of the variability in our estimate of a given ROC or EER. To approximate this variability, we have introduced a bootstrap methodology, the *two-instance bootstrap*, in Sect. 5.2. This methodology is the same as the *two-instance bootstrap* that we used in Chap. 4 except that we are bootstrapping match scores instead of decisions and that we include both genuine and imposter match scores here. We have used the two-instance bootstrap to estimate the variability for both confidence regions/intervals and hypothesis tests throughout this chapter. Previous work, Schuckers et al. [88], had proposed a confidence interval for a single ROC and a single EER using the same methodology given here. This methodology was extended throughout this chapter to allow inference about two or more performance metrics. These tools for comparing ROC's and EER's are an addition to the literature on these topics.

Inference for the ROC is complicated because an ROC is not a scalar quantity as the EER is. We address this by representing an ROC as a collection of angles and radii using polar coordinates. Since we make inference for the entire ROC at a series of angles given by the set, Θ^*, we must ensure that our methodology holds broadly across all of these angles. Because of this, the approaches given here are necessarily statistically conservative. That is, it takes extensive evidence to conclude that there is a difference between two or more ROC curves.

Several choices are required for making inference regarding an ROC following the methodology of this chapter. The first important choice when using the polar coordinates based approach here is the center of the radii. We have chosen $(1, 0)$. Other choices are possible and can be accommodated by this methodology. Another critical choice is to determine the angles in Θ^*. This requires consideration of both the number of angles, T, and the location of those angles. Certainly, if there is interest in estimation of the EER, χ, then one of the angles should be $\theta = 3\pi/4$ (assuming that we are using $(1, 0)$ as the center). Generally, we recommend that users of these methods choose an evenly-spaced range of angles unless there is a compelling reason to do otherwise. Similarly, the default choice of the range of angles should be $(\pi/2, \pi)$. The *optimal* choice of T, the number of angles is an open question. We have used a range of T's in this chapter. Given the correlated nature of radii length, it seems unlikely that more than $T = 100$ angles over 90° would be necessary. On the other end of the spectrum, $T = 20$ is probably too few. The final choice for utilizing these statistical tools is the number of bootstrap replications, M. Our recommendation is that $M \geq 999$.

The confidence region/intervals and hypothesis tests given in this chapter can be extended to other representations of the ROC, particularly the DET curve. Doing so is a straightforward process, since there is a one-to-one transformation from the ROC to the DET.

As a final note, many areas that use classification methods have used the area under the ROC curve (AUC) as a measure of the overall performance of a classification or matching system. Zhou et al. [108] has multiple examples of medical applications for the AUC. Like the EER, the AUC is a single measure for a matching or classification systems performance. Consequently, it is possible for two or more systems to have drastically different performance but the same AUC. That said, we prefer the overall measure of the AUC to the EER. It should be straightforward to adapt the methods above for the ROC to the AUC. The ROC as a summary of performance is superior to both of these scalar measures.

Part III
Biometric Specific Measures

Chapter 6
Failure to Enrol

One of the important aspects of a biometrics system is the universality of the system. That is, the ability of all individuals to use the system. See Jain et al. [51] for more on universality. One measure of universality is the ability to get an individual enrolled in the system. This is not a trivial endeavor as some devices have difficulty enrolling an individual in their database. There are many reasons that this may occur including the environment in which the system is operating or some characteristic of the individual that prevents a quality sample of that individual from being collected. The failure to enrol (FTE) rate is the percent of individuals who are unable to be enrolled in a biometric system. Thus, failure to enrol rate or FTE is an important measure of a biometric authentication system's performance. This chapter lays out statistical methods for estimation and comparison of FTE's. Little formal work has been done on statistical methods specifically for the FTE, though the FTE has been recognized as one of the important metrics for evaluation of a biometrics system. See, for example, Mansfield and Wayman [63], or Li [56]. Schuckers [86] first presented the correlation structure that we will use below.

This chapter is organized in the following way. We begin by describing the correlation structure of failure to enrol data as well as the notation that we will use in the rest of the chapter. Moving to statistical methods, we start with tools for a single enrolment process. This is followed by a discussion of methods for comparing two or more FTE's. Both large sample and bootstrap methods are used for implementing these methodologies. Sample size and power calculations, along with prediction interval are presented next. We conclude with a discussion of the work presented here. For the methodologies that we present below, where possible we use 'real' observed data from biometric collections in our examples. In some cases, it is necessary to use synthetic, generated enrolment data in order to illustrate a particular methodology.

6.1 Notation and Correlation Structure

The FTE or failure to enrol is typically defined as the proportion of enrolments that do not succeed. For this chapter we assume that enrolment is a decision made on

M.E. Schuckers, *Computational Methods in Biometric Authentication*, Information Science and Statistics,
DOI 10.1007/978-1-84996-202-5_6,

Table 6.1 Example of FTE data

i	E_i
1	0
2	1
3	0
4	1
5	0
6	0
7	0
8	0
9	0
10	0
11	1
12	0
13	0
14	0
15	0
16	0
17	0
18	0
19	0
20	1

an individual. Note that we will not be concerned with the number of attempts or transactions required to enrol. Thus, we will only consider a single determination of enrolment per individual. Let

$$E_i = \begin{cases} 1 & \text{if individual } i \text{ is unable to enrol} \\ 0 & \text{otherwise.} \end{cases} \tag{6.1}$$

Then E_i is a decision about whether the ith individuals is able to enrol. If multiple attempts for enrolment are necessary for the ith individual, then the final outcome (able to enrol or unable to enrol) is used for E_i. An example set of data for an FTE is given in Table 6.1. Another measure of interest might be the average number of attempts to enrol or AAE which would be a measure on the enrolment attempts. The AAE would be a measure of difficulty or ease of enrolment. We note here that it is possible that there would be other measures of the enrolment process that, likewise, would be appropriate. The process FTE is denoted by ε and it will be our focus here rather than AAE. For the single process case, we use E_i to denote the success or failure of enrolment where $i = 1, \ldots, n_\varepsilon$ where n_ε is the total number of people who attempted to enrol in a particular device.

Our estimate of the process FTE, ε, will be the sample FTE,

$$\hat{\varepsilon} = \frac{\sum_{i=1}^{n_\varepsilon} E_i}{n_\varepsilon}. \tag{6.2}$$

Since this is a linear combination of random variables, we will use Result 2.7. We will assume that we have a stationary enrolment process for the FTE and, thus, we will use Result 2.8. In order for us to estimate the variance of $\hat{\varepsilon}$, it is necessary to specify the covariances of the E_i's or, if we assume stationarity, to specify the correlations between the E_i's. Since we will assume stationarity of the enrolment process, we will do the latter. The basis for our correlation is the idea that one individual's ability to enrol is uncorrelated with another individual's ability to enrol. We acknowledge that this uncorrelated nature is conditional on the FTE of the process. This correlation can then be written as

$$Corr(E_i, E_{i'}) = \begin{cases} 1 & \text{if } i = i' \\ 0 & \text{otherwise.} \end{cases} \tag{6.3}$$

Schuckers [86] first proposed this correlation structure for the FTE.

As an alternative to (6.2), we can write the estimated FTE in vector notation. This is

$$\hat{\varepsilon} = n_\varepsilon^{-1} \mathbf{1}^T \mathbf{E} \tag{6.4}$$

where $\mathbf{1} = (1, 1, \ldots, 1)^T$ and $\mathbf{E} = (E_1, E_2, \ldots, E_{n_\varepsilon})^T$. By doing this, we stress that our estimated FTE, $\hat{\varepsilon}$, is a linear combination of the E_i's. Consequently, we can make use of the statistical methods for linear combinations found in Chap. 2. Likewise, we can derive the variance of an estimated FTE, $\hat{\varepsilon}$, by

$$\begin{aligned} V[\hat{\varepsilon}] = V[n_\varepsilon^{-1} \mathbf{1}^T \mathbf{E}] = V\left[\frac{\sum_{i=1}^{n_\varepsilon} E_i}{n_\varepsilon}\right] = n_\varepsilon^{-2} \sum_{i=1}^{n_\varepsilon} V[E_i] = n_\varepsilon^{-2}[n_\varepsilon \varepsilon(1-\varepsilon)] \\ = n_\varepsilon^{-1}[\varepsilon(1-\varepsilon)]. \end{aligned} \tag{6.5}$$

This variance is straightforward to derive because of the uncorrelated nature of enrolment decisions. Note also that the effective sample size for FTE's is simply n_ε since the enrolment decisions are uncorrelated.

Thus, the correlation structure for FTE is similar to that for binomial data, cf. Result 2.14, and we use methodology here for Binomial random variables, see Definition 2.27. Another direct consequence of this correlation structure is that the methodologies that we will use for resampling, i.e. bootstrap and randomization methods, can use the standard *iid* bootstrap described in Definition 2.51. This is quite different from the other metrics that we are considering in this book. Those all require bootstrap methods that are more complicated than the *iid* approach. In the rest of this chapter, we will present statistical methods for estimation of FTE's as well as for the comparison of more than one FTE's.

6.2 Statistical Methods

In this part of this chapter, we will outline techniques for estimation and comparison of FTE's. Following (6.3), we will assume that whether a given individual is able to enrol is uncorrelated with the enrolment decisions of other individuals that are measured as part of the biometric identification process we are evaluating. Below we present basic statistical methods for a single FTE, for comparing two FTE's and for comparing more than two FTE's. The methods that we will present encompass both large sample, bootstrap and randomization methods.

6.2.1 One Sample Methods for FTE

We begin by looking at methodology for evaluating a single FTE. Below we provide confidence intervals as well as hypothesis test for a single FTE. Our assumptions throughout this section is that the data collection that produced our enrolment decisions is a single stationary on-going enrolment process which implies that the FTE is constant.

6.2.1.1 Confidence Interval for Single FTE

In this section we present confidence interval methodology for estimation of a process FTE. This section has two methods for making a confidence interval for ε. The first is a method based upon large sample theory and the central limit theorem, Result 2.3. The second is a bootstrap approach that does not rely upon the assumptions necessary for the central limit theorem to apply.

Large Sample Approach If n_ε is large meaning that $n_\varepsilon\hat{\varepsilon} \geq 10$ and $n_\varepsilon(1-\hat{\varepsilon}) \geq 10$, then we can make a $(1-\alpha)100\%$ confidence interval for ε using the following:

$$\hat{\varepsilon} \pm z_{\alpha/2}\sqrt{\frac{\hat{\varepsilon}(1-\hat{\varepsilon})}{n_\varepsilon}} \tag{6.6}$$

where

$$\hat{\varepsilon} = \frac{\sum_{i=1}^{n_\varepsilon} E_i}{n_\varepsilon}. \tag{6.7}$$

As mentioned above, $\hat{\varepsilon}$ is the proportion of the sample that were unable to enrol. The $1-\alpha/2$th percentile of a Gaussian distribution is given by $z_{\alpha/2}$.

Example 6.1 In this example, we will make a 95% confidence interval for ε, the process FTE. The data we will use to illustrate this methodology is from the International Biometrics Group Comparative Biometrics Testing Round 7 [41]. In that test, the IrisGuard device was unable to enrol 18 individuals from $n_\varepsilon = 1035$ enrolment transactions. Then, we have that $\hat{\varepsilon} = 0.0174$. Since $n_\varepsilon\hat{\varepsilon} = 18 \geq 10$ and

$n_\varepsilon(1-\hat{\varepsilon}) = 1017 \geq 10$, we can proceed to use the confidence interval in (6.6). Then a 95% confidence interval for the FTE of this process is

$$0.0174 \pm 1.96\sqrt{\frac{0.0174(1-0.0174)}{1035}}. \tag{6.8}$$

This calculation yields an interval of $(0.0094, 0.0254)$. Thus, we can be 95% confident that the FTE, ε, for this device is between 0.94% and 2.54%.

Bootstrap Approach If n_ε, the sample size, is not sufficiently large, then the large sample methodology is not appropriate. One alternative is to use a bootstrap approach. Here we present a bootstrap approach assuming that there is at least one individual who has failed to enrol, i.e. that $n_\varepsilon\hat{\varepsilon} \geq 1$. In that case we use the following algorithm to calculate our confidence interval.

1. Calculate $\hat{\varepsilon}$ based upon the observed data, the E_i's.
2. Sample *with replacement* n_ε individuals from the n_ε individuals from whom an enrolment decision was collected. Call these individuals, $b_1, \ldots, b_{n_\varepsilon}$.
3. Calculate and store $e_\varepsilon = \hat{\varepsilon}^b - \hat{\varepsilon}$ where

$$\hat{\varepsilon}^b = \frac{\sum_{i=1}^{n_\varepsilon} E_{b_i}}{n_\varepsilon} \tag{6.9}$$

   is the bootstrapped FTE.
4. Repeat the previous two steps some large number of times, say M.
5. Create a $(1-\alpha) \times 100\%$ confidence interval for ε by taking $\hat{\varepsilon} - \varepsilon_H, \hat{\varepsilon} - \varepsilon_L$ where ε_H and ε_L are the $1-\alpha/2$th and $\alpha/2$th percentiles for the distribution of e_ε intersected with the interval $(0, 1)$.

Note that this is a Hall-type or second-type confidence interval as discussed by Manly [62]. Also, see Hall [44] for a more mathematical development of the same ideas.

Example 6.2 Fairhurst et al. [31] considered and reported the FTE's for several devices. In this example, we are interested in creating a 90% confidence interval for the FTE for the fingerprint device that they studied. They recruited $n_\varepsilon = 221$ volunteers and they reported an FTE of $\hat{\varepsilon} = 0.0271$. We can surmise from their results that 6 individuals were unable to enrol. Since we cannot, for these data, meet the conditions for the large sample approach, we will use the bootstrap approach given above. Resampling the individuals here $M = 5000$ times, we obtained a distribution of the values for e_ε. That distribution is summarized in Fig. 6.1. Using the 95th and the 5th percentiles from that distribution, we obtain a 90% confidence interval for ε to be $(0.0091, 0.0452)$. Thus, we can be 90% confident that the FTE for this process is between 0.91% and 4.52%.

Example 6.3 In this example, we will use data from the International Biometrics Group's Comparative Biometric Testing Round 7 [41]. We previously investigated these data in Example 6.1. We will use the bootstrap methodology to make a 95%

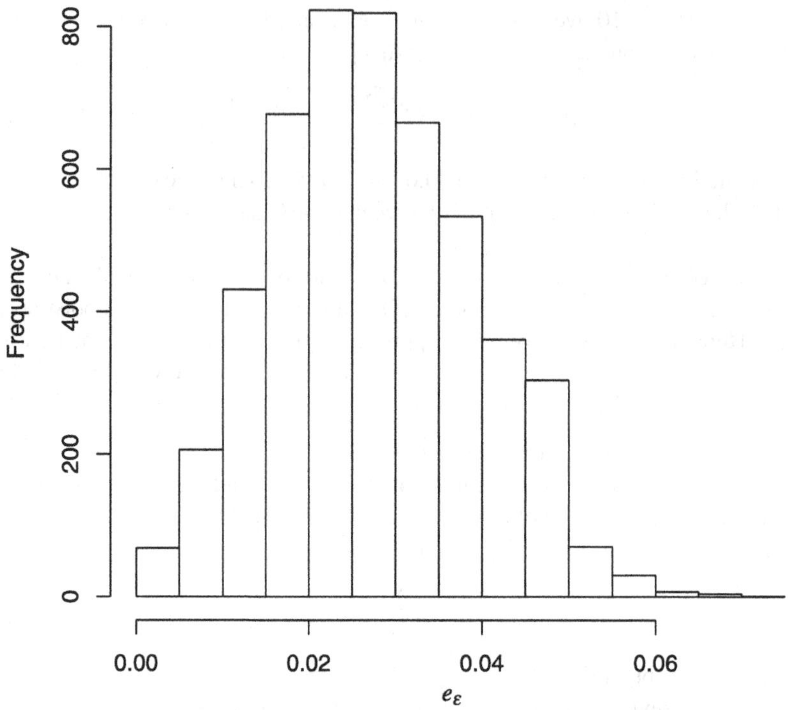

Fig. 6.1 Distribution of e_ε, the bootstrapped differences from the estimated failure to enrol using data from Fairhurst et al. [31]

confidence interval for the FTE for the IrisGuard device. For that device, there were $n_\varepsilon = 1035$ enrolment decisions there were 18 cases where the enrolment was denied. Following the algorithm above, we bootstrapped these data $M = 10000$ times. Figure 6.2 has the distribution of the resulting e_ε's. Using the 2.5th and 97.5th percentiles from that distribution, we obtain a 95% confidence interval for ε, the FTE of this process, to be 0.0097 and 0.0251. Thus, we can be 95% confident that the FTE for this process is between 0.97% and 2.51%. We note that this interval is very similar to the large sample interval that we obtained in Example 6.1 which was $(0.0094, 0.0254)$. The width of the bootstrapped interval is slightly smaller.

If $\sum_{i=1}^{n_\varepsilon} E_i = 0$ then neither of the two methods—the large sample method or the bootstrap method—above provide an interval. In this case, since the data is binomial, we can use the 'Rule of 3' discussed by Louis [57] and Jovanovic and Levy [52]. In the bioauthentication literature, this rule is considered by Mansfield and Wayman [63] among others. This rule suggests that when zero enrolment failures occur, it is possible to use $3/n_\varepsilon$ as an upper bound on a confidence interval for ε. Jovanovic and Levy note that $3/(n_\varepsilon + 1)$ performs better and so it is the recommendation that we use here. That interval is a one-sided CI meaning that bounds for the 95% CI can be written as $(0, 3/(n_\varepsilon + 1))$. The confidence interval methodology based upon

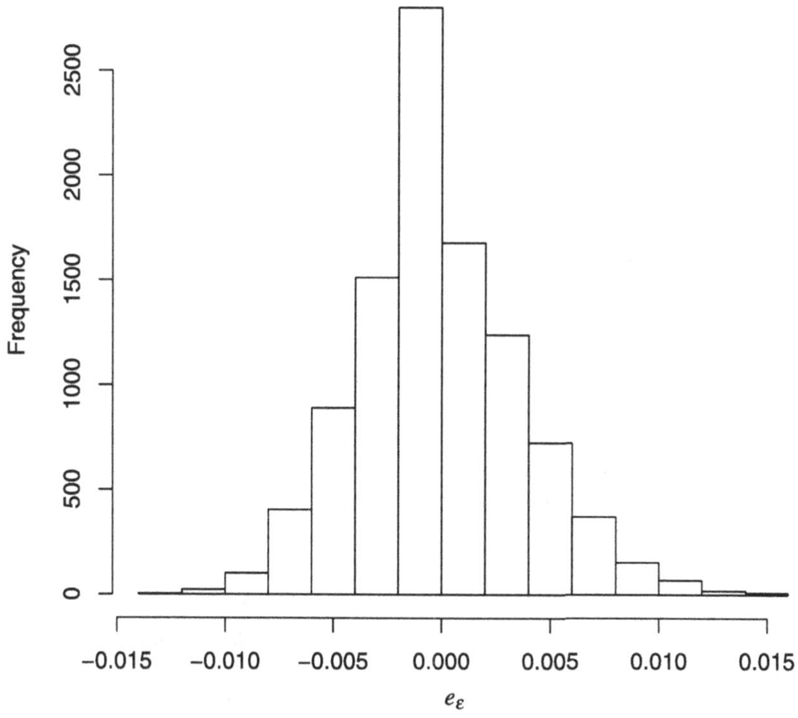

Fig. 6.2 Distribution of e_ε, the bootstrapped differences from the estimated failure to enrol using data from IBG [41]

a Gaussian distribution and upon the bootstrap can be modified to give one-sided confidence intervals. In each case the lower endpoints are replaced by zero and the upper endpoints are based upon a $(1 - \alpha)$th percentile.

6.2.1.2 Hypothesis Test for Single FTE

As mentioned in Chap. 2, hypothesis testing is about testing a specific value for a process parameter. As above, we will use ε to represent the FTE and ε_0 will represent the specific value that we want to test. The approach in this section and throughout this book is to assume that we want to test for a rate that is below some value. It is possible to test the other alternative hypotheses that were discussed in Chap. 2. In those cases, the calculation of the appropriate *p-value* would be different. Below we give both a large sample theory approach and a resampling approach to hypothesis testing.

Large Sample Approach For the large sample approach to testing whether the FTE is significant below a specified value, we must ensure that there are enough samples for the sampling distribution of $\hat{\varepsilon}$ to approximately follow a Gaussian distribution. For the following hypothesis test, we will decide that n_ε is sufficiently large if $n_\varepsilon \varepsilon_0 \geq 10$ and $n_\varepsilon(1 - \varepsilon_0) \geq 10$. If that is the case then, we can use the algorithm below to test that

$$H_0 : \varepsilon = \varepsilon_0$$
$$H_1 : \varepsilon < \varepsilon_0.$$

For this test we have the following:

Test Statistic:

$$z = \frac{\hat{\varepsilon} - \varepsilon_0}{\sqrt{\frac{\varepsilon_0(1-\varepsilon_0)}{n_\varepsilon}}}. \tag{6.10}$$

p-value: $p = P(Z < z)$ where Z is a standard Gaussian random variable and we can evaluate this probability using Table 9.1.

We will reject the null hypothesis if p is small or smaller than a prespecified significance level α.

Example 6.4 For this example, we will be testing whether a process FTE is less than 3%. The FTE that we will be testing is the IrisGuard data from the International Biometrics Group's Comparative Biometric Testing Round 7 [41]. The value of 0.03 or 3% is often one that is used by testing organizations to certify a product. For example, see the United States Transportation Security Administration's guidance document [2]. Thus, we will be testing the hypotheses

$$H_0 : \varepsilon = 0.03$$
$$H_1 : \varepsilon < 0.03.$$

In order to use the large sample approach for this test, we must verify that our sample size, n_ε, is sufficiently large. Since $1035(0.03) = 31.05$ and $1035(1 - 0.03) = 1003.95$, it is reasonable to use the Gaussian distribution as an approximation to the sampling distribution of $\hat{\varepsilon}$. The sample FTE from testing of the IrisGuard was $\hat{\varepsilon} = 0.0174$ based upon $n_\varepsilon = 1035$ enrolment attempts. Using the above calculations for our test statistic, we get

$$z = \frac{0.0174 - 0.0300}{\sqrt{\frac{0.03(1-0.03)}{1035}}} = -2.3779. \tag{6.11}$$

We can then get the *p-value* to be $p = P(Z < -2.38) \approx 0.0087$. Since this *p-value* is small we can reject the null hypothesis, H_0, and conclude that the IrisGuard FTE is significantly less than 3%.

Bootstrap Approach When the sample size is not sufficiently large or if one prefers to use randomization tests, the following test is available as an alternative using the same null and alternative hypotheses as above. The steps are as follows:

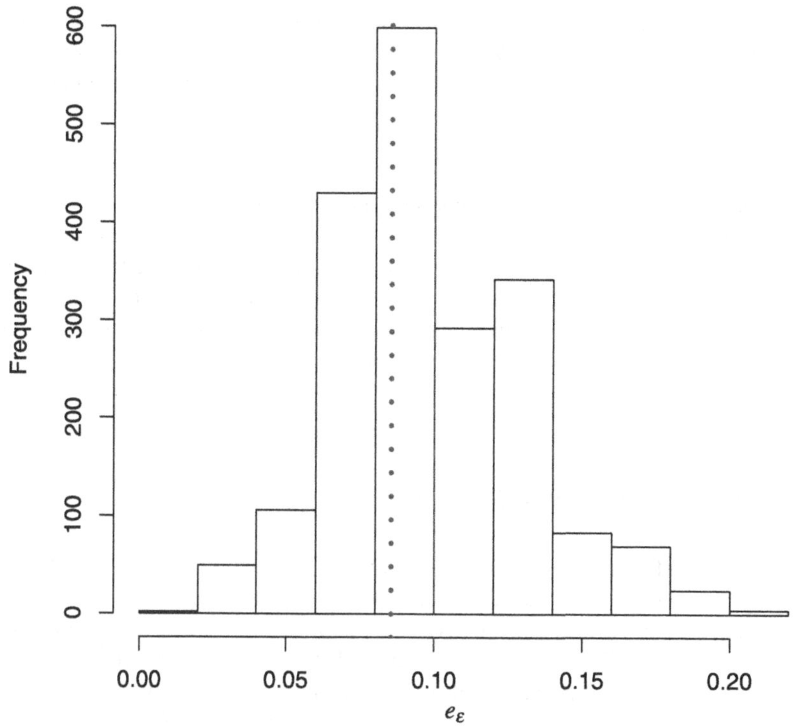

Fig. 6.3 Distribution of e_ε, the bootstrapped differences from the null hypothesis, for the FTE of a fingerprint device from Coventry et al. [16]

1. Calculate $\hat{\varepsilon}$ based upon the observed E_i's.
2. Sample *with replacement* n_ε from the n_ε individuals in the sample. Call the resampled individuals $b_1, \ldots, b_{n_\varepsilon}$
3. Calculate $e_\varepsilon = \hat{\varepsilon}^b - \hat{\varepsilon} + \varepsilon_0$ where

$$\hat{\varepsilon}^b = \frac{\sum_{i=1}^{n_\varepsilon} E_{b_i}}{n_\varepsilon}. \tag{6.12}$$

4. Repeat the previous two steps some large number of times, M.
5. Then our *p-value* can be found using

$$p = \frac{1 + \sum_{\varsigma=1}^{M} I_{\{e_\varepsilon \leq \hat{\varepsilon}\}}}{M+1}. \tag{6.13}$$

6. We will reject the null hypothesis, H_0, if the *p-value* is small or if $p < \alpha$ where α is a pre-specified significance level.

The subtraction of $\hat{\varepsilon}$ and the addition of ε in the third step of the algorithm above is to make the distribution of errors, e_ε, mimic the distribution of errors assuming that the null hypothesis that $\varepsilon = \varepsilon_0$ is true. See Sect. 3.10 of Manly [62] for more details.

Example 6.5 In this example, we will test whether a fingerprint device has an FTE that is significantly less than 10%. The fingerprint device that we are interested in testing comes from Coventry et al. [16]. Formally, we are testing the hypotheses

$$H_0 : \varepsilon = 0.10$$
$$H_1 : \varepsilon < 0.10.$$

In that paper they tested $n_\varepsilon = 82$ individuals and 7 were unable to enrol. The sample FTE is then $\hat{\varepsilon} = 0.0854$. We bootstrapped the individuals $M = 2000$ times, and Fig. 6.3 has a summary of the distribution of the e_ε's. The dashed vertical line there is $\hat{\varepsilon} = 0.0854$. Our *p-value* is 0.2940 which is the percentage of the e_ε's that are less than or equal to 0.0854. Since the *p-value* is large, we cannot reject the null hypothesis, H_0. Thus, our overall conclusion is that this FTE is not significantly less than 10%.

6.2.2 Two Sample Methods for FTE's

We now turn our focus to the comparison of two FTE's, denoted ε_1 and ε_2. Likewise, we'll use n_1 and n_2 to distinguish between the number of enrolment decisions being tested in each group. Specifically, our focus in this section is on how to determine if two methods have FTE's that are significantly different from each other. There are many possible cases where we might be interested in comparing two FTE's. These include determining if two demographic groups have different FTE's for the same device or determining if a group of individuals has the same or different FTE's for two different devices. There are two types of data collections that are common and need to be considered here. First, we consider the case where the data collections are 'independent.' Here that will mean that there is no overlap in the individuals from whom data was collected. An example of this might be two data collections of the same device in different locations. Second, we consider the case where the individuals involved in the data collection were the same for both FTE's. A laboratory test whether each subject had enrolment decisions for two or more devices. The former data is referred to as 'independent' data below and the latter is referred to as 'paired' data as each individual's observation has a 'mate' or 'pair' from a second process. Examples of 'independent' data include comparing different genders or comparing different age groups, while examples of 'paired' data include comparing FTE's on the same group between two different fingerprint devices or between two different sets of environmental conditions.

Confidence intervals and hypothesis tests for the difference of two FTE's are presented below. Both bootstrap methods and large sample theory methods are described for independent data collection and paired data collections.

6.2.2.1 Confidence Interval for Difference of Two Independent FTE's

Here we consider the case of making a confidence interval for the difference of two independent FTE's. We will denote this difference as $\varepsilon_1 - \varepsilon_2$. Here n_1 and n_2

represent the total number of enrolment decisions for the first and second process, respectively. The enrolment decisions will be denoted by $E_i^{(1)}$ and $E_i^{(2)}$ for the first and second FTE's.

Large Sample Approach If n_1 and n_2 are large, i.e. if $n_g\hat{\varepsilon}_g \geq 10$ and $n_g(1 - \hat{\varepsilon}_g) \geq 10$ for $g = 1, 2$ then we can (6.14) to make a $(1-\alpha)100\%$ confidence interval for $\varepsilon_1 - \varepsilon_2$ using (6.14) given below.

$$\hat{\varepsilon}_1 - \hat{\varepsilon}_2 \pm z_{\alpha/2}\sqrt{\frac{\hat{\varepsilon}_1(1-\hat{\varepsilon}_1)}{n_1} + \frac{\hat{\varepsilon}_2(1-\hat{\varepsilon}_2)}{n_2}} \tag{6.14}$$

where

$$\hat{\varepsilon}_g = \frac{\sum_{i=1}^{n_1} E_i^{(g)}}{n_1} \tag{6.15}$$

and $z_{\alpha/2}$ is the $1-\alpha/2$th percentile of a Gaussian distribution. We use the percentile of the Gaussian distribution since the sampling distribution of $\hat{\varepsilon}_1 - \hat{\varepsilon}_2$ is approximately Gaussian distribution when n_1 and n_2 are large, see, for example, Snedecor and Cochran [92].

Example 6.6 We illustrate the above procedure by deriving a confidence interval for the difference between two iris devices. The first is IrisGuard from the International Biometrics Group's Comparative Biometric Testing Round 7 [41] and the second is the iris device from set 1 of the FTE-Synth database. For the former, there were 18 enrolment attempts that were unsuccessful from among $n_1 = 1035$ attempts. For the latter, there were 73 unsuccessful enrolment attempts from $n_2 = 1214$ attempts. The resulting estimated FTE's are $\hat{\varepsilon}_1 = 18/1035 = 0.0174$ and $\hat{\varepsilon}_2 = 73/1214 = 0.0601$, respectively. In order to use the method above to create a 95% confidence interval, we first need to be assured that the sampling distribution of $\hat{\varepsilon}_1 - \hat{\varepsilon}_2$ approximately follows a Gaussian distribution. This is the case since $n_1\hat{\varepsilon}_1 = 18 \geq 10$, $n_1(1-\hat{\varepsilon}_1) = 1017 \geq 10$, $n_2\hat{\varepsilon}_2 = 73 \geq 10$, and $n_2(1-\hat{\varepsilon}_2) = 1141 \geq 10$. Then our 95% confidence interval is

$$0.0174 - 0.0601 \pm 1.960\sqrt{\frac{0.0174(1-0.0174)}{1035} + \frac{0.0601(1-0.0601)}{1214}} \tag{6.16}$$

which gives an interval of $(-0.0538, -0.0316)$. Thus, with 95% confidence, the difference in these two FTE's is between -5.38% and -3.16%.

Bootstrap Approach Next we present the bootstrap approach to making a confidence interval for the difference of two independent process FTE's. This methodology does not make assumptions about the sampling distribution of $\hat{\varepsilon}_1 - \hat{\varepsilon}_2$ as the large sample approach did. The basic outline here is that we bootstrap individuals from each set of FTE decisions separately to gain an approximation to the sampling distribution of $\hat{\varepsilon}_1 - \hat{\varepsilon}_2$. The algorithm for this confidence interval is given below.

1. Calculate $\hat{\varepsilon}_1 - \hat{\varepsilon}_2$.
2. Sample *with replacement* the n_1 individuals from the first group and call the selected individuals $b_1^{(1)}, \ldots, b_{n_1}^{(1)}$.
3. Sample *with replacement* the n_2 individuals from the second group and call the selected individuals $b_1^{(2)}, \ldots, b_{n_2}^{(2)}$.
4. Calculate and store

$$e_\varepsilon = \hat{\varepsilon}_1^b - \hat{\varepsilon}_2^b - (\hat{\varepsilon}_1 - \hat{\varepsilon}_2) \tag{6.17}$$

where

$$\hat{\varepsilon}_g^b = \frac{\sum_{i=1}^{n_g} E_{b_i^{(g)}}^{(g)}}{n_g} \tag{6.18}$$

for $g = 1, 2$.
5. Repeat the previous step some large number of times, M.
6. Create a $100(1-\alpha)\%$ confidence interval for $\varepsilon_1 - \varepsilon_2$ by taking $\hat{\varepsilon}_1 - \hat{\varepsilon}_2 - e_U$, $\hat{\varepsilon}_1 - \hat{\varepsilon}_2 - e_L$ where e_U and e_L are the $1-\alpha/2$th and $\alpha/2$th percentiles of the distribution of e_ε. Note that we expect e_L to be negative so that the upper bound for the confidence interval will be greater than $\hat{\varepsilon}_1 - \hat{\varepsilon}_2$.

Example 6.7 For this example, we will produce a 99% confidence interval for the difference between the FTE from the study done by Coventry et al. [16] and the FTE from the study done by Fairhurst et al. [31]. From the former, $n_1 = 82$ individuals attempted to enrol in a fingerprint system and 7 of those attempts were unsuccessful; from the latter, $n = 221$ individuals attempted to enrol in a voice system and 22 of those attempts were unsuccessful. From these data, we have that the estimated FTE's are $\hat{\varepsilon}_1 = 0.0854$ and $\hat{\varepsilon}_2 = 0.0995$. Following the bootstrap approach above, we generated $M = 2000$ values for e_ε. A histogram which summarizes this distribution is given in Fig. 6.4. Taking the 99.5th percentile and the 0.5th percentile, we obtain an 99% confidence interval to be $(-0.1204, 0.0740)$. Thus, we can be 99% confident that the difference between these two process is between -12.04% and 7.40%.

6.2.2.2 Hypothesis Test for Difference of Two Independent FTE's

This section focuses on testing the difference between two distinct data collections of FTE's. As above we use $\varepsilon_1 - \varepsilon_2$ as our parameter of interest. The hypotheses that we are testing are:

$$H_0 : \varepsilon_1 = \varepsilon_2$$
$$H_1 : \varepsilon_1 < \varepsilon_2.$$

These is equivalent to testing:

$$H_0 : \varepsilon_1 - \varepsilon_2 = 0$$
$$H_1 : \varepsilon_1 - \varepsilon_2 < 0.$$

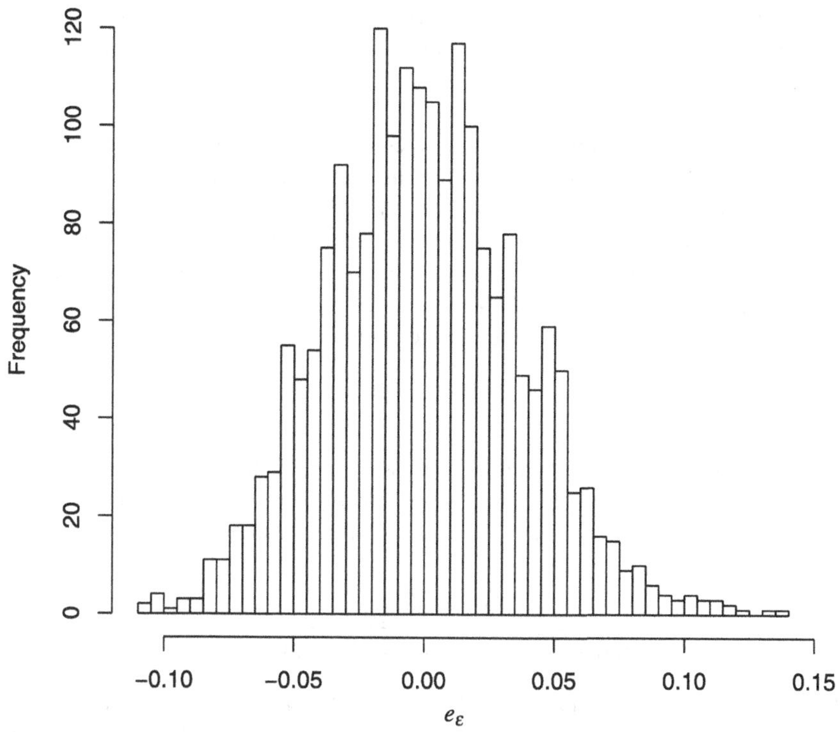

Fig. 6.4 Distribution of e_ε, the differences in the bootstrapped difference of two FTE's from the observed differences in those FTE's for enrolment decisions from Coventry et al. [16] and Fairhurst et al. [31]

We first provide a large sample methodology for this test and then provide a bootstrap methodology for the same test.

Large Sample Approach The following hypothesis test of equality of two FTE's when the data have been collected independently is to be used when the sample size are large. In this case that means that $n_g\hat{\varepsilon}_g \geq 10$ and $n_g(1-\hat{\varepsilon}_g) \geq 10$ for both groups, i.e. for $g = 1, 2$. To carry out this test we use the following elements.

Test Statistic:

$$z = \frac{\hat{\varepsilon}_1 - \hat{\varepsilon}_2}{\sqrt{\hat{\varepsilon}(1-\hat{\varepsilon})\left(\frac{1}{n_1} + \frac{1}{n_2}\right)}} \tag{6.19}$$

where

$$\hat{\varepsilon} = \frac{n_1\hat{\varepsilon}_1 + n_2\hat{\varepsilon}_2}{n_1 + n_2}. \tag{6.20}$$

p-value: From our test statistic we can calculate that $p = P(Z < z)$ where Z is a standard Gaussian random variable. Table 9.1 can be used to evaluate this probability.

Rejection of the null hypothesis will only result if we obtain a small *p-value*. If a significance level α is given *a priori* the analysis of the data, then we can reject the null hypothesis when $p < \alpha$.

Example 6.8 For this example, we will be comparing two iris FTE's. The first of these is for the IrisGuard device from the International Biometrics Group's Comparative Biometric Testing Round 7 [41]. From $n_1 = 1035$ enrolment attempts, that device had 18 individuals who were unable to enrol. The second device is from the FTE-Synth database that was discussed in Sect. 1.5.2. There were 73 unsuccessful enrolment attempts out of $n_2 = 1214$ total enrolment attempts for this iris device. Here, we would like to use the large sample approach given above to test

$$\begin{aligned} H_0 &: \varepsilon_1 = \varepsilon_2 \\ H_1 &: \varepsilon_1 < \varepsilon_2. \end{aligned}$$

Before we fully evaluate the quantities necessary for this test, we must first ensure that our sample sizes are sufficiently large. Since $n_1\hat{\varepsilon}_1 = 18 \geq 10$, $n_1(1 - \hat{\varepsilon}_1) = 1017 \geq 10$, $n_2\hat{\varepsilon}_2 = 73 \geq 10$ and $n_2(1 - \hat{\varepsilon}_2) = 1141$, we can proceed. Our test statistic is

$$z = \frac{0.0174 - 0.0601}{\sqrt{0.0405(1 - 0.0405)(\frac{1}{1035} + \frac{1}{1214})}} = -5.127. \tag{6.21}$$

We then get a *p-value* of $P(Z < -5.127) < 0.0001$. Since this *p-value* is small, we reject the null hypothesis, H_0, and conclude that the process FTE for the IrisGuard is significantly less than the process FTE for the FTE-Synth Iris device.

Bootstrap Approach The next methodology we present for testing whether two independent FTE's are significantly different is based upon the *iid* bootstrap. This test will use the same null and alternative hypotheses, H_0 and H_1, as the large sample test above. This procedure is given below.

1. Calculate $\hat{\varepsilon}_1 - \hat{\varepsilon}_2$ using the observed enrolment decisions from each group.
2. Sample *with replacement* the n_1 individuals from the first group and call the selected individuals $b_1^{(1)}, \ldots, b_{n_1}^{(1)}$.
3. Sample *with replacement* the n_2 individuals from the second group and call the selected individuals $b_1^{(2)}, \ldots, b_{n_2}^{(2)}$.
4. Calculate and store
$$e_\varepsilon = \hat{\varepsilon}_1^b - \hat{\varepsilon}_2^b - (\hat{\varepsilon}_1 - \hat{\varepsilon}_2) \tag{6.22}$$
where
$$\hat{\varepsilon}_g^b = \frac{\sum_{i=1}^{n_g} E_{b_i^{(g)}}^{(g)}}{n_g} \tag{6.23}$$
for $g = 1, 2$.
5. Repeat the previous step some large number of times, M.

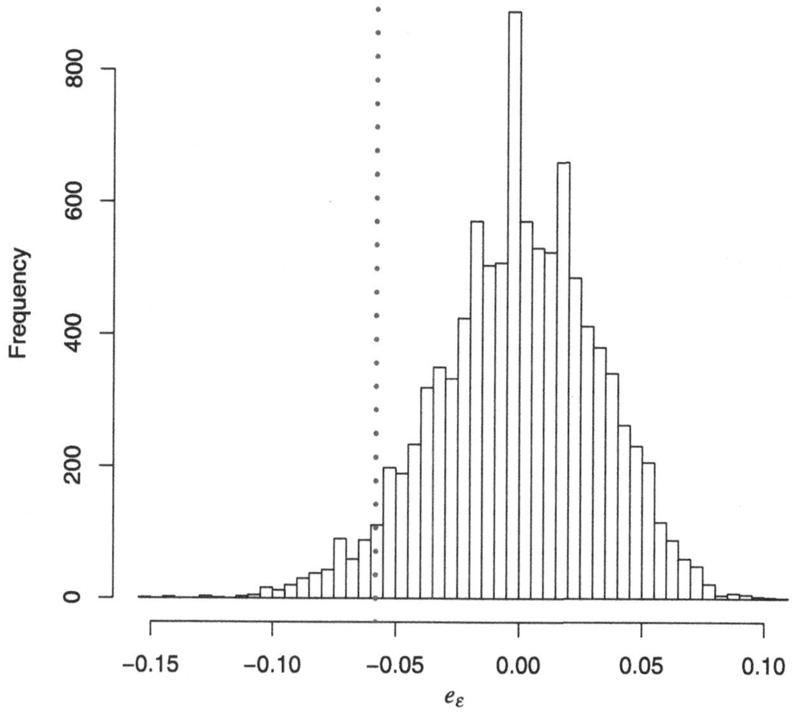

Fig. 6.5 Distribution of e_ε, the differences in the bootstrapped difference of two FTE's from the observed differences in FTE's from Fairhurst et al. [31] and Coventry et al. [16]

6. Then the *p-value* for this test is

$$p = \frac{1 + \sum_{\varsigma=1}^{M} I_{\{e_\varepsilon \leq \hat{\varepsilon}_1 - \hat{\varepsilon}_2\}}}{M+1}. \tag{6.24}$$

7. We will reject the null hypothesis, the equality of the two FTE's, if the *p-value* is small. For a given significance level α, we will reject that null hypothesis when $p < \alpha$.

The procedure outlined above does not depend on any distributional assumptions regarding the shape of the sampling distribution of the test statistic and, hence, it does not require n_1 and n_2 to be sufficiently large for the test to hold. As in the previous section, if both error rates are zero then we are unable to conclude that there is a significant difference between the FTE's regardless of sample size.

Example 6.9 In this example, we are testing the equality of two fingerprint FTE's. The first fingerprint FTE is from a study by Fairhurst et al. [31] and the second is from a study by Coventry et al. [16]. The first FTE from the sample collection of $n_1 = 221$ enrolment attempts is $\hat{\varepsilon}_1 = 6/221 = 0.0271$. The estimated FTE from

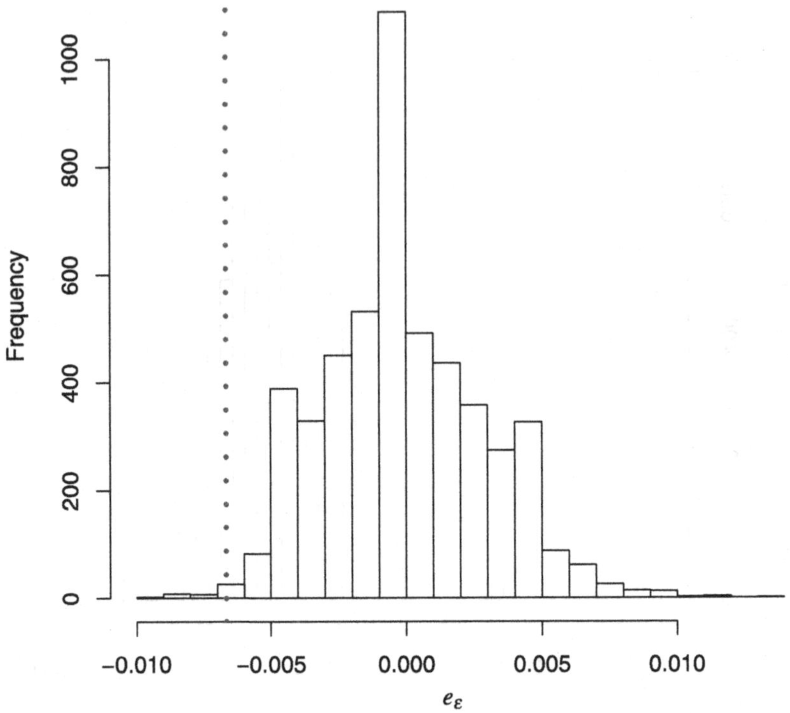

Fig. 6.6 Distribution of e_ε, the differences in the bootstrapped difference of the FTE's from the observed differences in FTE's from the FP-optical system in the FTE-Synth database and the Iris-Guard device from the International Biometrics Group's Comparative Biometric Testing Round 7 [41]

the Coventry study is $\hat{\varepsilon}_2 = 7/82 = 0.0854$ which is based upon $n_2 = 82$ enrolment decisions. We will formally test that

$$H_0 : \varepsilon_1 = \varepsilon_2$$
$$H_1 : \varepsilon_1 < \varepsilon_2.$$

This is testing whether the Fairhurst FTE is significantly less than the Coventry FTE given that the difference between these two FTE's is $\hat{\varepsilon}_1 - \hat{\varepsilon}_2 = 0.0271 - 0.0854 = -0.0582$. To test this, we will use the bootstrap algorithm given above. We repeated this methodology $M = 9999$ times. A summary of the distribution of the e_ε's is given in Fig. 6.5. The dashed vertical line in that figure is the observed difference of -0.0582. We can calculate the *p-value* of the proportion of the e_ε's that are less than that value. Here $p = 0.0434$. This particular *p-value* falls into a gray area for which we must consider our decision in context. If we had specified a significance level, α, then we could compare this *p-value* to α and make our decision. Instead, we have to make a decisions based purely on the value $p = 0.0434$. If there are serious consequences of choosing the Fairhurst device over the Coventry device, then we might require a smaller *p-value* than the one here. Otherwise, it is reasonable to conclude that the Fairhurst device has a significantly smaller FTE than the Coventry device.

Example 6.10 In this example, we are considering the difference between FTE's for an iris device and a fingerprint device with quite small FTE's. The fingerprint device is taken from the FTE-Synth database; it is the optical fingerprint device, FP-optical, reported there. That database is described in Sect. 1.5.2. That device had 13 of $n_1 = 1214$ individuals who were unable to enrol. The iris device is the IrisGuard device reported in the International Biometrics Group's Comparative Biometric Testing Round 7 [41]. From $n_2 = 1035$ enrolment attempts, that device had 18 individuals who were unable to enrol. Both of these devices exhibited fairly small FTE's, $\hat{\varepsilon}_1 = 0.0107$ and $\hat{\varepsilon}_2 = 0.0174$, respectively. Although this difference is small, both devices were tested extensively and so it is possible that the difference of -0.0067 is a significant one. To test the hypotheses

$$H_0 : \varepsilon_1 = \varepsilon_2$$
$$H_1 : \varepsilon_1 < \varepsilon_2,$$

we followed the bootstrap methodology above with $M = 4999$. A summary of the distribution of the resulting e_ε's is given in Fig. 6.6 where the dashed vertical line is $\hat{\varepsilon}_1 = \hat{\varepsilon}_2 = -0.0067$. From that distribution, we can calculate that the *p-value* is $p = 0.0036$. Since this *p-value* is small we reject the null hypothesis, H_0, that these two devices have the same FTE and conclude that the FP-optical device has a significantly small FTE than the IrisGuard device.

6.2.2.3 Confidence Interval for Difference of Two Paired FTE's

We now move from the comparison of independent FTE's to the comparison of paired FTE's. Paired enrolment decisions occur when each individual has two enrolment decisions, one for each device or one for each environment. The sample size for both groups is identical for the paired case, so that we will only worry about n_ε for the paired comparison. Thus we do not need to distinguish between n_1 and n_2 since they are equal. We do, however, distinguish between the estimated FTE's which we will denote $\hat{\varepsilon}_1$ and $\hat{\varepsilon}_2$, as above. Below we present both a large sample approach and a bootstrap approach.

Large Sample Approach To appropriately use the large sample approach for making a confidence interval for the difference of two paired FTE's, we must first determine that our sample size is sufficiently large. We will conclude this if $n_\varepsilon \hat{\varepsilon}_g \geq 10$ and $n_\varepsilon(1 - \hat{\varepsilon}_g) \geq 10$). Then, to make a $(1 - \alpha)100\%$ confidence interval for $\varepsilon_1 - \varepsilon_2$, we will use the following equation.

$$\hat{\varepsilon}_1 - \hat{\varepsilon}_2 \pm z_{\alpha/2}\sqrt{\frac{\hat{\varepsilon}_1(1 - \hat{\varepsilon}_1)}{n_\varepsilon} + \frac{\hat{\varepsilon}_2(1 - \hat{\varepsilon}_2)}{n_\varepsilon} - 2Cov(\hat{\varepsilon}_1, \hat{\varepsilon}_2)} \tag{6.25}$$

where

$$\hat{\varepsilon}_g = \frac{\sum_{i=1}^{n_\varepsilon} E_i^{(g)}}{n_\varepsilon} \tag{6.26}$$

and

$$Cov(\hat{\varepsilon}_1, \hat{\varepsilon}_2) = \frac{1}{n_\varepsilon^2} \sum_{i=1}^{n_\varepsilon} (E_i^{(1)} - \hat{\varepsilon}_1)(E_i^{(2)} - \hat{\varepsilon}_2). \tag{6.27}$$

Example 6.11 Our goal here is to create a 95% confidence interval for the difference of two paired FTE's. The two FTE's we will consider are from two fingerprint devices—FP-optical and FP-capacitive—from the FTE-Synth database, cf. Sect. 1.5.2. For each device, the same $n_\varepsilon = 1214$ individuals tried to enrol. We can use the methodology above for these two devices since the FP-optical device had a sample FTE of $\hat{\varepsilon}_1 = 0.0107$ and for the FP-capacitive device $\hat{\varepsilon}_2 = 0.0173$. Using these values, we can verify that the conditions for using the large sample confidence interval above are met. That is, $n_\varepsilon \hat{\varepsilon}_1 = 13 \geq 10$, $n_\varepsilon(1 - \hat{\varepsilon}_1) = 1201 \geq 10$, $n_\varepsilon \hat{\varepsilon}_2 = 21 \geq 10$, $n_\varepsilon(1 - \hat{\varepsilon}_2) = 1193 \geq 10$. Then our 95% confidence interval is $(-0.0157, 0.0025)$ using $Cov(\hat{\varepsilon}_1, \hat{\varepsilon}_2) = 0.000000526$. We can be 95% confident that the difference in these two process FTE's is between -1.57% and 0.25%.

Bootstrap Approach We next provide a bootstrap approach to making a confidence interval for the difference of two FTE's. The primary difference between this bootstrap approach and then one done for the difference of two independent FTE's is that we will only bootstrap once to replicate the data. We do this since there is a built in correlation between the decisions on a given individual and, therefore, at each iteration of the bootstrap, we will resample all of the individuals and take both decisions associated with the selected individuals. The algorithm for carrying this out is given below.

1. Calculate $\hat{\varepsilon}_1 - \hat{\varepsilon}_2$.
2. Sample *with replacement* n_ε individuals from the n_ε individuals. Call these selected individuals $b_1, \ldots, b_{n_1}$. For each selected individual, b_i, retain both decisions, $E_{b_i}^{(1)}$ and $E_{b_i}^{(2)}$, for that individual.
3. Calculate and store

$$e_\varepsilon = \hat{\varepsilon}_1^b - \hat{\varepsilon}_2^b - (\hat{\varepsilon}_1 - \hat{\varepsilon}_2) \tag{6.28}$$

   where

$$\hat{\varepsilon}_1^b - \hat{\varepsilon}_2^b = \frac{\sum_{i=1}^{n_\varepsilon} E_{b_i}^{(1)} - E_{b_i}^{(2)}}{n_\varepsilon} = \frac{\sum_{i=1}^{n_\varepsilon} E_{b_i}^{(1)}}{n_\varepsilon} - \frac{\sum_{i=1}^{n_\varepsilon} E_{b_i}^{(2)}}{n_\varepsilon}. \tag{6.29}$$

4. Repeat the two previous steps some large number of times, M.
5. Make a $(1 - \alpha) \times 100\%$ confidence interval for $\varepsilon_1 - \varepsilon_2$ using:

$$(\hat{\varepsilon}_1 - \hat{\varepsilon}_2 - e_U, \hat{\varepsilon}_1 - \hat{\varepsilon}_2 - e_L) \tag{6.30}$$

   where e_U and e_L represent the $1-\alpha/2$th and $\alpha/2$th percentiles of the distribution of e_ε, respectively.

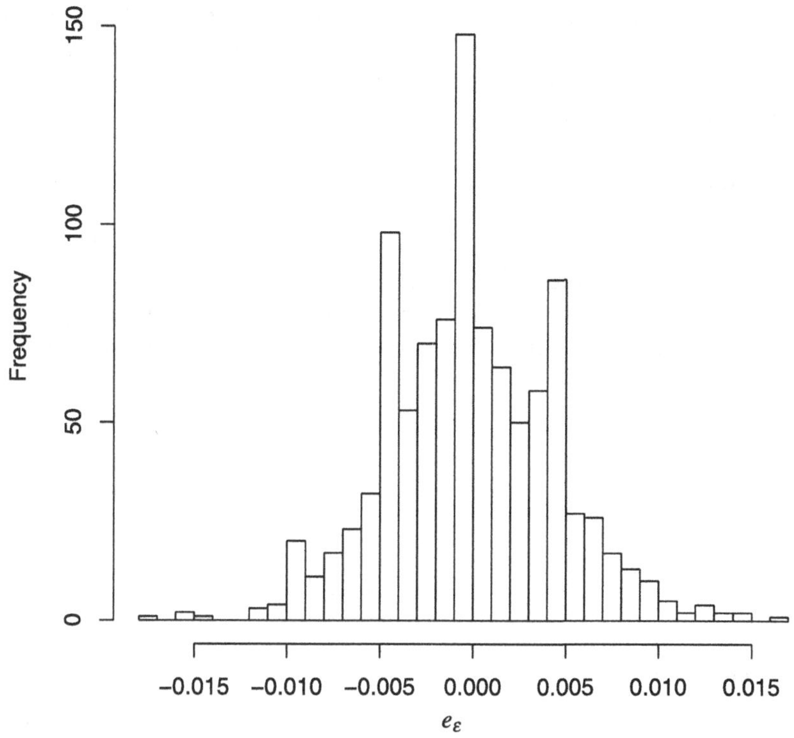

Fig. 6.7 Distribution of e_ε, the differences in the bootstrapped difference of two FTE's—FP-optical and FP-capacitive from the FTE-Synth database—from the observed differences in those FTE's

Example 6.12 In this example, we will use the bootstrap methodology above to make a 95% confidence interval for the difference of two FTE's. The two paired FTE's that we will use are the FP-optical and FP-capacitive fingerprint devices from the FTE-Synth database. The estimated FTE for the FP-optical device is $\hat{\varepsilon}_1 = 0.0107$ and for the FP-capacitive device is $\hat{\varepsilon}_2 = 0.0173$. We previously considered these paired enrolment decisions in Example 6.7. The large sample confidence interval from that example was $(-0.0157, 0.0025)$. Here we followed the bootstrap above and generated $M = 1000$ values of e_ε. Figure 6.7 has a summary of this distribution. Using the 97.5th and 2.5th percentile of that distribution, we obtain a confidence interval of $(-0.0157, 0.0025)$. Thus, we can be 95% confident that the difference between these two process FTE's is between -1.57% and 0.25%. We note that this interval is the same as that for the large sample confidence interval.

6.2.2.4 Hypothesis Test for Difference of Two Paired FTE's

We now move to the hypothesis test for comparing two paired FTE's. Our large sample test here is conducted in a slightly different manner than previously discussed tests. We will use a Chi-squared distribution to carry our test. The Chi-squared distribution was introduced in Definition 2.39 and Table 9.4 has some percentiles of the Chi-squared distribution. We will use McNemar's Test here since it is a common approach that is taken to testing paired binomial proportions.

Large Sample Approach The test that we will use is known as McNemar's Test. More details and a discussion on this test can be found in Snedecor and Cochran [92, pp. 121–124]. Here we are interested in the number of individuals, n_{01} who were enroled on the first test but not enroled on the second and in the number of individuals n_{10} who were not enroled on the first test and enroled on the second. This test should only be used when $n_{01}+n_{10}$ is more than 10. Formally, the hypotheses for this test are:

$$H_0: \varepsilon_1 = \varepsilon_2$$
$$H_1: \varepsilon_1 \neq \varepsilon_2.$$

These counts provide the summaries needed for the test we describe below.

Test Statistic:

$$X^2 = \frac{(|n_{01} - n_{10}| - 1)^2}{n_{01} + n_{10}}. \tag{6.31}$$

p-value:

$$p = P(\mathbb{X} > X^2) \tag{6.32}$$

where $\mathbb{X}$ is a Chi-squared random variable with 1 degree of freedom.

We will reject the null hypothesis if the *p-value* is small or if $p < \alpha$ where α is the significance level for the test.

Example 6.13 The focus of this example is two fingerprint devices, FP2 and FP4, from the FTE-Synth database, cf. 1.5.2. Each of those devices has enrolment attempts from the same $n = 628$ individuals. FP2 had a sample FTE of $\hat{\varepsilon}_1 = 0.0239$ and FP4 had a sample FTE of $\hat{\varepsilon} = 0.0318$. Our goal here is to test whether we can conclude that these two devices have different FTE's. We will use a significance level of $\alpha = 0.10$ for this test. In order to use the large sample McNemar's Test above, we need to check that $n_{01} + n_{10} \geq 10$. For these data, we have that $n_{01} = 18$ and $n_{10} = 31$. Consequently, we can proceed to analyze the differences between the FTE's of these two devices. Our test statistic is

$$X^2 = \frac{(|18 - 31| - 1)^2}{18 + 31} = 2.9388. \tag{6.33}$$

Using a Chi-squared table such as Table 9.4, we can calculate that $p = P(\mathbb{X} > 2.9388) \approx 0.0865$. Since that *p-value* is smaller than our significance level $\alpha = 0.10$, we reject the null hypothesis, H_0, that these two FTE's are the same. Our conclusion is then there is a significant difference between the process FTE's for these two fingerprint devices.

It may seem unusual that the test above does not *explicitly* take into account the estimates of ε_1 and ε_2. However, n_{01} and n_{10} are inherently part of the variability in the difference between these two estimates, $\varepsilon_1 - \varepsilon_2$. Thus, they are integral to the test. Their expected size will be proportional to the size of both FTE's. A discussion of this phenomena can found in the aforementioned Snedecor and Cochran [92].

Bootstrap Approach The bootstrap approach here is similar to the approach that we took with the bootstrap confidence interval for the difference of two paired FTE's.

1. Calculate $\hat{\varepsilon}_1 - \hat{\varepsilon}_2$.
2. Sample *with replacement* n_ε individuals from the n_ε individuals in the data. Call these selected individuals $b_1, \dots, b_{n_1}$. For each selected individual, b_i, retain both decisions, $E_{b_i}^{(1)}$ and $E_{b_i}^{(2)}$, for that individual.
3. Calculate and store

$$e_\varepsilon = \hat{\varepsilon}_1^b - \hat{\varepsilon}_2^b - (\hat{\varepsilon}_1 - \hat{\varepsilon}_2) \tag{6.34}$$

where

$$\hat{\varepsilon}_1^b - \hat{\varepsilon}_2^b = \frac{\sum_{i=1}^{n_\varepsilon}\{E_{b_i}^{(1)} - E_{b_i}^{(2)}\}}{n_\varepsilon} \tag{6.35}$$

which is equivalent to

$$\frac{\sum_{i=1}^{n_\varepsilon} E_{b_i}^{(1)}}{n_\varepsilon} - \frac{\sum_{i=1}^{n_\varepsilon} E_{b_i}^{(2)}}{n_\varepsilon}. \tag{6.36}$$

4. Repeat the two previous steps some large number of times M.
5. Our *p-value* can then be calculated as

$$p = \frac{1 + \sum_{\varsigma=1}^{M} I_{\{e_\varepsilon \le \hat{\varepsilon}_1 - \hat{\varepsilon}_2\}}}{M+1}. \tag{6.37}$$

6. We will then reject our null hypothesis if the *p-value* for this test is small.

Example 6.14 In this example, we consider two paired devices from the PURDUE-FTE database. In particular we will consider the left index finger enrolment decisions from the samples labeled '00' on device 4 and device 1. There were $n_\varepsilon = 85$ individuals that attempted to enrol in these two devices. Device 4 had a sample FTE of $\hat{\varepsilon} = 0.0118$ from these 85 individuals while device 1 had a sample FTE of $\hat{\varepsilon} = 0.0588$ from the same set of individuals. Bootstrapping these individuals $M = 4000$ times, we obtained the distribution of e_ε's that is summarized in Fig. 6.8. The dashed vertical line in that graphic occurs at the observed difference in the FTE's of $\hat{\varepsilon}_1 - \hat{\varepsilon}_2 = -0.0471$. We get the *p-value* associated with this value to be $p = 0.0735$. This is a relatively large *p-value*, so there is not enough evidence to reject the null hypothesis, H_0, that these two devices have the same FTE.

Finally, we recognize at the end of this section on comparing two FTE's that there has been some recent work in the statistical literature on the comparison of

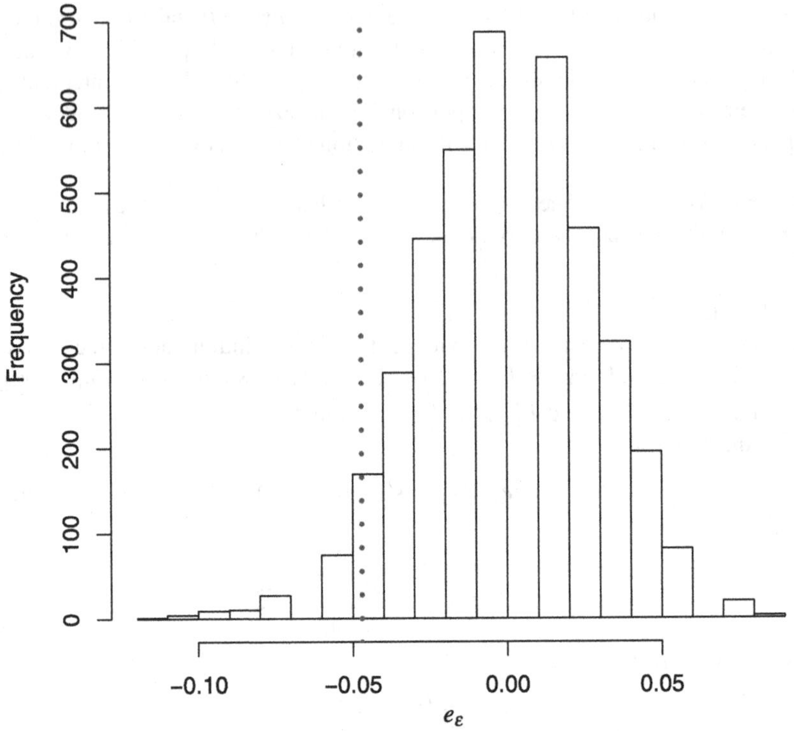

Fig. 6.8 Distribution of differences in the bootstrapped difference of two FTE's from the observed differences between device 4 and device 1 from the PURDUE-FTE database

two binomial proportions. The interested reader is directed to Agresti and Caffo [3] and a series of papers by Newcombe [70–72].

6.2.3 Multiple Sample Methods for FTE's

This section focuses on the comparison of three or more FTE's. Methods are presented for comparing independent FTE's and paired FTE's. Our notation is as follows. Let ε_g be the FTE rate for the gth biometric enrolment process. Let n_g be the number of individuals tested from a sample of the gth group and let $E_i^{(g)}$ be the ith enrolment decision from the gth group and then our estimate of the gth process FTE, ε_g, is

$$\hat{\varepsilon}_g = \frac{\sum_{i=1}^{n_g} E_i^{(g)}}{n_g}. \tag{6.38}$$

6.2.3.1 Hypothesis Test for Multiple Independent FTE's

Our interest here is in testing whether three or more process FTE's are the same or if there is enough evidence to conclude that at least one is different from the rest. For example, we might be interested in comparing the FTE rate for four different facial recognition systems implemented at four different buildings in four different cities with the assumption that the attempted enrolees are distinct among the four locations. Here the test we will use is a Chi-squared test of homogeneity. The hypotheses for this test are:

$$H_0 : \varepsilon_1 = \varepsilon_2 = \cdots = \varepsilon_G$$
$$H_1 : \text{ not } H_0.$$

Large Sample Approach The following test should be carried out if at least 80% of the $n_g\bar{\varepsilon}$'s and $n_g(1-\bar{\varepsilon})$'s are more than 5 and all of them are more than 1 where

$$E^{(g)} = \sum_{i=1}^{n_g} E_i^{(g)} \tag{6.39}$$

is the total number of enrolment rejections for the gth group.

Test Statistic:

$$X^2 = \sum_{g=1}^{n} \left\{ \frac{(E^{(g)} - n_g\bar{\varepsilon})^2}{n_g\bar{\varepsilon}} + \frac{(n_g - E^{(g)} - n_g(1-\bar{\varepsilon}))^2}{n_g(1-\bar{\varepsilon})} \right\} \tag{6.40}$$

where

$$\bar{\varepsilon} = \frac{\sum_{g=1}^{G}\sum_{i=1}^{n_g} E_i^{(g)}}{\sum_{g=1}^{G} n_g}. \tag{6.41}$$

p-value: $p = P(\mathbb{X} < X^2)$ where $\mathbb{X}$ is a chi-squared RV with $G-1$ degrees of freedom

We will reject the null hypothesis if $p < \alpha$ for a given significance level, α, or if the *p-value* is small.

Example 6.15 In this example, we are applying the method above to the FTE's from three different devices where the enrolment decision data collections are independent. The first is the fingerprint device from Fairhurst et al. [31]. The second is the FP-optical fingerprint device from the FTE-Synth database and the final one is IrisGuard device from the IrisGuard device from the from the International Biometrics Group Comparative Biometrics Testing Round 7 [41]. These devices have sample FTE's of $\hat{\varepsilon}_1 = 0.0271$, $\hat{\varepsilon}_2 = 0.0107$ and $\hat{\varepsilon}_3 = 0.0174$ based upon $n_1 = 221$, $n_2 = 1214$ and $n_3 = 1035$ enrolment decisions, respectively. We are using a large sample approach here, so we must ensure that our samples sizes are sufficiently large. Before we do that, we note that $\bar{\varepsilon} = 0.0150$. Table 6.2 contains the relevant values. Since 5/6th or 83% of the values in that table are larger than 5 and all of them are larger than 1, we can proceed. Using the values in Table 6.2, we can

Table 6.2 Calculations for Example 6.15

Device	$n_g\bar{\varepsilon}$	$n_g(1-\bar{\varepsilon})$
Fairhurst	3.31	217.69
FP-optical	18.19	1195.81
IrisGuard	15.50	1019.50

calculate that $X^2 = 4.1271$. Then using 2 degrees of freedom for the Chi-squared distribution, we find that $p = P(\chi^2 > 4.1271) \approx 0.1270$. Therefore, since that *p-value* is large, we do not reject the null hypothesis, H_0, of equality of these three independent process FTE's.

Randomization Approach The randomization approach that we take here uses the same hypotheses given in the large sample approach above. We also use the same test statistic. However, instead of using a Chi-squared distribution as our reference distribution we create a randomized distribution for the test statistic.

1. Calculate

$$X^2 = \sum_{g=1}^{G} \left\{ \frac{(E^{(g)} - n_g\bar{\varepsilon})^2}{n_g\bar{\varepsilon}} + \frac{(n_g - E^{(g)} - n_g(1-\bar{\varepsilon}))^2}{n_g(1-\bar{\varepsilon})} \right\} \tag{6.42}$$

   where

$$\bar{\varepsilon} = \frac{\sum_{g=1}^{G}\sum_{i=1}^{n_g} E_i^{(g)}}{\sum_{g=1}^{G} n_g}. \tag{6.43}$$

2. Combine all of the $E_i^{(g)}$'s and randomly select without replacement n_1 of them. Let their sum be $E^{(1)*}$.
3. Continue this for $g = 2, \ldots, G-1$. For the gth process, select n_g enrolment decisions without replacement from the remaining unselected enrolment decisions. Denote the sum of these selected enrolment decisions $E^{(g)*}$.
4. Call the sum of the remaining n_G enrolment decisions $E^{(G)*}$.
5. We then calculate

$$X_{\varsigma}^2 = \sum_{g=1}^{G} \left\{ \frac{(E^{(g)*} - n_g\bar{\varepsilon})^2}{n_g\bar{\varepsilon}} + \frac{(n_g - E^{(g)*} - n_g(1-\bar{\varepsilon}))^2}{n_g(1-\bar{\varepsilon})} \right\}. \tag{6.44}$$

6. Repeat steps 2 to 5 some large number of times, say M times.
7. Our p-value is then

$$p = \frac{1 + \sum_{\varsigma=1}^{M} 1_{\{X_{\varsigma}^2 \geq X^2\}}}{M+1}. \tag{6.45}$$

8. Reject the null hypothesis, H_0, if the *p-value* is small.

Example 6.16 For this example, we will test whether three devices whose enrolment decisions were collected independently of one another have the same FTE's.

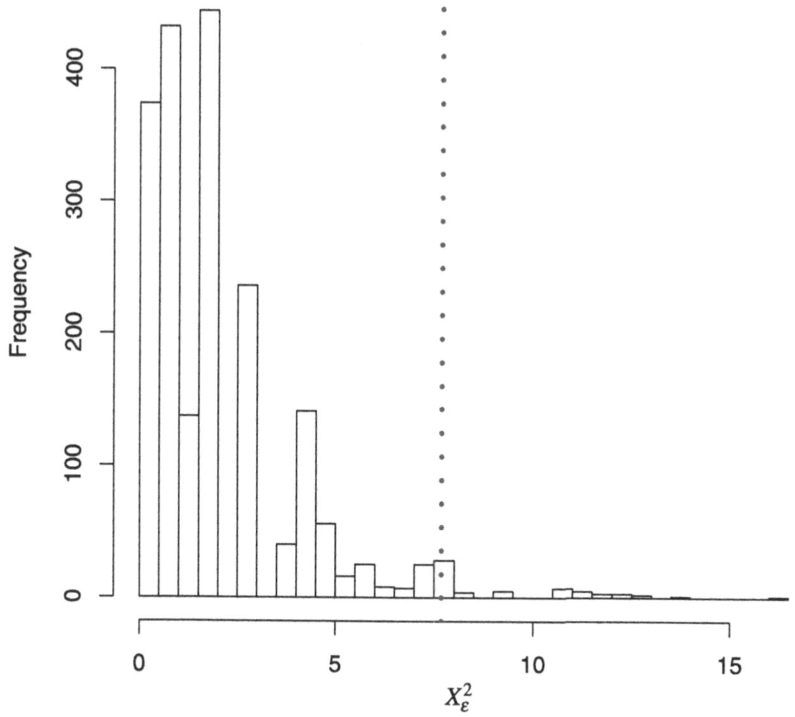

Fig. 6.9 Distribution of the randomized test statistic X_ε^2 for comparing the FTE's of a fingerprint device from Fairhurst et al. [31], the fingerprint device from Coventry et al. [16] and left index finger enrolments from fingerprint device 4 from the PURDUE-FTE database

The three devices are: the fingerprint device from Fairhurst et al. [31], the fingerprint device from Coventry et al. [16] and fingerprint device 4 from the PURDUE-FTE database, cf. Sect. 1.5.1. For device 4 from the PURDUE-FTE database, we used the left index finger enrolment decisions. These devices had sample FTE's of $\hat{\varepsilon}_1 = 0.0271$, $\hat{\varepsilon}_2 = 0.0854$ and $\hat{\varepsilon}_3 = 0.0118$, respectively. There were $n_1 = 221$, $n_2 = 82$ and $n_3 = 85$ enrolment decisions for each device, respectively. We randomized the enrolment decisions from these devices following the algorithm give above $M = 1999$ times. Figure 6.9 gives a histogram of the distribution of the resulting X_ε^2's. The value of $X^2 = 7.6786$ is represented by the dashed vertical line in that figure. The resulting *p-value* is $p = 0.0280$. As a consequence, we can reject the null hypothesis, H_0, that these three devices have the same process FTE.

Example 6.17 In this example, we will compare the equality of three FTE's from three distinct data collections. With a significance level of $\alpha = 0.01$, we will test the hypothesis that

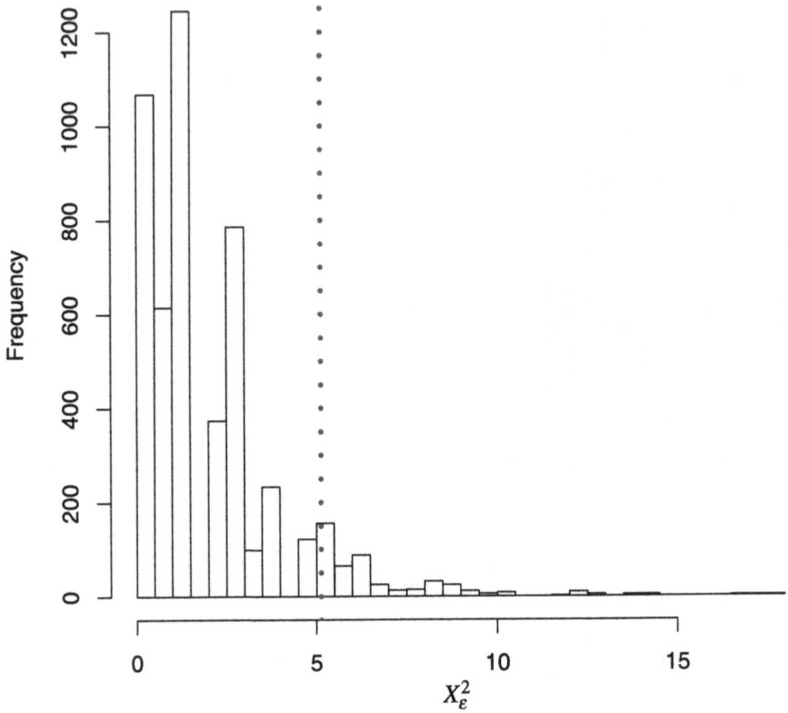

Fig. 6.10 Distribution of the randomized test statistic, X_ε^2, for comparing three fingerprint FTE's: the fingerprint device in Fairhurst et al. [31], device 2 from the PURDUE-FTE database and the fingerprint device in Coventry et al. [16]

$$H_0 : \varepsilon_1 = \varepsilon_2 = \varepsilon_3$$
$$H_1 : \text{ not } H_0.$$

The first FTE is from the fingerprint device discussed in Fairhurst et al. [31]. There were $n_1 = 221$ enrolment attempts and the sample FTE was $\hat{\varepsilon}_1 = 6/221 = 0.0272$. The second FTE is from the PURDUE-FTE database. From that database, we use the $n_2 = 81$ sample '00' enrolment decisions on the left index finger applied to device 2. The sample FTE for that device was $\hat{\varepsilon}_2 = 3/81 = 0.0370$. The final device that we will consider here is the fingerprint device discussed in Coventry et al. [16]. In their testing, of the $n_3 = 82$ enrolment decisions, there were 7 individuals who were unable to enrol which gives an FTE of $\hat{\varepsilon}_3 = 0.0854$. Applying the randomization algorithm given above to these data $M = 4999$ times, we obtained a *p-value* of $p = 0.0842$. The distribution of the X_ε^2's is given in Fig. 6.10. The value of the test statistic, $X^2 = 5.1314$ is represented by the dashed vertical line on the histogram in that figure. Since our *p-value* from this test is larger than the significance level $\alpha = 0.01$, then we fail to reject H_0, the null hypothesis that all three of these FTE's are identical.

6.2.3.2 Hypothesis Test for Multiple Paired FTE's

Paired FTE's are possible when every individuals in the sample has enrolment decisions on G different devices. We note here that by definition we must have that $n_1 = n_2 = \cdots = n_G = n_\varepsilon$ for this hypothesis test. As above $\hat{\varepsilon}_g$ is the estimated FTE for the gth group. Here we introduce q_i as the total number of (out of G) devices on which the ith individual is unable to enrol. That is,

$$q_i = \sum_{g=1}^{G} E_i^{(g)}. \tag{6.46}$$

Formally, we are testing the equality of G different process FTE's. This is

$$H_0 : \varepsilon_1 = \varepsilon_2 = \cdots = \varepsilon_G$$
$$H_1 : \text{ not } H_0.$$

We will have both a large sample test for these hypotheses as well as a randomization test for these hypotheses. One situation where the methods below would apply is the case where we are testing multiple devices on the same group of individuals. We would have each individual try to enrol in each system. Thus, for every enrolment decision by the ith individual on a given device, we would have $G - 1$ additional enrolment decisions for that individual.

Large Sample Approach For this large sample approach, we will use the test known as Cochran's Q Test [15]. Vitaliano [98] suggests that this test works well when $n_\varepsilon = 40$ or more. While that may be true for relatively large values of $\bar{\varepsilon}$—the average FTE across the G groups—it is likely not the case when the $\hat{\varepsilon}_g$'s are small. A reasonable approach might be to use the following test when $n_\varepsilon \hat{\varepsilon}_g \geq 10$ for all $g = 1, \ldots, G$.

Test Statistic:

$$Q = \frac{G(G-1)n_\varepsilon^2 \sum_{g=1}^{G} (\hat{\varepsilon}_g - \bar{\varepsilon})^2}{G(\sum_{i=1}^{n_\varepsilon} q_i) - (\sum_{i=1}^{n_\varepsilon} q_i^2)} \tag{6.47}$$

where

$$\bar{\varepsilon} = \frac{1}{G} \sum_{g=1}^{G} \hat{\varepsilon}_g. \tag{6.48}$$

p-value: $p = P(\mathbb{X} > Q)$ where $\mathbb{X}$ is a Chi-squared random variable with $G - 1$ degrees of freedom.

We will reject the null hypothesis, H_0, if the *p-value* is small or if $p < \alpha$ for a significance level α.

Example 6.18 In this example we are comparing the paired FTE's for three devices from the FTE-Synth database. See Sect. 1.5.2 for details about this database. The three devices that have paired data collections are the FP-optical, the FP-capacitive and the Iris device. As we are doing a large sample test, we need to confirm that the samples are of sufficient size. For these data, we have that $n_\varepsilon = 1214$ and that the sample FTE's are $\hat{\varepsilon}_1 = 0.0107$, $\hat{\varepsilon}_2 = 0.0173$, and $\hat{\varepsilon}_3 = 0.0601$. Then based upon the criteria given above, we can proceed with this large sample test since $n_\varepsilon \hat{\varepsilon}_g$ equals 13, 21, and 73, respectively for $g = 1, 2, 3$. Based upon these data, we get $Q = 135.1698$. For a Chi-squared distribution with $G - 1 = 2$ degrees of freedom, we get a *p-value* of $p = P(\chi^2 > 135.1698) < 0.0001$. Therefore, we can reject the null hypothesis of equality of the FTE's, and conclude that at least one of these FTE's is significantly different from the others.

Randomization Approach As with the previous test, we are testing whether G paired process FTE's are equal. We use the same test statistic here for our randomization approach as the large sample approach. The difference here is that we do not assume that are test statistic follows a χ^2 distribution. Instead, we generate a distribution based upon randomizing the failure to enrol decisions between the G groups. This procedure is given below.

1. Calculate

$$Q = \frac{G(G-1)n_\varepsilon^2 \sum_{g=1}^{G} (\hat{\varepsilon}_g - \bar{\varepsilon})^2}{G(\sum_{i=1}^{n_\varepsilon} q_i) - (\sum_{i=1}^{n_\varepsilon} q_i^2)} \tag{6.49}$$

   where

$$\bar{\varepsilon} = \frac{1}{G} \sum_{g=1}^{G} \hat{\varepsilon}_g. \tag{6.50}$$

2. For each individual i, permute, i.e. sample without replacement, their G enrolment decisions. Repeat this process for all n_ε individuals. The permuted decisions are denoted by $E_i^{(b_g)}$ which is the decision of the ith individual that is randomly assigned to the gth group.
3. Calculate and store

$$Q_\varepsilon = \frac{G(G-1)n_\varepsilon^2 \sum_{g=1}^{G} (\hat{\varepsilon}_g^b - \bar{\varepsilon})^2}{G(\sum_{i=1}^{n} q_i) - (\sum_{i=1}^{n} q_i^2)} \tag{6.51}$$

   where

$$\hat{\varepsilon}_g^b = \frac{\sum_{i=1}^{n} E_i^{(b_g)}}{n_\varepsilon}. \tag{6.52}$$

   We note that the q_i's and $\bar{\varepsilon}$ are unchanged by this randomization of the enrolment decisions between groups within an individual.
4. Repeat the two previous steps some large number of times M.
5. Then the *p-value* is

$$p = \frac{1 + \sum_{\varsigma=1}^{M} 1_{\{Q \geq Q_\varepsilon\}}}{M+1}. \tag{6.53}$$

6. If the *p-value* is small, then we will reject the null hypothesis.

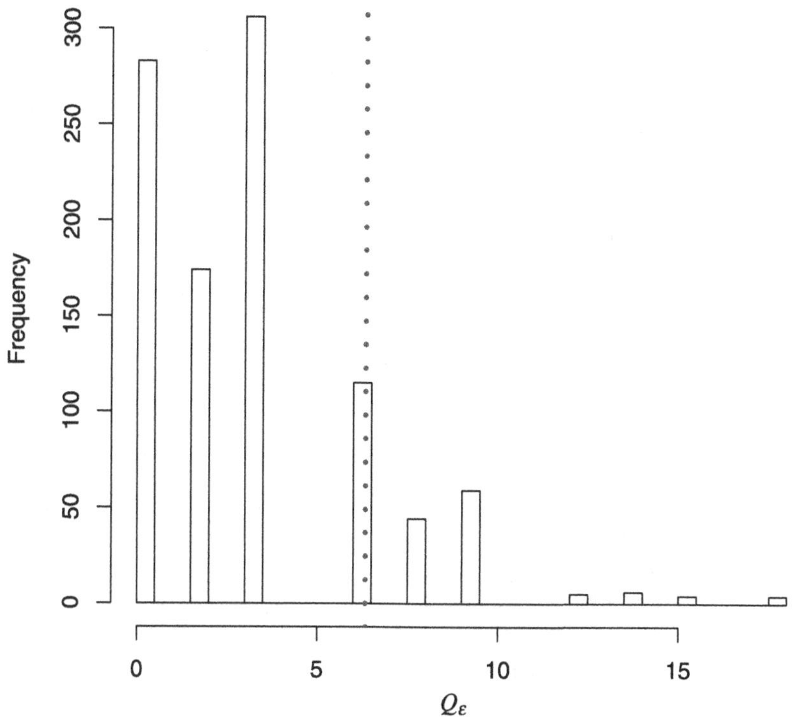

Fig. 6.11 Distribution of the randomized test statistic Q_ε for comparing the FTE's of three devices from the PURDUE-FTE database

Example 6.19 To illustrate the randomization approach for testing paired FTE's given above, we will use three devices from the PURDUE-FTE database. In particular, we will use devices 1, 3 and 4 from that database. The enrolment decisions that we will use are the ones from sample '00' of the enrolment of the left index finger on these three fingerprint devices. For these data, device 1 had an FTE of $\hat{\varepsilon}_1 = 0.0588$, while devices 3 and 4 had FTE's of $\hat{\varepsilon}_2 = 0.0471$ and $\hat{\varepsilon}_3 = 0.0118$. All of these FTE's are based upon enrolment decisions for the same $n_\varepsilon = 85$ individuals. We next applied the randomization approach described above to these data $M = 999$ times. Figure 6.11 gives a summary of the distribution of the randomized test statistic Q_ε. The value $Q = 6.3243$ is represented by the dashed vertical line on that figure. The resulting *p-value* is 0.237. Thus, since this *p-value* is large, we do not have enough evidence to reject the null hypothesis, H_0, that these three devices have the same process FTE.

We note here that the distribution of Q_ε in the above example takes a small set of values. This results from having a few errors for each of the devices, and so, there are only a small number of values that Q_ε can take.

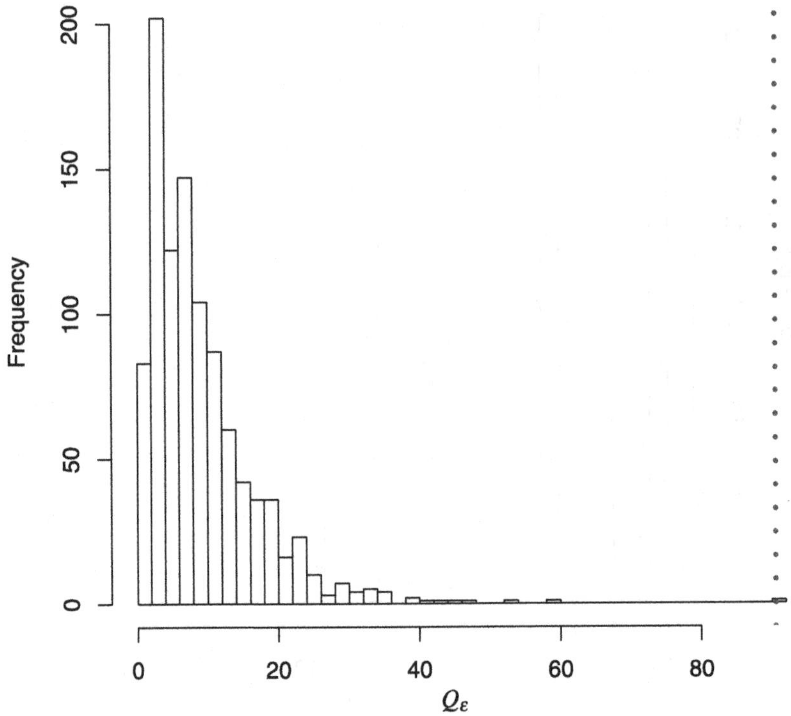

Fig. 6.12 Distribution of the randomized test statistic Q_ε for comparing four fingerprint devices from the FTE-Synth database

Example 6.20 To illustrate the randomization approach above, we will utilize data from the FTE-Synth database which was described in Sect. 1.5.2. For this test we will consider the four fingerprint devices: FP1, FP2, FP3, and FP4 from that database. Then our hypothesis test is

$$H_0 : \varepsilon_1 = \varepsilon_2 = \varepsilon_3 = \varepsilon_4$$
$$H_1 : \text{ not } H_0.$$

For those four devices, there were $n_\varepsilon = 628$ individuals who had enrolment attempts on each device. The sample FTE's for each is given by $\hat{\varepsilon}_1 = 0.0239$, $\hat{\varepsilon}_2 = 0.0318$, $\hat{\varepsilon}_3 = 0.0032$ and $\hat{\varepsilon}_4 = 0.0525$. Following the randomization algorithm above for testing paired FTE's, we generated $M = 999$ replicate data sets. Figure 6.12 has a histogram of the values for Q_ε generated from this process. The value for Q that we obtain from analyzing the original data is $Q = 90.4933$ which is represented by the dashed vertical line on the histogram in Fig. 6.12. The *p-value* that we obtain for that distribution is $p = 0.001$. Therefore, since that *p-value* is small, we will reject the null hypothesis and conclude that there are significant differences between the FTE's of these devices.

6.3 Sample Size and Power Calculations

One of the common questions asked by testers and evaluators of biometric identification devices is 'How large of a sample do I need?' Statistically, this is a question that involves the inversion of the confidence intervals and hypothesis tests described in previous sections. That is, it is necessary to take the formulae given above and convert them to calculations in terms of a sample size. In this section, we will derive the sample size calculations needed for two types of outcomes: a confidence interval for a single FTE and a hypothesis test for a single FTE. As is the case for all sample size calculations, it is necessary to posit *a priori* estimates for some quantities that comprise the confidence interval. These estimates can be obtained in several ways and these are discussed in Sect. 2.3.4. It is possible (and straightforward) to derive sample size calculations for the comparison of two FTE's but we will not do so in this book. We begin with the sample size calculation for a confidence interval.

6.3.1 Sample Size Calculations

In this section, we would like to determine the number of individuals that are needed to create a $100(1-\alpha)\%$ confidence interval for an FTE with a margin of error of B. To derive the sample size needed to obtain such a confidence interval, we must first recall the formula for the desired confidence interval. Here a $100(1-\alpha)\%$ confidence interval for an FTE is

$$\hat{\varepsilon} \pm z_{\alpha/2}\sqrt{\frac{\hat{\varepsilon}(1-\hat{\varepsilon})}{n_{\varepsilon}}}. \tag{6.54}$$

Setting the margin of error for this interval to be B, we get

$$B = z_{\alpha/2}\sqrt{\frac{\hat{\varepsilon}(1-\hat{\varepsilon})}{n_{\varepsilon}}}. \tag{6.55}$$

Next, we solve the previous equation for n_{ε} or, simply, n,

$$n = \left\lceil \frac{z^2_{\alpha/2}\varepsilon(1-\varepsilon)}{B^2} \right\rceil. \tag{6.56}$$

Here $\lceil\,\rceil$ represent the ceiling function or the 'next largest integer' function. In order to evaluate the right hand side of (6.56), we must specify a value for ε before the data is collected. As discussed in Chap. 2, there are several ways to proceed. The best way is to take a small sample of enrolment decisions from the system of interest and use the sample FTE obtained from the pilot study in (6.56). Alternatively, an estimate for ε from another similar data collection would be reasonable choice.

6.3.2 Power Calculations

Power calculations correspond to the sample size needed to obtain a certain power, $1-\beta$ for a fixed significance level, α, assuming a particular alternative hypothesis. More concretely, this means that a given sample size will ensure that we reject the null hypothesis (in favor of the alternative hypothesis) with probability $1-\beta$ for a particular alternative hypothesis. In order to carry out a power calculation, it is necessary to specify an exact alternative to the null hypothesis. This means that rather than stating that we would like to test a hypothesis of $\varepsilon = 0.05$ against an alternative of less than $\varepsilon < 0.05$ we must determine an exact value for the alternative, say $\varepsilon_a = 0.035$ or $\varepsilon_a = 0.02$. Additionally, the desired power of the test $1-\beta$ or the of rejecting the null hypothesis when $\varepsilon = \varepsilon_a$ is true needs to be specified.

Recall that for a large sample hypothesis test of a single FTE, we reject when

$$\frac{\hat{\varepsilon} - \varepsilon_0}{\sqrt{\frac{\varepsilon_0(1-\varepsilon_0)}{n}}} < -z_\alpha \tag{6.57}$$

where z_α is the $1-\alpha$th percentile of a Gaussian distribution. We would like to reject the null hypothesis, H_0, with probability $1-\beta$ when $\varepsilon = \varepsilon_a$. That is,

$$P\left(\frac{\hat{\varepsilon} - \varepsilon_0}{\sqrt{\frac{\varepsilon_0(1-\varepsilon_0)}{n}}} < -z_\alpha \middle| \varepsilon = \varepsilon_a\right) = 1-\beta \tag{6.58}$$

where $\varepsilon_a < \varepsilon_0$. We can write this probability statement as

$$P\left(\hat{\varepsilon} < \varepsilon_0 - z_\alpha\sqrt{\frac{\varepsilon_0(1-\varepsilon_0)}{n}} \middle| \varepsilon = \varepsilon_a\right) = 1-\beta. \tag{6.59}$$

Then,

$$P\left(\frac{\hat{\varepsilon} - \varepsilon_a}{\sqrt{\frac{\varepsilon_a(1-\varepsilon_a)}{n}}} < \frac{\varepsilon_0 - z_\alpha\sqrt{\frac{\varepsilon_0(1-\varepsilon_0)}{n}} - \varepsilon_a}{\sqrt{\frac{\varepsilon_a(1-\varepsilon_a)}{n}}}\right) = 1-\beta. \tag{6.60}$$

With $\varepsilon = \varepsilon_a$, the left of the inequality in (6.60) is a standard Gaussian random variable and we can then equate the right hand side of that equation to the relevant percentile of that distributions. That is,

$$z_\beta = \frac{\varepsilon_0 - z_\alpha\sqrt{\frac{\varepsilon_0(1-\varepsilon_0)}{n}} - \varepsilon_a}{\sqrt{\frac{\varepsilon_a(1-\varepsilon_a)}{n}}}. \tag{6.61}$$

We can then solve the equation above for n, which gives us

$$n = \left\lceil \left(\frac{z_\alpha\sqrt{\varepsilon_0(1-\varepsilon_0)} + z_\beta\sqrt{\varepsilon_A(1-\varepsilon_A)}}{\varepsilon_0 - \varepsilon_A}\right)^2 \right\rceil. \tag{6.62}$$

6.4 Prediction Intervals

Prediction intervals are a different type of statistical interval from confidence intervals. The focus of confidence intervals is the estimation of a parameter for an entire process, while the focus of prediction intervals is future observations. The variability in a prediction interval is larger because it includes variability in estimating the process parameter as well as the variability for the future observations from that process. We will denote unobserved future observations using the symbol $^\diamond$. Because most tests of biometric devices are meant to be analytic rather than enumerative in the sense of Hahn and Meeker [43], prediction intervals can be a very useful tool. Suppose that we have collected failure to enrol data on $n_\varepsilon = 100$ individuals and we would like to be able to estimate the FTE on some future set of $n_\varepsilon^\diamond = 500$ individuals collected under similar circumstances. In that case, a prediction interval would be an appropriate tool. For the FTE, ε, of a biometric system, we can find the variance of the FTE of $n_\varepsilon^\diamond$ future observations by using the following:

$$V[\hat{\varepsilon}^\diamond] = \hat{\varepsilon}(1-\hat{\varepsilon})\left(\frac{1}{n_\varepsilon^\diamond} + \frac{1}{n_\varepsilon}\right), \tag{6.63}$$

where

$$\varepsilon^\diamond = \frac{\sum_{i=1}^{n_\varepsilon^\diamond} E_i^\diamond}{n_\varepsilon^\diamond} \tag{6.64}$$

is the future FTE based upon $n_\varepsilon^\diamond$ enrolment decisions. Each future enrolment decision is denoted by $E_i^\diamond$.

Having calculated the variance for a predicted interval, it is possible to calculate the prediction interval, a confidence interval for the predicted sample FTE, $\hat{\varepsilon}^\diamond$. Thus a $100(1-\alpha)\%$ prediction interval for a future FTE with a sample size of $n^\diamond$ based upon an observed sample of size n_ε is the following:

$$\hat{\varepsilon} \pm z_{\alpha/2}\sqrt{\hat{\varepsilon}(1-\hat{\varepsilon})\left(\frac{1}{n_\varepsilon^\diamond} + \frac{1}{n_\varepsilon}\right)}. \tag{6.65}$$

The interval above is appropriate to use when we have a sufficiently large sample. The conditions for the use of the above formula as an appropriate prediction interval are based upon having a sufficiently large sample size. This means that we need to ensure that n_ε is sufficiently large. As with the large sample *confidence interval*, we need $n_\varepsilon\hat{\varepsilon} \geq 10$ and $n_\varepsilon(1-\hat{\varepsilon}) \geq 10$. With those conditions met, the sampling distribution for $\hat{\varepsilon}$ will approximately follow a Gaussian distribution. We also assume here that the enrolment process is stationary across both data collections.

Example 6.21 Our goal in this example is to create a 95% prediction interval for the IrisGuard device from the from the International Biometrics Group Comparative Biometrics Testing Round 7 [41]. We previously made a large sample 95% confidence interval for this device in Example 6.1. Suppose we would like to create a 95% prediction interval for $n_\varepsilon^\diamond = 250$ future enrolments. In the original test, the IrisGuard device had an estimated FTE of $\hat{\varepsilon} = 0.0174$, based upon was 18 individuals that were unable to enrol from $n_{\varepsilon} = 1035$ enrolment decisions. Since we checked the conditions for using the large sample confidence interval for a single FTE in Example 6.1, we will not redo those calculations here. Then our 95% prediction interval is

$$0.0174 \pm 1.960\sqrt{0.0174(1-0.0174)\left(\frac{1}{1035}+\frac{1}{250}\right)} \tag{6.66}$$

which yields a 95% prediction interval of $(-0.0007, 0.0354)$. Truncating this interval to only plausible, positive values, we can be 95% confident that the future FTE, $\hat{\varepsilon}^{\diamond}$, from this process will be between 0.0% and 3.54%.

6.5 Discussion

The focus of this chapter has been statistical methods for failure to enrol rates. We have presented confidence intervals and hypotheses tests for estimation and testing of one or more FTE's. Using large sample, bootstrap and randomization approaches, we have proffered a range of methodologies. In addition to these, we have introduced a sample size calculation, a power calculation and a prediction interval for a single FTE. All of this work has been done assuming that enrolment decisions between individuals are uncorrelated conditional on the process FTE. As is the case throughout this book, we have aimed to provide a basic foundation for statistical decision making rather than to be exhaustive in our methods.

We prefer the use of confidence intervals for inference regarding FTE's rather than hypothesis tests since more information is gained from knowing the range of values that is the output of a confidence interval than from knowing whether a hypothesis is rejected, even if the *p-value* is reported. Further, we recommend using at least $M \geq 999$ replicates when applying a bootstrap or randomization approach to the analysis of FTE's.

Finally, we note that there has been a series of recent articles in the statistics literature on estimation of binomial proportions or rates. These new methods suggest some adjustments to the methods described here that may be appropriate for FTE's given an uncorrelated structure for enrolment decisions. The interested reader can find more details in work by Agresti and Coull [4] and Newcombe [72].

Chapter 7
Failure to Acquire

In this chapter we are focusing on the failure to acquire rate or FTA. The FTA is a measure of the ability of a given sensor to collect an appropriate image or biometric signal in a given environment. The FTA is often used to quantify how well a particular system does at the job of collecting quality images. As a consequence of this, our measurements here are made on a particular presentation of a biometric. These measurements are in contrast to the false match and false non-match rates—FMR and FNMR—which are based upon comparisons of two signals. If a system is unable to acquire images of sufficient quality to match, it does not matter how well the matching component of the system performs. Thus, the FTA is an important measure of the performance of a biometric system.

There has been little research on statistical methodology for FTA. It has certainly not received the attention that FMR's and FNMR's have. However, there have been some recent attempts to study and analyze failure to acquire rates. Elliot and Kukula [30] have created a framework for parsing failure to acquire errors. Kukula [53] applied advanced statistical methods, a multi-way analysis of variance, to compare different FTA's in a variety of circumstances. The approach taken in these articles is to treat the acquisition decisions as though they are uncorrelated. Schuckers [85] proposed a correlation structure for failure to acquire decisions that is more general and is the approach that we will follow. The basics of the structure are that we will treat acquisition attempts by the same person as correlated but acquisition attempts between different individuals as uncorrelated. All of that structure is conditional on the FTA process being stationary in the sense given by Definition 2.23. We will let γ represent the process FTA for a given system. Therefore, inference about γ will be the focus of our interest in this chapter. As we do throughout this book, we will treat each sample as coming from an ongoing biometric process rather than from a fixed population. Consequently, the methods and the examples below will be about trying to make inference and to make decisions about such processes from a sample or from samples of those processes.

In the next section, we present the notation that we will use in this chapter as well as the basic correlation structure for acquisition decisions. We then discuss statistical methods for a single FTA, γ. Both confidence intervals and hypothesis

M.E. Schuckers, *Computational Methods in Biometric Authentication*,
Information Science and Statistics,
DOI 10.1007/978-1-84996-202-5_7, © Springer-Verlag London Limited 2010

tests are given. Methods for comparing two and then three or more FTA's are given in thereafter. We present large sample, bootstrap and randomization approaches to inference about multiple FTA's. Sample size calculations and power calculations are presented next. We end this chapter with a section on prediction intervals for an FTA and a discussion section reiterating the results for this chapter.

7.1 Notation and Correlation Structure

The fundamental unit for a given FTA is the acquisition decision that come from the biometric system. Each acquisition attempt is either successful or not. Therefore, each acquisition decision is binary. We will denote each acquisition attempt by $A_{i\kappa}$ for the κth attempt by the ith individual where $\kappa = 1, \ldots, K_i$ and $i = 1, \ldots, n$. We will assume that there are n_A individuals who attempt acquisition and that each of these individuals makes K_i attempts. Formally then,

$$A_{i\kappa} = \begin{cases} 1 & \text{if the } \kappa\text{th acquisition attempt by} \\ & \text{individual } i \text{ is not acquired} \\ 0 & \text{otherwise.} \end{cases} \tag{7.1}$$

We estimate γ, the FTA, in the following way:

$$\hat{\gamma} = \frac{\sum_{i=1}^{n_A} \sum_{\kappa=1}^{K_i} A_{i\kappa}}{\sum_{i=1}^{n_A} K_i}. \tag{7.2}$$

Our estimated FTA from (7.2) is an average of the acquisition attempts and, hence, it is a linear combination of those same attempts. In order to perform statistical estimation regarding γ, we need to derive the estimated variance based upon observed data.

The correlations structure described above can then be given by

$$Corr(A_{i\kappa}, A_{i'\kappa'}) = \begin{cases} 1 & \text{if } i = i', \kappa = \kappa' \\ \psi & \text{if } i = i', \kappa \neq \kappa' \\ 0 & \text{otherwise.} \end{cases} \tag{7.3}$$

This is the correlation structure proposed by Schuckers [85]. Table 7.1 contains an example of failure to acquire decision data. The intra-individual correlation is represented by the parameter ψ. ψ plays an equivalent role to the one played by ϱ for the false non-match rate, FNMR, cf. (3.3). This correlation structure is known as a 'repeated measures' correlation structure.

In order to estimate γ appropriately, we need to find the variance of our estimate of γ, $V[\hat{\gamma}]$. We can write the variance of the estimated FTA, $\hat{\gamma}$ as

$$V[\hat{\gamma}] = V\left[\frac{\sum_{i=1}^{n_A} \sum_{\kappa=1}^{K_i} A_{i\kappa}}{\sum_{i=1}^{n_A} K_i}\right] = V[N_\gamma^{-1}\mathbf{1}^T\mathbf{A}] = N_\gamma^{-2}V[\mathbf{1}^T\mathbf{D}]$$

Table 7.1 Example of FTA data

i	κ	$A_{i\kappa}$
1	1	0
1	2	0
1	3	0
1	4	1
2	1	0
2	2	0
2	3	0
3	1	0
3	2	1
3	3	1
3	4	0
3	5	0
4	1	0
4	2	0
5	1	0
5	2	0
5	3	0
6	1	0
6	2	0

$$= N_\gamma^{-2}\mathbf{1}^T\boldsymbol{\Sigma}_\gamma\mathbf{1} = N_\gamma^{-2}\gamma(1-\gamma)\mathbf{1}^T\boldsymbol{\Phi}_\gamma\mathbf{1}$$

$$= N_\gamma^{-2}\gamma(1-\gamma)\left[N_\gamma + \psi\sum_{i=1}^{n} K_i(K_i-1)\right] \tag{7.4}$$

where $\boldsymbol{\Sigma}_\gamma = Var[\mathbf{A}]$, $N_\gamma = \sum_{i=1}^{n_A} K_i$, $\boldsymbol{\Phi}_\gamma = Corr(\mathbf{A})$, $\mathbf{1} = (1, 1, \ldots, 1)^T$, $\hat{\gamma}$ can be written as $(N_\gamma)^{-1}\mathbf{1}^T\mathbf{A}$ and $\mathbf{A} = (A_{11}, \ldots, A_{1m_1}, A_{21}, \ldots, A_{2m_2}, \ldots, A_{n1}, \ldots, A_{n_A m_{n_A}})^T$. The covariance matrix is denoted by $\boldsymbol{\Sigma}_\gamma$ and the correlation matrix is denoted by $\boldsymbol{\Phi}_\gamma$. Writing the estimated FTA in this manner is done as a reminder that the FTA is a linear combination and that statistical methods for linear combinations apply to the FTA. An estimator for ψ is given by

$$\hat{\psi} = \frac{\sum_{i=1}^{n_A}\sum_{\kappa=1}^{K_i}\sum_{\kappa'=1,\kappa'\neq\kappa}^{K_i}(A_{i\kappa}-\hat{\gamma})(A_{i\kappa'}-\hat{\gamma})}{\hat{\gamma}(1-\hat{\gamma})\sum_{i=1}^{n_A}K_i(K_i-1)}. \tag{7.5}$$

Following the correlation structure given in (7.3), we can derive the explicit equation given above. An estimated variance, $\hat{V}[\hat{\gamma}]$, can be calculated by substituting estimates of the parameters γ and ψ into (7.4). The variance above, (7.4), can be simplified to

$$(n_A K)^{-1}\gamma(1-\gamma)[1+\psi(K-1)] \tag{7.6}$$

if $K_i = K$ for all i, i.e. if all individuals have the same number of attempted acquisitions. Several authors including Fleiss et al. [33] have suggested a simplified version of the result in (7.4) that is similar in structure to (7.6) when the K_i's are unequal. This is

$$(n_A\bar{K})^{-1}\gamma(1-\gamma)[1+\psi(K_0-1)] \tag{7.7}$$

where

$$K_0 = \frac{\sum_{i=1}^{n_A} K_i^2}{\sum_{i=1}^{n_A} K_i} \quad \text{and} \quad \bar{K} = \frac{1}{n_A}\sum_{i=1}^{n_A} K_i. \tag{7.8}$$

In addition to γ, we need to be able to estimate ψ in order to use the confidence intervals and hypothesis tests given below. ψ is a measure of similarity within an individual's acquisition attempts. This sort of binary *repeated measures* correlation structure has been studied extensively by statisticians since it is one that is common. We will thus use an estimator proposed by Fleiss et al. [33]. This is

$$\hat{\psi} = \left(\hat{\gamma}(1-\hat{\gamma})\sum_{i=1}^{n_A} K_i(K_i-1)\right)^{-1} \sum_{i=1}^{n_A}\sum_{\kappa=1}^{K_i}\sum_{\substack{\kappa'=1\\ \kappa'\neq\kappa}}^{K_i}(A_{i\kappa}-\hat{\gamma})(A_{i\kappa'}-\hat{\gamma}). \tag{7.9}$$

Due to the correlation system here, we do not have independence of each acquisition attempt. To quantify the amount of information in a particular set of FTA observations, we use the effective sample size. For a single FTA, the *effective sample size*, $N_\gamma^\dagger$, can be written as

$$N_\gamma^\dagger = \left\lfloor \frac{N_\gamma}{1+\hat{\psi}(K_0-1)} \right\rfloor \tag{7.10}$$

where

$$N_\gamma = \sum_{i=1}^{n_A} K_i. \tag{7.11}$$

This follows from Definition 2.34. The amount of 'independent' information, $N_\gamma^\dagger$, is dependent upon the intra-individual correlation, ψ, and upon K_0. The effective sample size will play a prominent role below in ascertaining whether a sample is large enough for the use of a Gaussian distribution to approximate a sampling distribution. Note that for data with this type of correlation framework, although it is not independent, there is still a version of the Central Limit Theorem which applies. See Theorem 2.4.

7.2 Statistical Methods

Having established calculations for estimating a process FTA, γ, and for estimating the variance of our estimate, we turn to presenting statistical methodology for a single FTA or for multiple FTA's. We will assume that for a given group the acquisition process is stationary. Below we give methods for comparing multiple processes or even for comparing whether the same FTA applies to a particular set of processes. We start with methodology for a single FTA. Subsequently, we propose methodology for comparing two or more FTA's. Large sample methods as well as bootstrap and randomization methods are given.

7.2.1 One Sample Methods for FTA

We begin by investigating approaches when the focus is a single failure to acquire rate. Assuming that we are dealing with a single stationary acquisition process, then we can apply the methods described below. For a single FTA, we will consider both large sample approaches and bootstrap resampling approaches for confidence intervals and hypothesis tests. The bootstrap methodology we use is the *subsets bootstrap* that was proposed by Bolle et al. [9]; however, they did not propose it for failure to acquire data. It is the appropriate choice given that we have the same repeated measures correlation structure here as we had for the false non-match rate covered in Chap. 3.

7.2.1.1 Confidence Interval for Single FTA

Two methods are presented in this section for making a confidence interval for a single process FTA. The first is a large sample approach. This method depends upon having a sample that is large enough to assume that the sampling distribution of $\hat{\gamma}$ is approximately Gaussian. The second approach is a bootstrap one.

Large Sample Approach If $N_\gamma^\dagger$ is large (generally $N_\gamma^\dagger \hat{\gamma} \geq 10$ and $N_\gamma^\dagger(1-\hat{\gamma}) \geq 10$, then we can use the following equation to make a $(1-\alpha)100\%$ confidence interval for γ.

$$\hat{\gamma} \pm z_{\frac{\alpha}{2}} \sqrt{\frac{\hat{\gamma}(1-\hat{\gamma})[1+(K_0-1)\hat{\psi}]}{n_A \bar{K}}}, \tag{7.12}$$

where

$$\hat{\gamma} = \frac{\sum_{i=1}^{n_A} \sum_{\kappa=1}^{K_i} A_{i\kappa}}{\sum_{i=1}^{n_A} K_i} \tag{7.13}$$

which is the fraction of observed acquisition errors to total acquisition attempts in the sample.

Example 7.1 We would like to make a 90% confidence interval for the failure to acquire on attempted acquisitions for the left index finger of Device 1 from the

PURDUE-FTA database. Attempts were obtained six times each for 186 individuals. Thus, there were $N_\gamma = 1116$ acquisition attempts. Accounting for the intra-individual correlation of $\hat{\psi} = 0.6771$, we have $N_\gamma^\dagger = 254.47$. The estimated FTA is $\hat{\gamma} = 0.0090$. In order to use the large sample approach above, we must have a sufficiently large sample. To check this, we considered $N_\gamma^\dagger \hat{\gamma} = 254.47(0.0090) = 2.37$ and $N_\gamma^\dagger(1 - \hat{\gamma}) = 254.47(1 - 0.0090) = 252.18$. Since the former value is not sufficiently large, i.e. $\ngeq 10$, we cannot use the above methodology to create our confidence interval. We will reconsider this set of acquisition decisions using a bootstrap approach in Example 7.5.

Example 7.2 In this example our focus is on Device 3 from the PURDUE-FTA database and the failure to acquire for the left middle finger. We will use the methods given above to make a 95% confidence interval for the FTA, γ. For this device, we have $N_\gamma = 1116$ acquisition decisions, six each from the $n_A = 186$ individuals. The estimated intra-individual correlation for these data is $\hat{\psi} = 0.6799$. Calculating the effective sample size, we find that $N_\gamma^\dagger = 253.66$. To ensure that we have a sufficient sample size to conclude that the sampling distribution of $\hat{\gamma}$ is approximately Gaussian, we need to have enough sample decisions so that $N_\gamma^\dagger \hat{\gamma} \geq 10$ and $N_\gamma^\dagger(1 - \hat{\gamma}) \geq 10$. In this case the former calculation gives us $253.66 * (0.0520) = 13.19$ and the latter gives us $253.66 * (1 - 0.0520) = 240.47$ where $\hat{\gamma} = 0.0520$. Both are clearly larger than 10; therefore, we can use the methods given above. Our 95% confidence interval is then $0.0520 \pm 1.960(0.0139)$ which gives us an interval of $(0.0247, 0.0793)$. With 95% confidence, the process FTA for the left middle finger for this device is between 2.47% and 7.93%.

Example 7.3 In this example, we will use information from Device 5 from the Purdue FTA database. In this case, we will be interested in making a 90% confidence interval for the failure to acquire, γ, for attempted acquisition on the middle finger of the left hand. There are $n_A = 186$ individuals and $N_\gamma = 1116$ acquisition decisions from those individuals. Each individual made six acquisition attempts. The relevant summary statistics here are $\hat{\gamma} = 0.0556$, $K_0 = 6$, $\hat{\psi} = 0.8361$ and $N_\gamma^\dagger = 215.43$. Since $N_\gamma^\dagger \hat{\gamma} = 11.97$ and $N_\gamma^\dagger(1 - \hat{\gamma}) = 203.47$, we have a sufficient number of samples to infer that the sampling distribution of $\hat{\gamma}$ is approximately Gaussian. Then, following (7.12), we obtain $0.0556 \pm (1.645)0.0156$ which yields an interval of $(0.0299, 0.0812)$. We can then be 90% confident that the process FTA for this device applied to the left middle finger is between 2.99% and 8.12%.

Bootstrap Approach Here we propose a bootstrap approach to making a confidence interval for an FTA. The bootstrap approach does not depend upon the large sample size for making confidence intervals. This is in marked contrast to the large sample methods given above. Our methodology follows the algorithm given below.

1. Calculate $\hat{\gamma}$ from the observed complete data.
2. Sample n_A individuals *with replacement* from the n_A individuals from which there are acquisition attempts. Denote these selected individuals by b_1, $b_2, \ldots, b_{n_A}$.

3. For each selected individual, b_i, take all K_{b_i} acquisition attempts. Call these selected acquisition attempts, $A_{b_i 1}, A_{b_i 2}, \ldots, A_{b_i K_{b_i}}$'s.
4. Calculate and store $e_\gamma = \hat{\gamma}^b - \hat{\gamma}$ where

$$\hat{\gamma}^b = \frac{\sum_{i=1}^{n_A} \sum_{\kappa=1}^{K_{b_i}} A_{b_i \kappa}}{\sum_{i=1}^{n_A} K_{b_i}}. \tag{7.14}$$

5. Repeat the three previous steps some large number of times M, each time storing the differences, the e_γ's, from the estimated FTA, $\hat{\gamma}$.
6. A $100(1-\alpha)\%$ confidence interval for the FTA, γ, is then formed by taking the interval

$$(\hat{\gamma} - e_U, \hat{\gamma} - e_L) \tag{7.15}$$

where e_L and e_U represent the $\alpha/2$th and the $1-\alpha/2$th percentiles of the distribution of e_γ's.

Example 7.4 Here we apply the bootstrap approach to acquisitions for the left index finger attempts using Device 5 from the PURDUE-FTA database. We attempted to make a confidence interval for this FTA using a large sample approach in Example 7.1. However, under our criteria, that approach requires larger samples than were available. Consequently, we will use the bootstrap approach. We bootstrapped the FTA here $M = 5000$ times. Our estimated FTA is $\hat{\gamma} = 0.0547$. To make a 99% confidence interval for the process FTA, we take the 0.5th and 99.5th percentiles of the bootstrapped differences, the $e_\gamma = \hat{\gamma}^b - \hat{\gamma}$'s, and subtract them from our estimated FTA. Here that means that the endpoints for our confidence interval are 0.0099 and 0.0923. Figure 7.1 summarizes the sampling distribution of e_γ. We can then conclude with 99% confidence that the process FTA here is between 0.99% and 9.23%.

Example 7.5 Here we apply the bootstrap approach to acquisitions for the left index finger attempts using Device 5 from the PURDUE-FTA database. We attempted to make a confidence interval for this FTA using a large sample approach in Example 7.1. However, using our criteria, our sample size was insufficient for that methodology. Consequently, we have no choice but to use the bootstrap approach in this case. We bootstrapped the FTA here $M = 1000$ times. Our estimated FTA is $\hat{\gamma} = 0.0090$. Figure 7.2 gives a summary of the distribution of the bootstrapped differences, the e_γ's, from the sample FTA using the algorithm given above. For our 90% confidence interval, we find that the appropriate limits are -0.0036 and 0.0161. Since it is not possible to have a negative FTA, we will truncate this interval to be $(0.0000, 0.0161)$. Thus, we can be 90% confident that the FTA for Device 5 from the PURDUE-FTA database applied to the left index finger is below 1.61%.

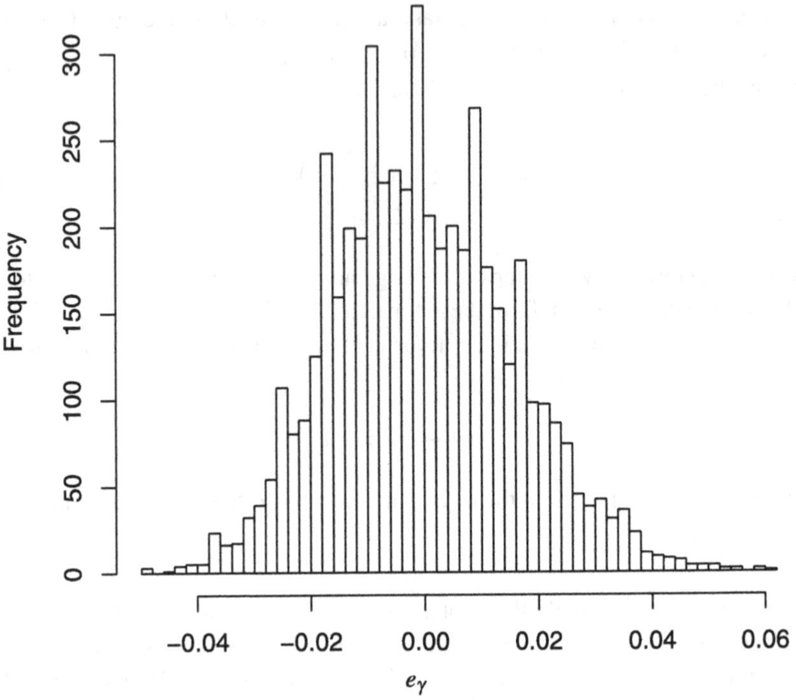

Fig. 7.1 Distribution of e_γ, the bootstrapped differences from the sample left index finger failure to acquire rate for Device 5 from the PURDUE-FTA database

7.2.1.2 Hypothesis Test for Single FTA

In this section, we are testing whether a particular device or a particular process has an FTA significantly below a set bound. This necessitates a less than alternative hypothesis, H_1. The other typical alternative hypotheses—greater than *and* not equal to—are possible and discussed in Sect. 2.3.3. However, the usual role of biometrics testing is to verify that an error rate is significantly below a particular value. Thus, we will focus on the following hypothesis test.

$$H_0 : \gamma = \gamma_0$$
$$H_1 : \gamma < \gamma_0.$$

First, we provide a large sample approach for testing these hypotheses. That approach is followed by a bootstrap method for testing the same hypotheses.

Large Sample Approach The large sample approach here is dependent upon having a sample of sufficient size. For this significance test, we need the $N_\gamma^\dagger \gamma_0 \geq 10$ and $N_\gamma^\dagger(1 - \gamma_0) \geq 10$. Having met those conditions, the test below allows us to evaluate whether a sample FTA is significantly below the value γ_0.

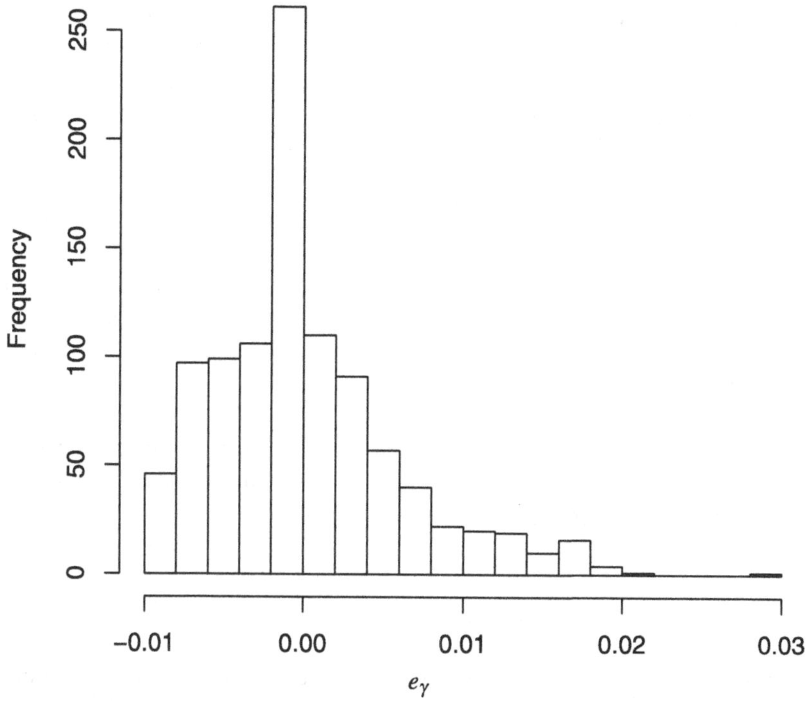

Fig. 7.2 Distribution of e_γ, bootstrapped differences from the right index finger failure to acquire rate for Device 1 from the PURDUE-FTA database

Test Statistic:

$$z = \frac{\hat{\gamma} - \gamma_0}{\sqrt{\frac{\gamma_0(1-\gamma_0)(1+\hat{\psi}(K_0-1))}{n_A \bar{K}}}}. \tag{7.16}$$

p-value: $p = P(Z < z)$ where Z is a standard Gaussian random variable and this probability can be calculated from Table 9.1.

We will reject the null hypothesis, H_0, that $\gamma = \gamma_0$ if the *p-value* is small or for a given significance level, α, if $p < \alpha$.

Example 7.6 The first case that we will use to illustrate our methodology is testing whether the FTA for left middle fingers from Device 3 of the PURDUE-FTA database is significantly less than 0.10. The sample FTA is $\hat{\gamma} = 0.0520$. We will test if that value is significantly less than 10% and this test is formally

$$H_0 : \gamma = 0.10$$

$$H_1 : \gamma < 0.10.$$

There are $n_A = 186$ individuals and each had $K_i = K_0 = 6$ acquisition attempts for all individuals i. We need to ensure that the total number of decisions here is sufficiently large for the use of the Gaussian distribution. The required criteria are that $N_\gamma^\dagger \gamma_0 \geq 10$ and $N_\gamma^\dagger(1-\gamma_0) \geq 10$. Here we have an effective sample size of $N_\gamma^\dagger = 253.66$. Using $\gamma_0 = 0.10$, we have that $N_\gamma^\dagger \gamma_0 = 25.37 \geq 10$ and $N_\gamma^\dagger(1-\gamma_0) = 228.29 \geq 10$. Thus, we can proceed with our test. The estimated intra-individual correlation is $\hat{\psi} = 0.6799$. Then our test statistics is

$$z = \frac{0.0520 - 0.10}{0.0188} = -2.548. \tag{7.17}$$

From this, our *p-value* is calculated to be $p = P(Z < -2.548) \approx 0.0054$. This *p-value* is small and, therefore, we will reject the null hypothesis that $\gamma = 0.10$. We can then conclude that the process FTA under consideration here is significantly less than 10%.

Example 7.7 We next wish to test whether the FTA for Device 5 from the PURDUE-FTA database is less than 0.05 when applied to middle fingers from the right hand at a significance level of $\alpha = 0.10$. Formally, our test is

$$H_0 : \gamma = 0.05$$
$$H_1 : \gamma < 0.05.$$

Our estimated FTA is $\hat{\gamma} = 0.0484$ and the estimated intra-individual correlation is $\hat{\psi} = 0.8755$. Six acquisition attempts—$K_i = K_0 = 6$ for all i—are made by each of the $n_A = 186$ individuals. Before we can carry out this test, we need to ensure that it is reasonable to suppose that the sampling distribution of $\hat{\gamma}$ approximately follows a Gaussian distribution. To do this, we need to check that there are a sufficient number of samples to carry out this calculation. Since we have $N_\gamma^\dagger = 207.54$ and our estimated FTA is 0.0484, then our criteria are met and we can proceed. The test statistic for this test is

$$z = \frac{0.0484 - 0.05}{0.0151} = -0.1058. \tag{7.18}$$

Using Table 9.1, we find the *p-value* to be $p = P(Z < -0.1058) \approx 0.4562$ which is larger than our significance level $\alpha = 0.10$. Consequently, we fail to reject the null hypothesis that the FTA here is 0.05 and conclude that this FTA is not significantly less than the 0.05.

Bootstrap Approach Since the process of FTA evaluation involves collecting individuals and then taking repeated samples from those individuals there are two sources of variability in this process. Both stages are captured by the correlation structure given in (7.3). To mirror these, we will use the *subsets bootstrap*. As part of any hypothesis test, we assume the null hypothesis is true until we have sufficient evidence otherwise. Thus, we must use a sampling distribution centered at the hypothesized value in the null hypothesis. We will adjust the bootstrapped proportions accordingly in the procedure given below.

1. For the observed data, calculate

$$\hat{\gamma} = \frac{\sum_{i=1}^{n_A} \sum_{\kappa=1}^{K_i} A_{i\kappa}}{\sum_{i=1}^{n_A} K_i}. \tag{7.19}$$

2. Randomly select *with replacement* n_A individuals from the list of individuals $\{1, 2, \ldots, n_A\}$ and call the selected individuals $b_1, b_2, \ldots, b_{n_A}$'s.
3. For each selected individual b_i, bootstrap take all K_{b_i} acquisition attempts for that individual. We will refer to these acquisition attempts as $A_{b_i\kappa}$'s where $\kappa = 1, \ldots, K_{b_i}$.
4. Calculate and store

$$e_\gamma = \hat{\gamma}^b - \hat{\gamma} + \gamma_0 \tag{7.20}$$

where

$$\hat{\gamma}^b = \frac{\sum_{i=1}^{n_A} \sum_{\kappa=1}^{K_{b_i}} A_{b_i\kappa}}{\sum_{i=1}^{n_A} K_{b_i}}. \tag{7.21}$$

5. Repeat the previous three steps some large number of times M.
6. The *p-value* for this test is then the percentage of times that $e_\gamma \leq \hat{\gamma}$ out of $M+1$, which is the total number of $\hat{\gamma}$'s that we have observed or generated. This *p-value* is then

$$p = \frac{1 + \sum_{\varsigma=1}^{M} 1_{\{e_\gamma \leq \hat{\gamma}\}}}{M+1}. \tag{7.22}$$

7. We will reject the null hypothesis, $H_0 : \gamma = \gamma_0$, if the *p-value* is small. Small can be defined as $p < \alpha$ if a particular significance level α is specified.

The intuition for this test is that the distribution of e_γ is an approximation to the sampling distribution for the FTA, $\hat{\gamma}$, if we assume that the null hypothesis is true. This is the reason that we take the resampled FTA, $\hat{\gamma}^b$ and recenter it by subtracting the estimated FTA, $\hat{\gamma}$, and adding the hypothesized FTA, γ_0.

Example 7.8 In this example, we are testing whether the FTA for Device 5 applied to the right index finger of the users in the PURDUE-FTA database is significantly less than 0.05. The observed FTA is $\hat{\gamma} = 0.0341$ based upon 1116 acquisition decisions from $n_A = 186$ individuals. Following the algorithm given above, we bootstrapped $M = 1000$ times these failure to acquire decisions. The distribution of the e_γ's is given in Fig. 7.3. In that figure, the vertical dashed line represents the estimated FTA, $\hat{\gamma} = 0.0341$. The *p-value* is $p = 0.1030$ which is the percent of the e_γ's that are equal to or below $\hat{\gamma}$. Since that *p-value* is large, we can conclude that there is not enough evidence to reject the null hypothesis that the FTA is 0.0500. Thus, the FTA for this device is not significantly less than 5%.

Example 7.9 To illustrate the methodology here, we will test whether Device 1 from the PURDUE-FTA database applied to fingerprint acquisition decisions from

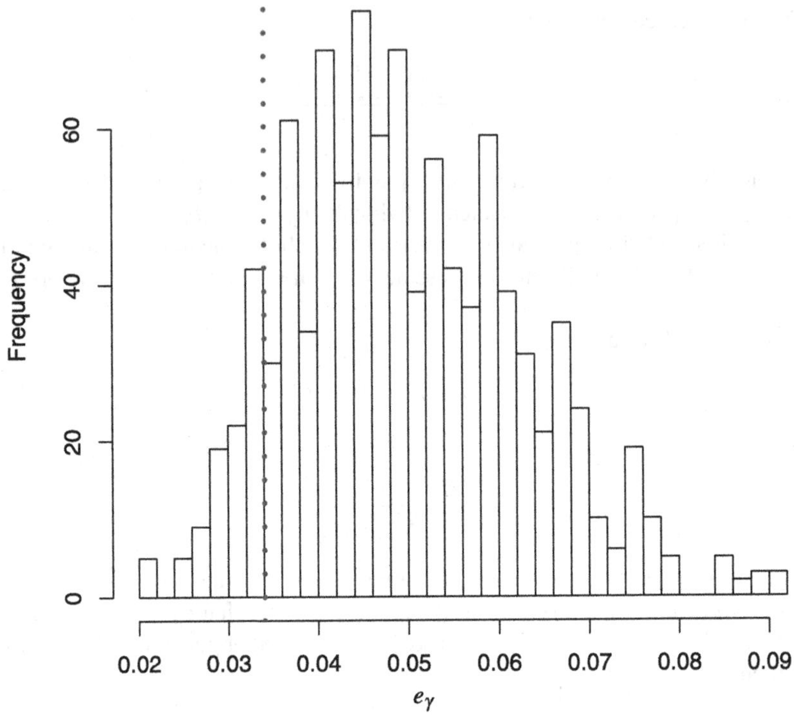

Fig. 7.3 Distribution of e_γ, the bootstrapped failure to acquire rate for Device 5 applied to the right index finger for individuals in the PURDUE-FTA database

attempts on the left middle finger has an FTA that this less than 5%. Formally, that hypothesis is

$$H_0 : \gamma = 0.05$$
$$H_1 : \gamma < 0.05.$$

We will test this hypothesis using a significance level of $\alpha = 0.02$. Our estimated FTA from this sample is $\hat{\gamma} = 0.0211$. Generating the bootstrapped approximation to the sampling distribution of $\hat{\gamma}$ assuming that $\gamma = 0.05$ yields the distribution of e_γ that is summarized in Fig. 7.4. We calculate the *p-value* as the percent of those bootstrapped e_γ's that are less than the estimated FTA, $\hat{\gamma}$, from the original data. The vertical dashed line at 0.0211 on this figure represents this value. Of the $M = 4999$ bootstrapped values, $p = 0.0002$ are less than or equal to our estimated FTA, 0.0211. Since $p < \alpha$, we reject the null hypothesis and conclude that the FTA here is significantly less than 5%.

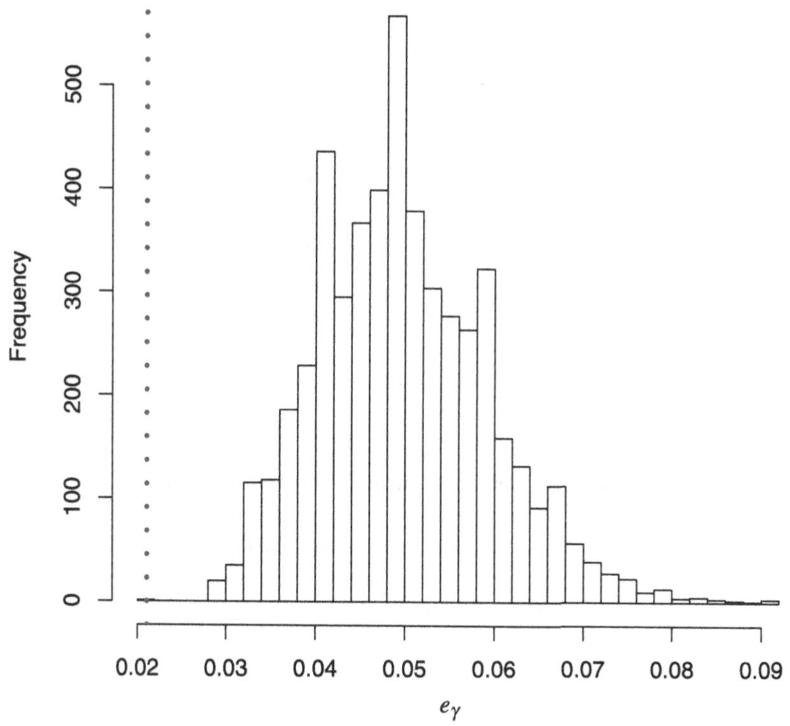

Fig. 7.4 Distribution of e_γ, the bootstrapped left middle finger failure to acquire rate for Device 1 from the PURDUE-FTA database

7.2.2 Two Sample Methods for FTA's

In this section, we present methodology for the statistical comparison of two FTA's. As in the previous chapters, we must consider how the data was collected. Focusing on data that are either paired or independent, we will present inferential methods for those two cases. More complicated data structures are beyond the scope of this text. Notationally, we will let γ_1 and γ_2 represent the two failure to acquire rates. Each is assumed to be mean stationary for that data considered to be analyzed. We use a subscript, $g = 1, 2$, to denote the different process FTA's. We will distinguish between other quantities for the two groups by a superscript of (g) for $g = 1, 2$ when a subscript is already in use, e.g. $N_\gamma^{(2)}$. Thus, the κth decision for the ith individual in group 1 is $A_{i\kappa}^{(1)}$ and the estimated intra-class correlation for the second group is $\hat{\psi}_2$. We propose methodology for both confidence interval and hypothesis tests for the difference of two FTA's, $\gamma_1 - \gamma_2$, below.

Table 7.2 Correlation parameter estimates for Example 7.10

Measure	Device 3 (Left Middle)	Device 5 (Left Index)
g	1	2
$n_A^{(g)}$	186	186
$N_\gamma^{(g)}$	1116	1116
K_0	6	6
$N_\gamma^{\dagger(g)}$	274.58	201.74
$\hat{\gamma}_g$	0.0520	0.0547
$\hat{\psi}$	0.6799	0.9064
$\sqrt{\hat{V}[\hat{\upsilon}_g]}$	0.0139	0.0160

7.2.2.1 Confidence Interval for Difference of Two Independent FTA's

In this section, we provide a confidence interval for the difference of two independent FTA's, $\gamma_1 - \gamma_2$. For example, if we want to compare the FTA's for men and women for the same iris system. The individuals in those two groups are distinct and, thus, fit the profile for using an approach for two independent FTA's. Below we present two methods: a large sample one and a bootstrapped based one.

Large Sample Approach If $N_\gamma^{(1)}$ and $N_\gamma^{(2)}$ are large meaning that $N_\gamma^{\dagger(g)}\hat{\gamma}_g \geq 10$ and $N_\gamma^{\dagger(g)}(1-\hat{\gamma}_g) \geq 10$ for $g = 1, 2$, then we can use (7.23) to make a $(1-\alpha)100\%$ confidence interval for the difference of two FTA's, $\gamma_1 - \gamma_2$.

$$\hat{\gamma}_1 - \hat{\gamma}_2 \pm z_{\frac{\alpha}{2}} \sqrt{\frac{\hat{\gamma}_1(1-\hat{\gamma}_1)[1+(K_0^{(1)}-1)\hat{\psi}_1]}{n_1\bar{K}_1} + \frac{\hat{\gamma}_2(1-\hat{\gamma}_2)[1+(K_0^{(2)}-1)\hat{\psi}_2]}{n_2\bar{K}_2}}, \tag{7.23}$$

where

$$\hat{\gamma}_g = \frac{\sum_{i=1}^{n_g}\sum_{\kappa=1}^{K_i^{(g)}} A_{i\kappa}^{(g)}}{\sum_{i=1}^{n_g} K_i^{(g)}}. \tag{7.24}$$

$K_0^{(g)}$ is the value of K_0 for the gth group, n_g is the number of individuals in the gth group and $A_{i\kappa}^{(g)}$ is the κth decision from the ith individual from the gth group.

Example 7.10 In this example, we want to make a confidence interval for the difference between the FTA's of two devices from the PURDUE-FTA database. We are comparing the FTA's of the left middle finger acquisition attempts for Device 3 to the left index finger acquisition attempts for Device 5. For the purposes of illustrating this example, we must assume that these decisions are independent, although they come from the same individuals. (We note that the correlation between FTA's here is not large, $Corr(\hat{\gamma}_1, \hat{\gamma}_2) = 0.0569$, so that this assumption is reasonably tenable.) Table 7.2 gives the summary statistics for these two devices. Using these summaries to create our 95% confidence interval for the difference here, we find that $\hat{\gamma}_1 - \hat{\gamma}_2 = -0.0027$. Our confidence interval is then calculated as

$$-0.0027 \pm 1.96\sqrt{\frac{0.0520(1-0.0520)[1+(6-1)0.6799]}{1116} + \frac{0.0547(1-0.0547)[1+(6-1)0.9064]}{1116}} \tag{7.25}$$

which gives an interval of $(-0.0443, 0.0389)$. Thus we can be 95% confident that the difference between the FTA's is between -4.43% and 3.89%.

Bootstrap Approach A bootstrap approach to making a confidence interval for the difference of two independent FTA's, $\gamma_1 - \gamma_2$ is found by separately resampling from each group and then taking the difference of the two bootstrap estimated FTA's. The bootstrap procedure that we describe is the same as we used for bootstrapping a single FTA but here we apply it to the two FTA's separately.

More formally, the procedure is

1. Calculate $\hat{\gamma}_1 - \hat{\gamma}_2$ from the observed data.
2. Sample n_1 individuals *with replacement* from the n_1 individuals in the first group from which there are acquisition decisions. Denote these selected individuals by $b_1^{(1)}, b_2^{(1)}, \ldots, b_{n_1}^{(1)}$.
3. For the selected individual, $b_i^{(1)}$, take all $K_{b_i^{(1)}}$ acquisition decisions from that selected individual. Call these selected acquisition decisions, $A_{b_i^{(1)}\kappa}$'s for $\kappa = 1, \ldots, K_{b_i}$. Do this for all individuals selected in the previous step.
4. Sample n_2 individuals *with replacement* from the n_2 individuals in the second group from which there are acquisition decisions. Denote these selected individuals by $b_1^{(2)}, b_2^{(2)}, \ldots, b_{n_2}^{(2)}$.
5. For each selected individual, $b_i^{(2)}$, take all $K_{b_i^{(2)}}$ acquisition decisions from that individual. Call these selected acquisition decisions, $A_{b_i^{(2)}\kappa}$'s for $\kappa = 1, \ldots, K_{b_i}^{(2)}$.
6. Calculate

$$e_\gamma = \hat{\gamma}_1^b - \hat{\gamma}_2^b - (\hat{\gamma}_1 - \hat{\gamma}_2) \tag{7.26}$$

where

$$\hat{\gamma}_g^b = \frac{\sum_{i=1}^{n_g} \sum_{\kappa=1}^{K_{b_i^{(g)}}} A_{b_i^{(g)}}}{\sum_{i=1}^{n_g} K_{b_i^{(g)}}} \tag{7.27}$$

for $g = 1, 2$.

7. Repeat the five previous steps some large number of times M, each time storing the differences, e_γ's.
8. A $100(1-\alpha)\%$ confidence interval for the difference of two FTA's, $\gamma_1 - \gamma_2$ is then formed by taking the interval

$$(\hat{\gamma}_1 - \hat{\gamma}_2 - e_U, \hat{\gamma}_1 - \hat{\gamma}_2 - e_L) \tag{7.28}$$

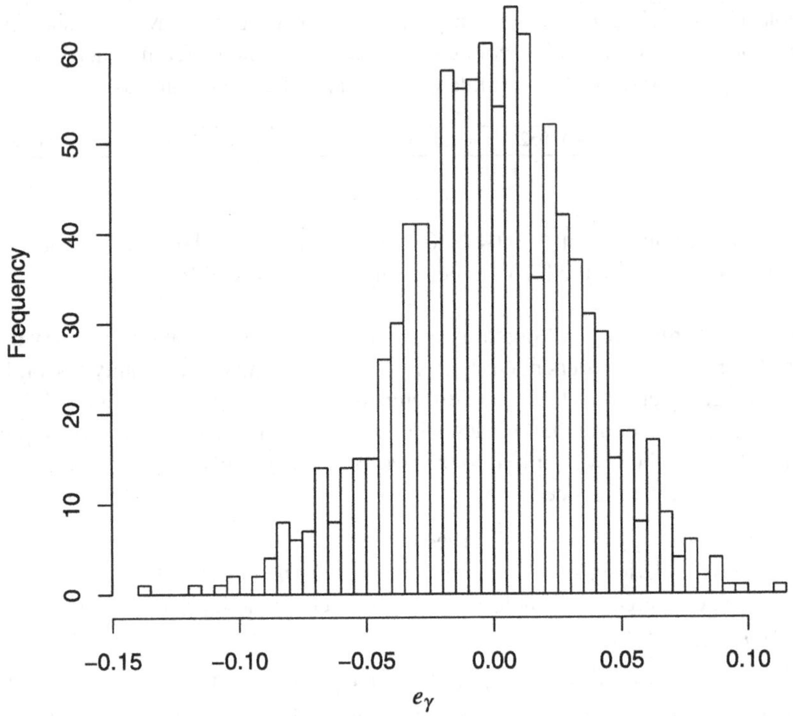

Fig. 7.5 Distribution of e_γ, the bootstrapped differences in the failure to acquire rate for those 25 years of age and older and for those 24 years of age and younger for the left index finger for Device 5 in the PURDUE-FTA database

where e_L and e_U represent the $\alpha/2$th and the $1-\alpha/2$th percentiles of the distribution of e_γ.

Example 7.11 For this example we are creating a 95% confidence interval for the difference in two failure to acquire rate for Device 5 from the PURDUE-FTA database. The acquisition attempts come from the left index finger of the users. The groups we will consider here are those aged 24 or younger—$g = 1$—and those 25 and older—$g = 2$. There cannot be overlap in these two groups and, therefore, it is appropriate to use the independent bootstrap methodology. The first group has an FTA of 0.0494 and the second has an FTA of 0.0637. There are $n_1 = 118$ individuals who are 24 or younger in this data collection and $n = 2 = 68$ individuals who are 25 or older. Each of the individuals in both groups had $K_i = K_0 = 6$ acquisition decisions. The distribution of bootstrapped differences from the observed sample difference, the e_γ's, is given in Fig.7.5. Using the 97.5th and 2.5th percentiles of this distribution, we obtain a confidence interval of $(-0.0816, 0.0598)$. Thus, we can infer that the difference between the process FTA's for these two groups is between -8.16% and 5.98% with 95% confidence. Given that this interval includes

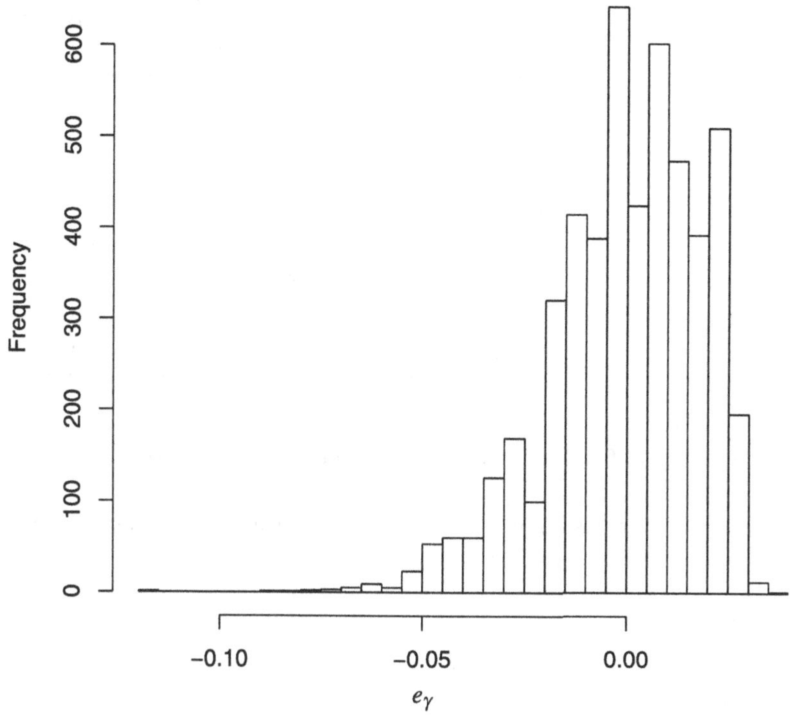

Fig. 7.6 Distribution of e_γ, the differences in the bootstrapped FTA's from the observed FTA's for males and for females for the right index finger for Device 1 from the PURDUE-FTA database

zero, there is enough variability in the estimates that we cannot conclude that the younger group's FTA is significantly less than the older group's FTA.

Example 7.12 Suppose that we would like to consider differences between men and women on their respective FTA's. We will illustrate our methodology by using acquisition decisions from Device 1 of the PURDUE-FTA database to make a 99% confidence interval for the difference in FTA's between men and women. We note that the methodology given above for two independent groups is satisfied, since these groups are clearly distinct. For this device there are 128 males and 58 females each of which recorded $K_i = 6$ acquisition attempts for all i. The estimated FTA for the males is $\hat{\gamma}_1 = 0.0026$ and for the females is $\hat{\gamma}_2 = 0.0230$. Following the algorithm above, we bootstrapped each group $M = 5000$ times and calculated e_γ for each of those bootstrapped data sets. Figure 7.6 summarizes this distribution. Using the 99.5th and 0.5th percentiles from the distribution of e_γ, we create a confidence interval that is $(-0.0486, 0.0371)$. Thus, we can be 99% confident that the difference between these process FTA's is between -4.86% and 3.71%. This suggests that for this particular data there is not a significant difference in the FTA's of men and women.

Table 7.3 Correlation parameter estimates for Example 7.13

Measure	Device 7 (Left Middle)	Device 5 (Left Index)
g	1	2
$n_A^{(g)}$	186	186
$N_\gamma^{(g)}$	1116	1116
K_0	6	6
$N_\gamma^{\dagger(g)}$	210.86	201.74
$\hat{\gamma}_g$	0.0493	0.0547
$\hat{\psi}_g$	0.8585	0.9064
$\sqrt{\hat{V}[\hat{v}_g]}$	0.0149	0.0160

7.2.2.2 Hypothesis Test for Difference of Two Independent FTA's

In this section, we are testing whether there are significant differences between two independent FTA's. Formally, the hypotheses that we are testing are

$$H_0 : \gamma_1 = \gamma_2$$
$$H_1 : \gamma_1 < \gamma_2.$$

This is testing whether the first FTA, γ_1 is significantly less than the second, γ_2. The data collection mechanism assumed here is that there is not an overlap in the individuals between the two groups. If there is overlap, then the methods given below are *not* appropriate for this analysis. We will present both a large sample approach as well as a bootstrap approach for carrying out this test.

Large Sample Approach If $N_\gamma^{(1)}$ and $N_\gamma^{(2)}$, the total number of decisions from the first and second groups, respectively, are large meaning that $N_\gamma^{\dagger(g)}\hat{\gamma}_g \geq 10$ and $N_\gamma^{\dagger(g)}(1-\hat{\gamma}_g) \geq 10$ for $g = 1, 2$, then we can use the following procedure for the difference of two independent FTA's.

Test Statistic:

$$z = \frac{\hat{\gamma}_1 - \hat{\gamma}_2}{\sqrt{\left[\hat{\gamma}_1(1-\hat{\gamma}_1)\frac{1+\hat{\psi}_1(K_0^{(1)}-1)}{n_1\bar{K}_1}\right] + \left[\hat{\gamma}_2(1-\hat{\gamma}_2)\frac{1+\hat{\psi}_2(K_0^{(2)}-1)}{n_2\bar{K}_2}\right]}} \tag{7.29}$$

where $K_0^{(g)}$ is K_0 calculated on the acquisition decisions from group g. Similarly, $\hat{\psi}_g$ is the estimated intra-individual correlation for the gth group.

p-value: $p = P(Z < z)$ where Z is a standard Gaussian random variable. This probability can be evaluated using Table 9.1.

We will reject the null hypothesis of the equality of these two FTA's, that is, reject $H_0 : \gamma_1 = \gamma_2$, if $p < \alpha$ or if the *p-value* is small.

Example 7.13 This example illustrates the method for a hypothesis test of two independent FTA's. We will be testing the differences in the FTA's for the left middle finger of Device 7 and the left index finger of Device 5 from the PURDUE-FTA database. In order to use this particular set of data, the acquisition decisions for each device were made on the same set of individuals. Thus, to use the large sample approach for testing the difference of two independent FTA's we must *assume* that acquisition attempts from the left middle finger and the left index finger of the same individual are uncorrelated. We note that the correlation between these two sets of decision is relatively small, 0.0740. Table 7.3 gives the relevant summary quantities for each of the two data collections here. Our goal in this example is to test if the FTA for Device 7 is significantly less than the FTA for Device 5, the former for the left middle finger and the latter for the left index finger. From Table 7.3, we can surmise that our sample sizes are sufficiently large to carry out the hypothesis test

$$H_0 : \gamma_1 = \gamma_2$$
$$H_1 : \gamma_1 < \gamma_2.$$

Our test statistic for this test is

$$z = \frac{0.0493 - 0.0547}{\sqrt{[0.0493(1-0.0493)\frac{1+0.8585(6-1)}{1116}] + [0.0547(1-0.0547)\frac{1+0.9064(6-1)}{1116}]}}$$
$$= -0.2458. \tag{7.30}$$

From this *p-value*, we can calculate the *p-value* to be $p = P(Z < -0.2458) \approx 0.4013$ using Table 9.1. This is a large *p-value* and as a consequence we cannot reject the null hypothesis that the FTA for these two devices applied to these two particular fingers are the same.

Bootstrap Approach To test the hypotheses of interest using a resampling approach, we assume the same null and alternative hypotheses H_0 and H_1, respectively, as in the large sample approach above. Here we will bootstrap the two groups separately using the subsets bootstrap to approximate the sampling distribution for $\hat{\gamma}_1 - \hat{\gamma}_2$ assuming that the null hypothesis is true, i.e. $\gamma_1 = \gamma_2$. Our algorithm for carrying out this test is the following one.

1. Calculate $\hat{\gamma}_1 - \hat{\gamma}_2$ from the observed data.
2. Sample n_1 individuals *with replacement* from the n_1 individuals in the first group from which there are acquisition decisions. Denote these selected individuals by $b_1^{(1)}, b_2^{(1)}, \ldots, b_{n_1}^{(1)}$.
3. For each selected individual, $b_i^{(1)}$, take the $K_{b_i^{(1)}}$ acquisition attempts. Repeat this for each individuals selected in the previous step. Call these selected acquisition attempts, $A_{b_i^{(1)}}$'s.
4. Sample n_2 individuals *with replacement* from the n_2 individuals in the second group from which there are acquisition attempts. Denote these selected individuals by $b_1^{(2)}, b_2^{(2)}, \ldots, b_{n_2}^{(2)}$.

5. For each selected individual in the second group, $b_i^{(2)}$, take the $K_{b_i^{(2)}}$ acquisition attempts. Call these selected acquisition attempts, $A_{b_i^{(2)}}$'s.
6. Calculate and store

$$e_\gamma = (\hat{\gamma}_1^b - \hat{\gamma}_2^b) - (\hat{\gamma}_1 - \hat{\gamma}_2) \tag{7.31}$$

where

$$\hat{\gamma}_g^b = \frac{\sum_{i=1}^{n_g} \sum_{\kappa=1}^{K_{b_i^{(g)}}} A_{b_i^{(g)}\kappa}}{\sum_{i=1}^{n_g} K_{b_i^{(g)}}}. \tag{7.32}$$

7. Repeat the previous five steps a large number of times M.
8. The *p-value* for this test is then

$$p = \frac{1 + \sum_{\varsigma=1}^{M} 1_{\{e_\gamma \le \hat{\gamma}_1 - \hat{\gamma}_2\}}}{M+1}. \tag{7.33}$$

9. We then will reject the null hypothesis if the *p-value* is small. If a significance level α has been specified *a priori*, then we will reject that same null hypothesis if $p < \alpha$.

Example 7.14 In this example, we are using the bootstrap methodology given above to test whether there is a significant difference between the FTA's of two groups of individuals: those in their 20's—between 20 and 29 years of age—and those in their 30's—between 30 and 39 years of age. There are $n_1 = 114$ individuals in this database in their 20's and $n_2 = 19$ individuals in their 30's. We have $K_i^{(g)} = 6$ acquisition attempts from each of those individuals in the database. We will compare these two groups from acquisition attempts on the left middle finger of Device 1 from the PURDUE-FTA database. The FTA for the younger group is 0.0194 and the FTA for the older group is 0.0526. The goal here is to answer the question of whether that difference of 3.33% is large enough for us to conclude that the difference is not due to chance alone. Following the algorithm above, we resampled the two groups and Fig. 7.7 has a summary of the distribution of the e_γ's. The dashed vertical line on that graph is the observed difference in the FTA's, −0.0333. By calculating the *p-value* based upon the observed difference and our distribution of bootstrapped differences, we find that $p = 0.2800$. This *p-value* is large and, hence, we conclude that we cannot reject the null hypothesis that the FTA of these two difference age groups is the same.

Example 7.15 Our next example compares the FTA for males and females on the right index finger for Device 7 from the PURDUE-FTA database. Six, $K_i^{(g)} = 6$, acquisition decisions were taken for each individual i in this collection and there were $n_1 = 128$ males and $n_2 = 58$ females in this data collection. Figure 7.8 shows the distribution of the e_γ's, the bootstrapped differences from the observed differences between the FTA's of these two groups. This gives us an approximation to

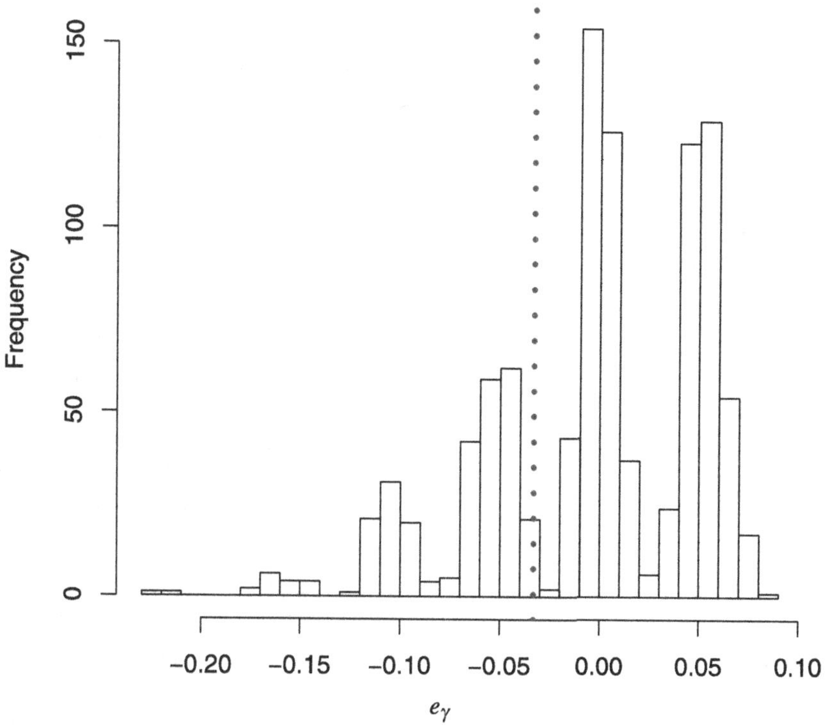

Fig. 7.7 Distribution of e_γ, the bootstrapped differences from the observed differences for comparing the Device 1 FTA's of individuals aged 20 to 29 to the FTA's individuals of aged 30 to 39 from the PURDUE-FTA database

the distribution of $\hat{\gamma}_1 - \hat{\gamma}_2$ assuming that the null hypothesis of equality of the two FTA's holds. Here our observed difference is $0.0026 - 0.0201 = -0.0175$—male FTA minus female FTA. That difference is represented in Fig. 7.8 as the dashed vertical line. The *p-value* is the percent of the bootstrapped observations that fall below that value. Thus, we have $p = 0.144$. Since that *p-value* is large, we cannot reject the null hypothesis that males and females have the same FTA for acquisition of the right index finger on this device (Device 7).

7.2.2.3 Confidence Interval for Difference of Two Paired FTA's

We move now to paired data and a confidence interval when acquisition attempts are paired. Specifically, we are making a $100(1-\alpha)\%$ confidence interval for the difference of two process FTA's, $\gamma_1 - \gamma_2$. In the case of FTA, this scenario is likely to occur less often than it does for false match or false non-match rates . For acquisition attempts to be paired, we need to have some way of associating each decision from each individual in the first group with an accompanying decision from that same individual in the second group. One circumstance where paired data can occur is

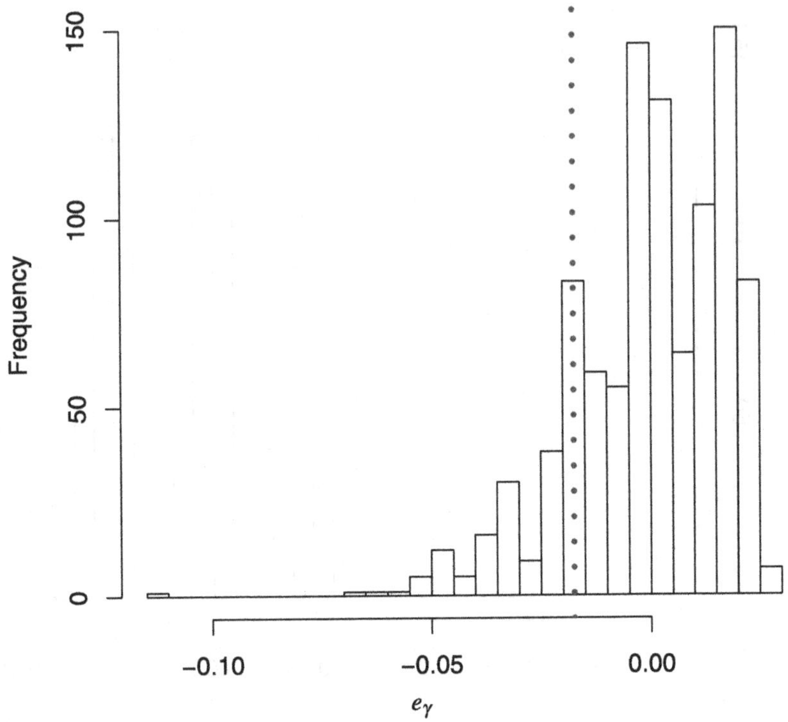

Fig. 7.8 Distribution of e_γ, the bootstrapped differences from the observed difference in the right index finger FTA's of males and females for Device 7 from the PURDUE-FTA database

when the same number of acquisition attempts from each individual were collected on two devices at every visit. For example, we might want to contrast the FTA's for different gait recognition systems. Each time that an individual involved in this test records an acquisition decision on one device, they do the same for the other device.

Below we present methodology that depends on having a sufficiently large sample size and methodology that does *not*. Because of the paired nature of this data, we will resample in a slightly different way than we did for the independent case above. This pairing also means that the number of acquisition decisions for each individual must be the same in each group, i.e. $K_i^{(1)} = K_i^{(2)}$ for every individual i. Consequently, we will drop the superscripts and subscripts that denote the different groups for K_i, n and N_γ. Note that the *effective sample size*, $N_\gamma^{\dagger(g)}$, $g = 1, 2$ is different in each group since that is dependent upon the intra-class correlation within that group.

Large Sample Approach If N_γ, the number of acquisition decisions from each group, is large, then we can use the following procedure to create a $100(1-\alpha)\%$ confidence interval for the difference of two paired FTA's, $\gamma_1 - \gamma_2$. Sufficiently large in this context will mean that $N_\gamma^{\dagger(g)}\hat{\gamma}_g \geq 10$ and $N_\gamma^{\dagger(g)}(1-\hat{\gamma}_g) \geq 10$ for $g = 1, 2$.

Our equation for making a $100(1-\alpha)\%$ confidence interval is then:

$$\hat{\gamma}_1 - \hat{\gamma}_2 \pm z_{\frac{\alpha}{2}} \sqrt{\hat{\gamma}_1(1-\hat{\gamma}_1)\left[\frac{1+(K_0-1)\hat{\psi}_1}{N_\gamma}\right] + \hat{\gamma}_2(1-\hat{\gamma}_2)\left[\frac{1+(K_0-1)\hat{\psi}_2}{N_\gamma}\right] - 2Cov(\hat{\gamma}_1, \hat{\gamma}_2)}, \tag{7.34}$$

where

$$\hat{\gamma}_g = \frac{\sum_{i=1}^{n}\sum_{\kappa=1}^{K_i} A_{i\kappa}^{(g)}}{\sum_{i=1}^{n} K_i} \tag{7.35}$$

and

$$Cov(\hat{\gamma}_1, \hat{\gamma}_2) = \frac{1}{N_\gamma^2}\sum_{i=1}^{n}\sum_{\kappa=1}^{K_i}(A_{i\kappa}^{(1)} - \hat{\gamma}_1)(A_{i\kappa}^{(2)} - \hat{\gamma}_2). \tag{7.36}$$

There are no superscripts on K_0 and N_γ. Due to the paired nature of the data, those quantities are the same for each group.

Example 7.16 To make a 95% confidence interval for the difference of the FTA's for two different fingers on the left hand: the index and the middle is the goal of this example. The observations we will be using to illustrate the method above come from Device 5 of the PURDUE-FTA database. For the left index finger, the observed FTA is $\hat{\gamma}_1 = 0.0547$ and the intra-individual correlation is $\hat{\psi} = 0.9064$. For the right index finger, the observed FTA is $\hat{\gamma}_2 = 0.0556$ and the intra-individual correlation is $\hat{\psi} = 0.8361$. There were $n = 186$ individuals tested as part of this data collection and each of those individuals gave $K_i = 6$ acquisition decisions. From the above information, we can calculate that $N_\gamma^{\dagger(1)} = 201.74$ and $N_\gamma^{\dagger(2)} = 215.43$. Thus, we have that $N_\gamma^{\dagger(1)} \times \hat{\gamma}_1 = 11.03$, $N_\gamma^{\dagger(1)} \times (1-\hat{\gamma}_1) = 190.72$, $N_\gamma^{\dagger(2)} \times \hat{\gamma}_2 = 11.97$, $N_\gamma^{\dagger(2)} \times \hat{\gamma}_2 = 203.46$. Since all of those quantities are at least 10, we can proceed to use the methodology above to calculate our confidence interval. To that end, we note that the $Cov(\hat{\gamma}_1, \hat{\gamma}_2) = 0.0000374$. Our confidence interval is then $(-0.0413, 0.0395)$. So we can be 95% confident that the difference in the process FTA's for these two fingers using Device 5 is between -4.13% and 3.95%.

Example 7.17 This example considers comparing two devices for the same modality, the left middle finger. We will do this by making a 90% confidence interval for the difference of the FTA's for these devices. The acquisition decisions we will use are from Device 5 and Device 3 from the PURDUE-FTA database. For both devices, we have $n = 186$ individuals and $N_\gamma = 1116$ total decisions. There are $K_i = 6$ acquisition decisions for each individual i. Device 5 has an FTA from these data of $\hat{\gamma}_1 = 0.0556$, while Device 3 has an FTA from these data of $\hat{\gamma}_2 = 0.0520$. To use the large sample methodology given above, we must verify that there are a sufficient number of decisions so that we can approximate the sampling distribution of $\hat{\gamma}_1 - \hat{\gamma}_2$ by a Gaussian distribution. Given that $\hat{\psi}_1 = 0.8361$ and $\hat{\psi}_2 = 0.6799$, we find that $N_\gamma^{\dagger(1)} = 215.43$ and $N_\gamma^{\dagger(2)} = 253.66$. Using these figures, we note that $N_\gamma^{\dagger(1)}\hat{\gamma}_1 = 11.97$, $N_\gamma^{\dagger(1)}(1-\hat{\gamma}_1) = 203.46$, $N_\gamma^{\dagger(2)}\hat{\gamma}_2 = 13.18$, and $N_\gamma^{\dagger(2)}\hat{\gamma}_2 = 240.48$.

All of these quantities are large enough that we can proceed. For making our interval, we will also need to know the covariance between these two sample FTA's, $Cov(\hat{\gamma}_1, \hat{\gamma}_2) = 0.0000187$. Then following (7.34), we have our 90% confidence interval to be $(-0.0299, 0.0371)$. Thus, we can be 90% confident that the difference between the process FTA's for these two devices is between -2.99% and 3.71%.

Bootstrap Approach The bootstrap approach for making a confidence interval for $\gamma_1 - \gamma_2$, the difference between two paired process FTA's, does not depend upon having a sample size that is sufficiently large as the previous large sample approach did. Instead, we use resampling of the acquisition decisions to approximate the sampling distribution for $\hat{\gamma}_1 - \hat{\gamma}_2$. We then use this approximate distribution to derive our confidence interval. This algorithm for creating such a $100(1-\alpha)\%$ confidence interval for $\gamma_1 - \gamma_2$ is given below.

1. Calculate $\hat{\gamma}_1 - \hat{\gamma}_2$ where
$$\hat{\gamma}_g = \frac{\sum_{i=1}^{n} \sum_{\kappa=1}^{K_i} A_{i\kappa}^{(g)}}{\sum_{i=1}^{n} K_i} \tag{7.37}$$
for $g = 1, 2$.
2. Resample *with replacement* the n individuals and call the resampled individuals $b_1, \ldots, b_n$.
3. For the b_ith selected individual in the previous step, obtain all K_{b_i} decisions from both group 1 and group 2. Call the decisions from the first group $A_{b_i 1}^{(1)}, \ldots, A_{b_i K_{b_i}}^{(1)}$ and the decisions from the second group $A_{b_i 1}^{(1)}, \ldots, A_{b_i K_{b_i}}^{(1)}$. Do this for each of the n resampled individuals.
4. Calculate and store
$$e_\gamma = \hat{\gamma}_1^b - \hat{\gamma}_2^b - (\hat{\gamma}_1 - \hat{\gamma}_2) \tag{7.38}$$
where
$$\hat{\gamma}_g^b = \frac{\sum_{i=1}^{n} \sum_{\kappa=1}^{K_{b_i}} A_{b_i\kappa}^{(g)}}{\sum_{i=1}^{n} K_{b_i}} \tag{7.39}$$
for $g = 1, 2$.
5. Repeat the previous three steps some large number of times, M each time storing e_γ.
6. Find the $\alpha/2$th and the $1-\alpha/2$th percentiles for the distribution of e_γ and denote those by e_L and e_U, respectively.
7. Then a $100(1-\alpha)\%$ confidence interval for the difference of two paired FTA's $\gamma_1 - \gamma_2$ is given by
$$(\hat{\gamma}_1 - \hat{\gamma}_2 - e_U, \hat{\gamma}_1 - \hat{\gamma}_2 - e_L). \tag{7.40}$$

Example 7.18 In this example, we are comparing the same modality applied to two different devices and we would like to make a 90% confidence interval for this

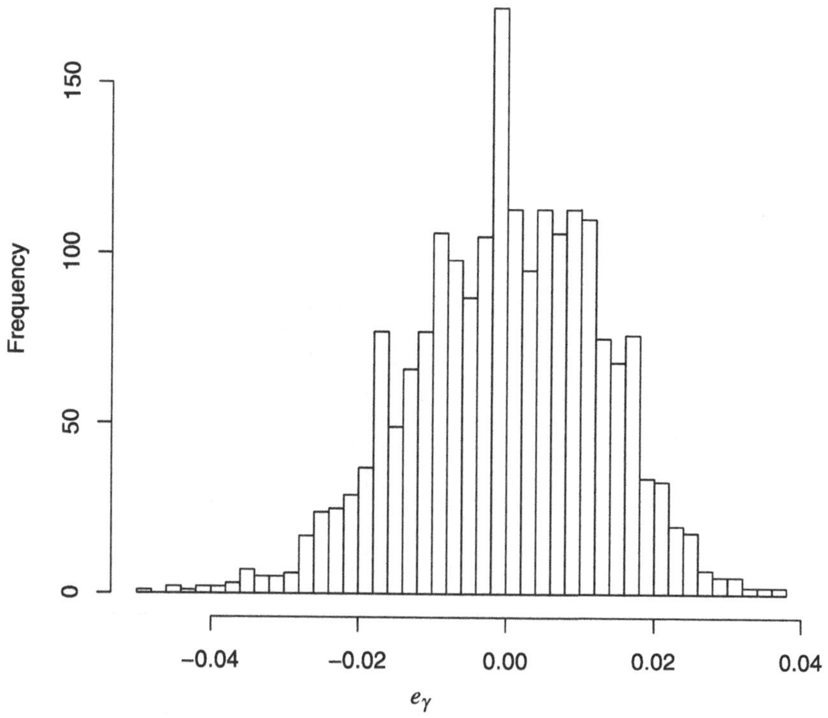

Fig. 7.9 Distribution of e_γ, the bootstrapped differences in the FTA's from the observed differences in the FTA's, for the right middle finger acquisition decisions from Device 1 and Device 3 of the PURDUE-FTA database

difference. Specifically those are the right middle finger for fingerprint devices Device 1 and Device 3 from the PURDUE-FTA database. We use the methodology for the bootstrap given above and create $M = 2000$ e_γ's. The distribution of e_γ is summarized in Fig. 7.9. The individual FTA's are $\hat{\gamma}_1 = 0.0108$ for Device 1 and $\hat{\gamma}_2 = 0.0403$ for Device 3. The difference is then $\hat{\gamma}_1 - \hat{\gamma}_2 = -0.0296$. Finding the 95th and 5th percentiles from the distribution of the e_γ's and subtracting them from that difference, we get a 90% confidence interval for the difference in the FTA for these two devices to be from -0.0493 to -0.0080. Thus, we can be 90% confident that the difference in the right middle finger FTA between Device 1 and Device 3 is between -4.93% and -0.80%. Since this interval does not contain zero, we can conclude with 90% confidence that there is a significant difference between the process FTA's of these two devices.

Example 7.19 We are going to consider making a 90% confidence interval for the difference in the FTA's of Device 5 from the PURDUE-FTAdatabase applied to the right middle finger and the left middle finger in this example. As noted above, our approach is to bootstrap the individuals and to take the acquisition decisions for those selected individuals in both groups. The right middle finger has an FTA from

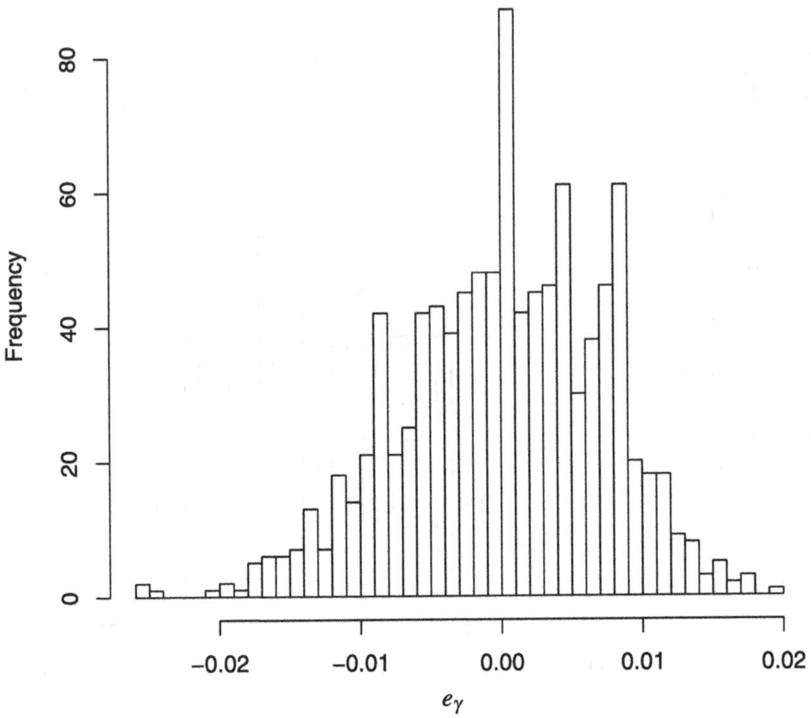

Fig. 7.10 Distribution of e_γ, the bootstrapped differences from the observed differences for comparing the FTA's of the right middle finger to the left middle finger for Device 5 from the PURDUE-FTAdatabase

the sample of $\hat{\gamma}_1 = 0.0484$ and the left middle finger has an FTA from the sample of $\hat{\gamma}_2 = 0.0556$. The estimated difference is then $\hat{\gamma}_1 - \hat{\gamma}_2 = -0.0072$. This will be the center of our 90% confidence interval. We then resampled $M = 1000$ times to get the distribution of the e_γ's given in the algorithm above. Figure 7.10 summarizes this distribution. Taking the 95th and 5th percentiles from that distribution and subtracting them from the observed difference of -0.0072, we get a confidence interval from -0.0179 to 0.0054. Thus, we can be 90% confident that the difference in the two process FTA's for these fingers and this device is between -1.79% and 0.54%.

7.2.2.4 Hypothesis Test for Difference of Two Paired FTA's

The focus of this section is testing for significant differences between two process FTA's where the attempts, the $A_{i\kappa}$'s, are paired. This means that there is some association between the κth acquisition attempts for the ith individual in both processes, i.e. between $A^{(1)}_{i\kappa}$ and $A^{(2)}_{i\kappa}$. This relationship can be, for example, that each acquisition attempt occurred at approximately the same time on two different devices or two different systems. One characteristic of paired data is that the K_i's will be exactly the same for each i in each of the two groups. Data that does not have this structure cannot be paired data and the methodology given here will not be appropriate. We will use the same notation as above for the case of making a confidence interval for the difference between two paired FTA's. The hypotheses that we will be testing are:

$$H_0: \gamma_1 = \gamma_2$$
$$H_1: \gamma_1 < \gamma_2.$$

We will focus on the test above which can be restated as testing whether the first process has a significantly smaller FTA than the second. Below, both a large sample approach and a bootstrap approach are proffered for this hypothesis test.

Large Sample Approach If the sample size, N_γ, is sufficiently large, meaning that $N_\gamma^{\dagger(g)}\hat{\gamma}_g \geq 10$ and $N_\gamma^{\dagger(g)}(1-\hat{\gamma}_g) \geq 10$ for $g = 1, 2$, then we can use the following hypothesis test for the difference of two independent FTA's. Note that while the number of acquisition decisions from each group is the same in a paired data analysis, the *effective sample size*, $N_\gamma^{\dagger(g)}$, will be different since it is dependent upon the actual decisions in each group, particularly the intra-class correlation. The methodology for this test is the following.

Test Statistic:

$$z = \frac{\hat{\gamma}_1 - \hat{\gamma}_2}{\sqrt{\hat{\gamma}_1(1-\hat{\gamma}_1)[\frac{1+\hat{\psi}_1(K_0-1)}{N_\gamma}] + \hat{\gamma}_2(1-\hat{\gamma}_2)[\frac{1+\hat{\psi}_2(K_0-1)}{N_\gamma}] - 2Cov(\hat{\gamma}_1, \hat{\gamma}_2)}} \tag{7.41}$$

where

$$Cov(\hat{\gamma}_1, \hat{\gamma}_2) = \frac{1}{N_\gamma^2}\sum_{i=1}^{n}\sum_{\kappa=1}^{K_i}(A_{i\kappa}^{(1)} - \hat{\gamma})(A_{i\kappa}^{(2)} - \hat{\gamma}), \tag{7.42}$$

and

$$\hat{\gamma} = \frac{\hat{\gamma}_1 + \hat{\gamma}_2}{2}. \tag{7.43}$$

p-value: $p = P(Z < z)$ where Z is a standard Gaussian random variable and this probability can be evaluated using Table 9.1.

We will reject the null hypothesis, H_0 if $p < \alpha$ for some specified value of the significance level α or if the *p-value* is small.

Example 7.20 In this example, we are testing to see if there is a difference between the process FTA for Device 5 applied to the left middle finger and the process FTA for Device 5 applied to the right middle finger. Specifically, we are testing whether the right middle finger has a significantly lower FTA than the left middle finger. There are $N_\gamma = 1116$ paired acquisition decisions that we will use to make our inference about the significance of the difference between these two FTA's. Those

Table 7.4 Correlation parameter estimates for Example 7.21

Measure	Device 7 (Left Middle)	Device 3 (Left Middle)
g	1	2
$n_A^{(g)}$	186	186
$N_\gamma^{(g)}$	1116	1116
K_0	6	6
$N_\gamma^{\dagger(g)}$	210.87	253.66
$\hat{\gamma}_g$	0.0493	0.0520
$\hat{\psi}_g$	0.8585	0.6799
$\sqrt{\hat{V}[\hat{v}_g]}$	0.0149	0.0139

decisions come from $n = 186$ individuals who each had $K_0 = K_i = 6$ attempts resulting in an acquisition decision. The sample FTA for the right middle finger is $\hat{\gamma}_1 = 0.0484$ and the sample FTA for the left middle finger is $\hat{\gamma}_2 = 0.0556$. Since we intend to use the large sample approach above, it is necessary to ensure that we have samples of sufficient size. The effective sample size for the two processes here is $N_\gamma^{\dagger(1)} = 207.54$ and $N_\gamma^{\dagger(2)} = 207.54$ because $\hat{\psi}_1 = 0.8755$ and $\hat{\psi}_2 = 0.8361$, respectively. Then, $207.54 \times 0.0484 = 10.04 \geq 10$, $207.54 \times (1 - 0.0484) = 197.50 \geq 10$, $215.43 \times 0.0556 = 11.98 \geq 10$ and $215.43 \times (1 - 0.0556) = 203.45 \geq 10$. Consequently, we can proceed with the above test since our conditions are met. Using the summaries already presented we can calculate the test statistic z to be $z = -0.3608$. From this we find that the *p-value* can be approximated as $P(Z < -0.36) \approx 0.3594$ using Table 9.1. The *p-value* here is large and, therefore, we cannot reject the null hypothesis that the process FTA's that generated the data that we observed are the same.

Example 7.21 For this example, we are testing whether the process FTA for Device 7 applied to the left middle finger is the same or significantly less than the process FTA from the same finger for Device 3 using a significance level of $\alpha = 0.05$. Both of these devices come from the PURDUE-FTA database. Formally, the hypotheses we are testing are

$$H_0 : \gamma_1 = \gamma_2$$
$$H_1 : \gamma_1 < \gamma_2.$$

The relevant summaries for the acquisition decisions from these two devices are given in Table 7.4. Since our data is paired, we must also note that the $Cov(\hat{\gamma}_1, \hat{\gamma}_2) = 0.0000186$. Then we can calculate our test statistic to be

$$z = \frac{0.0493 - 0.0520}{\sqrt{0.0493(1-0.0493)[\frac{1+0.8585(6-1)}{1116}] + 0.0520(1-0.0520)[\frac{1+0.6799(6-1)}{1116}] - 2(0.0000186)}}$$
$$= -0.1380. \quad (7.44)$$

Then we can find our *p-value* to be $p = P(Z < -0.1380) \approx 0.4443$ using Table 9.1. Since $p > \alpha$, we conclude that there is not a significant difference between these two FTA's at the 5% significance level.

Bootstrap Approach In order to test the equality of two FTA's using a bootstrap approach, we assume the same null and alternative hypotheses as in the large sample approach. That is, we are testing:

$$H_0 : \gamma_1 = \gamma_2$$
$$H_1 : \gamma_1 < \gamma_2.$$

An algorithm for deriving the *p-value* necessary for assessing whether the first FTA is significantly less than the second FTA is given below.

1. Calculate $\hat{\gamma}_1 - \hat{\gamma}_2$ based upon the observed acquisition decisions, the $A_{i\kappa}$'s.
2. Sample n individuals *with replacement* from the n individuals from which there are acquisition attempts. Denote these selected individuals by $b_1, b_2, \ldots, b_{n_A}$.
3. For each selected individual, b_i, obtain all K_{b_i} acquisition attempts from the first group for that individual and call those attempts $A^{(1)}_{b_i 1}, \ldots, A^{(1)}_{b_i K_{b_i}}$'s. Similarly, take the K_{b_i} acquisition attempts from the second group for that same individual. Call those $A^{(2)}_{b_i 1}, \ldots, A^{(2)}_{b_i K_{b_i}}$.
4. Calculate and store

$$e_\gamma = \hat{\gamma}_1^b - \hat{\gamma}_2^b - (\hat{\gamma}_1 - \hat{\gamma}_2) \tag{7.45}$$

where

$$\hat{\gamma}_g^b = \frac{\sum_{i=1}^{n} \sum_{\kappa=1}^{K_{b_i}} A^{(g)}_{b_i \kappa}}{\sum_{i=1}^{n} K_{b_i}} \tag{7.46}$$

for $g = 1, 2$.
5. Repeat steps 2 through 4 some large number of times M.
6. Then the *p-value* for this test is

$$p = \frac{1 + \sum_{\varsigma=1}^{M} 1_{\{e_\gamma \leq \hat{\gamma}_1 - \hat{\gamma}_2\}}}{M + 1}. \tag{7.47}$$

7. We then will reject the null hypothesis of equality of the two FTA's if the *p-value* is small. For a prespecified value of the significance level α, we will reject the null hypothesis if $p < \alpha$.

Example 7.22 In this example, we are testing whether there is a significant difference in the process FTA for two devices from the PURDUE-FTA database applied to the right index finger. These devices are Device 3 and Device 1. The estimated FTA from Device 3 is $\hat{\gamma}_1 = 0.0090$ based upon $N_\gamma = 1116$ acquisition decisions. For Device 1, we have an estimated FTA of $\hat{\gamma}_2 = 0.0251$. There were six attempted

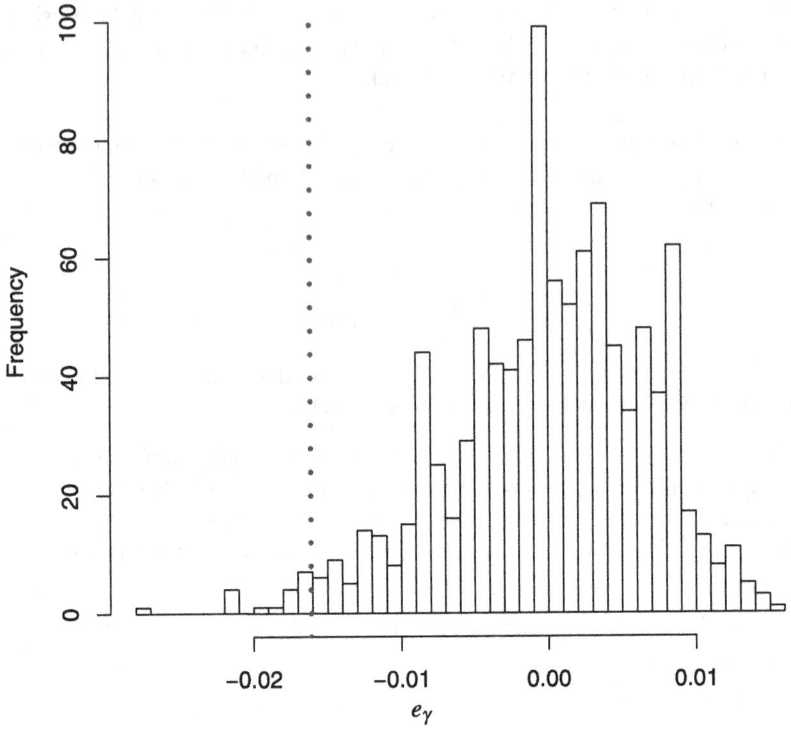

Fig. 7.11 Distribution of e_γ, the difference in the bootstrapped differences in the failure to acquire rate from the observed differences in those rates for right index finger FTA from Devices 1 and 3 in the PURDUE-FTAdatabase

acquisitions by each of the $n = 186$ individuals in this database. We would like to test the hypotheses that

$$H_0 : \gamma_1 = \gamma_2$$
$$H_1 : \gamma_1 < \gamma_2$$

using a significance level of $\alpha = 0.05$. Following the bootstrap methodology given above, we generated $M = 1000$ e_γ's and a summary of their distribution is given in Fig. 7.11. The dashed vertical line in that figure is the observed difference in the estimated FTA's of $\hat{\gamma}_1 - \hat{\gamma}_2 = -0.0161$. We can obtain our *p-value* by determining the percent of the e_γ's that are less than or equal to that estimated difference. Note that the distribution of the e_γ's is an approximation to the distribution of $\hat{\gamma}_1 - \hat{\gamma}_2$ when the null hypothesis is true. The *p-value*, 0.018, is less than the significance level $\alpha = 0.05$ and, as a consequence, we can reject the null hypothesis, H_0. Thus, we can conclude that Device 3 for the right index finger has a significantly smaller FTA than Device 1 for that same finger.

Example 7.23 Suppose that we would like to use the above methodology to test whether female users have different FTA's for two devices. We will use Device 5 and Device 7 from the PURDUE-FTA database as an example of using the bootstrap approach given above to carry out such a test. The estimated FTA for left middle finger attempts on Device 5 is $\hat{\gamma}_1 = 0.0288$ and for left middle finger attempts on Device 7 is $\hat{\gamma}_2 = 0.0661$. The data here is paired since each of the $n = 58$ women in this sample had $K_0 = K_i = 6$ acquisition attempts per device. Thus, the total number of acquisition decisions is $N_\gamma = 348$. Following the bootstrap methodology above, we resampled the individuals here $M = 1000$ times. A summary of the distribution of the e_γ's is given in Fig. 7.12. From this sample we get a *p-value* that is 0.1400. This number is the proportion of the e_γ's that are less than or equal to the dashed vertical line in Fig. 7.12 at $\hat{\gamma}_1 = \hat{\gamma}_2 = -0.0374$. Since this *p-value* is large, we cannot reject the null hypothesis that these two devices have the same FTA for females.

Note that the difference in the FTA's in the preceding example are smaller than the differences in the FTA's for this example. Those differences are -0.0161 and -0.0374, respectively. However, the first difference is significant while the second is not. Our differing conclusions are a result of the small number of observations in the latter example. As a consequence, there is much more variability in the second estimated difference and we do have enough evidence to conclude that the second difference is significant.

7.2.3 Multiple Sample Methods for FTA's

We move now to the case where there are three or more FTA's to consider. Confidence intervals when comparing more than two FTA's will be beyond the scope of this text. Linear combinations of estimates are often referred to as contrasts and the interested reader is directed to works such as Rao [78] for further details. The focus of this section then is hypothesis tests for the equality of G FTA's. The methods given below are akin to analysis of variance approaches since we are testing the equality of multiple quantities, i.e. multiple FTA's. We will assume that the FTA's within each group are stationary in the sense of Definition 2.23. As above, we will denote statistical summaries for the gth groups by a subscript or superscript of 'g' for $g = 1, \dots, G$. There are no large sample approaches given below, only resampling approaches. As we mentioned in Chap. 3, there are large sample approaches that can be applied to the types of tests given below, see e.g. Crowder [17], but they require an assumption on the distribution of acquisition decisions. The validity of such an assumption can be tested by a goodness-of-fit test like that given by Garren et al. [35]. Below we present our approaches for testing the equality of three or more FTA's for both independent and paired data collections.

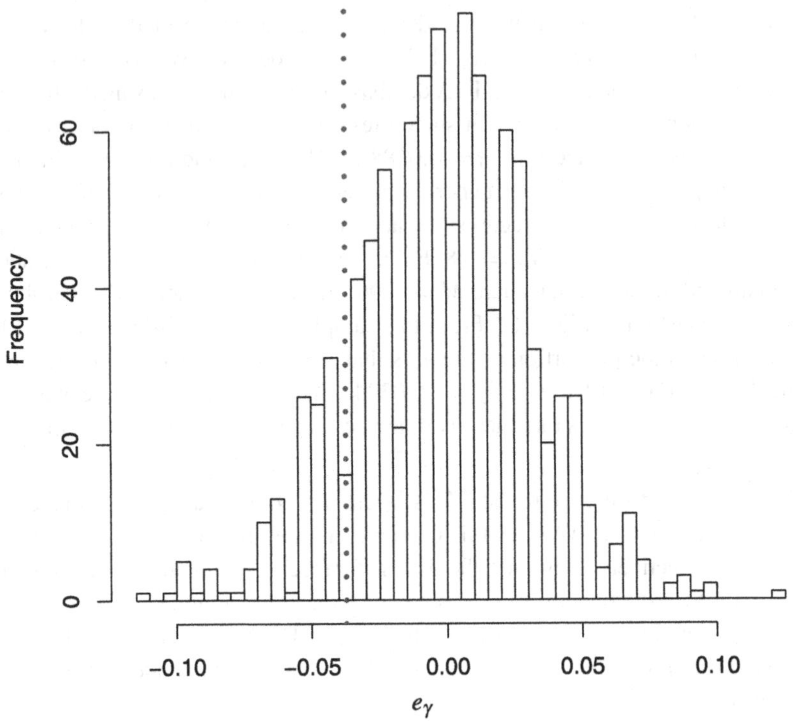

Fig. 7.12 Distribution of e_γ, the difference in the bootstrapped differences in the failure to acquire rate from the observed differences in the left middle finger FTA's of female users of Devices 5 and 7 in the PURDUE-FTA database

7.2.3.1 Hypothesis Test for Multiple Independent FTA's

This section focuses on testing the equality of G independent FTA's. Thus, we are assuming that the FTA's here are G different groups or for different processes. Independence here means that the data collections from which our acquisition attempts and decisions derive were distinct and separate. One instance where we might use this test is if we want to test whether four demographics groups have the same FTA for a single device. Likewise, we could use the methods below for comparing the FTA's of six different fingerprint devices implemented in six different countries. Formally, the hypotheses that we will be testing are:

$$H_0: \gamma_1 = \cdots = \gamma_G$$

$$H_1: \text{ not } H_0.$$

Below we give both a bootstrap approach for this test as well as a randomization approach.

Bootstrap Approach Our bootstrap method for testing the equality of G independent FTA's is similar in structure to the bootstrap method for testing the equality of two independent FTA's. We will resample each group separately to obtain a summary of the distribution for our test statistic F.

1. Calculate

$$F = \frac{[\sum_{g=1}^{G} N_\gamma^{(g)} (\hat{\gamma}_g - \bar{\gamma})^2]/(G-1)}{[\sum_{g=1}^{G} N_\gamma^{(g)} \hat{\gamma}_g (1-\hat{\gamma}_g)(1+(K_0^{(g)}-1)\hat{\psi}_g)]/(N-G)} \tag{7.48}$$

from the observed data where

$$\bar{\gamma} = \frac{\sum_{g=1}^{G} N_\gamma^{(g)} \hat{\gamma}_g}{\sum_{g=1}^{G} N_\gamma^{(g)}}, \tag{7.49}$$

$$\hat{\gamma}_g = \frac{\sum_{i=1}^{n_g} \sum_{\kappa=1}^{K_i^{(g)}} A_{i\kappa}^{(g)}}{\sum_{i=1}^{n_g} K_i^{(g)}}. \tag{7.50}$$

and $N = \sum_{g=1}^{G} N_\gamma^{(g)}$.
2. For each group g, sample n_g individuals *with replacement* from the n_g individuals in the gth group from which there are acquisition attempts. Denote these selected individuals by $b_1^{(g)}, b_2^{(g)}, \ldots, b_{n_g}^{(g)}$. For each selected individual, $b_i^{(g)}$, take all $K_{b_i}^{(g)}$ acquisition attempts for that individual. We will represent these attempts by $A_{b_i^{(g)}1}^{(g)}, \ldots, A_{b_i^{(g)}K_{b_i}^{(g)}}^{(g)}$. Do this for all g, $g = 1, \ldots, G$.
3. Repeat the previous two steps some large number of times M each time calculating and storing

$$F_\gamma = \frac{[\sum_{g=1}^{G} N_\gamma^{(g)b} (\hat{\gamma}_g^b - \hat{\gamma}_g)^2]/(G-1)}{[\sum_{g=1}^{G} N_\gamma^{(g)b} \hat{\gamma}_g^b (1-\hat{\gamma}_g^b)(1+(K_0^{(g)b}-1)\hat{\psi}_g^b)]/(N^b-G)} \tag{7.51}$$

where

$$\hat{\gamma}_g^b = \frac{\sum_{i=1}^{n_g} \sum_{\kappa=1}^{K_{b_i}^{(g)}} A_{b_i\kappa}^{(g)}}{\sum_{i=1}^{n_g} K_{b_i}^{(g)}}, \tag{7.52}$$

$$N_\gamma^{(g)b} = \sum_{i=1}^{n_g} K_{b_i}^{(g)} \tag{7.53}$$

and $N^b = \sum_{g=1}^{G} N_\gamma^{(g)b}$.
4. Then the *p-value* for this test is

$$p = \frac{1 + \sum_{\varsigma=1}^{M} 1_{\{F_\gamma \geq F\}}}{M+1}. \tag{7.54}$$

5. To reject the null hypothesis, we must have a *p-value* that is small. If a significance level α is specified *a priori* the data analysis, then we can reject the null hypothesis if $p < \alpha$.

Example 7.24 To illustrate the methodology above, we consider comparing three different age groups: individuals 29 and under, individuals between 30 and 39 and individuals 40 and over. Since these groups are clearly distinct, we can use the bootstrap methodology given above for comparing three or more independent groups. We will apply this approach to acquisition decisions from right middle finger attempts on Device 7 of the PURDUE-FTA database. The youngest age group has a sample FTA of $\hat{\gamma}_1 = 0.0075$, individuals in their 30's had a sample FTA of $\hat{\gamma}_2 = 0.0614$, and those 40 and over at a sample FTA of $\hat{\gamma}_3 = 0.2361$. We created $M = 999$ replicated bootstrap copies of the original data following the above approach. Figure 7.13 summarizes the distribution of the calculated F_γ's. (Note that 12 values of F_γ were very large,i.e. > 300, we have not displayed these values in Fig. 7.13 since a display of the full distribution on the original scale would not allow the reader to get a complete sense of the shape of the distribution.) The test statistic from the original data $F = 19.7785$ is represented on the graph by the dashed vertical line. Our *p-value* $= 0.183$ and is taken from the number of F_γ's that are larger than that value. Since that *p-value* is large, we fail to reject the null hypothesis, H_0, that the FTA's for these three groups are equal.

Randomization Approach Here we are testing the same hypothesis of equality of G FTA's as we did in the bootstrap approach. These hypotheses are

$$H_0 : \gamma_1 = \gamma_2 = \cdots = \gamma_G$$
$$H_1 : \text{not } H_0.$$

The randomization approach is similar to the bootstrap approach in that both try to estimate the sampling distribution of a statistic assuming that all of the FTA's, the γ_g's, are equal. To do this with a randomization approach, we will permute individuals (and their associated acquisition attempts) between the groups. The basic idea being that if the FTA's are all equal, then permuting the individuals between the groups should give us an idea of the distribution of our statistic if all of the FTA's are the same. We will use the same statistic that we did for the bootstrap approach above. We enumerate this approach below.

1. Calculate

$$F = \frac{[\sum_{g=1}^{G} N_\gamma^{(g)} (\hat{\gamma}_g - \bar{\gamma})^2]/(G-1)}{[\sum_{g=1}^{G} N_\gamma^{(g)} \hat{\gamma}_g (1-\hat{\gamma}_g)(1+(K_0^{(g)}-1)\hat{\psi}_g)]/(N-G)} \tag{7.55}$$

from the observed acquisition attempts where

$$\bar{\gamma} = \frac{\sum_{g=1}^{G} N_\gamma^{(g)} \hat{\gamma}_g}{\sum_{g=1}^{G} N_\gamma^{(g)}}, \tag{7.56}$$

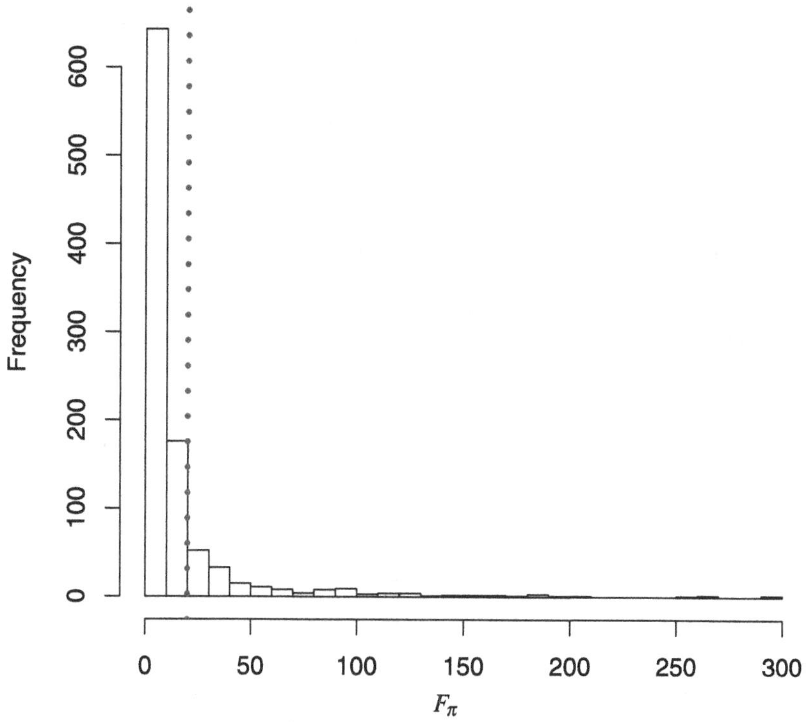

Fig. 7.13 Distribution of F_γ for testing the equality of the right middle finger failure to acquire rate for three different age groups using Device 7 of the PURDUE-FTA database

$$\hat{\gamma}_g = \frac{\sum_{i=1}^{n_g} \sum_{\kappa=1}^{K_i^{(g)}} A_{i\kappa}^{(g)}}{\sum_{i=1}^{n_g} K_i^{(g)}} \tag{7.57}$$

and $N = \sum_{g=1}^{G} N_\gamma^{(g)}$.

2. Our next step is to collect all of the individuals from each of the G groups into a single list and then, having permuted or reordered the list, reassign individuals to groups. Formally let $\mathscr{I}_g$ be the collection of individuals in the gth group where $n_g = |\mathscr{I}_g|$ is the number of individuals in the gth group and let $\mathscr{I} = \bigcup_{g=1}^{G} \mathscr{I}_g$ with $n = |\mathscr{I}|$. Sample *without* replacement n_1 individuals from $\mathscr{I}$. Call these individuals $b_1^{(1)}, b_2^{(1)}, \ldots, b_{n_1}^{(1)}$ and call the collection of these individuals $\mathscr{I}_1^b$.
3. For the gth group where $2 \leq g \leq G-1$, sample *without* replacement n_g individuals from the set of remaining individuals

$$\mathscr{I} \setminus \left\{ \bigcup_{t=1}^{g-1} \mathscr{I}_t^b \right\} \tag{7.58}$$

and call these individuals $b_1^{(g)}, \ldots, b_{n_g}^{(g)}$ and the collection of them $\mathscr{I}_g^b$.

4. Assign to the Gth group the remaining n_G individuals, the set of individuals defined by

$$\mathscr{I} \setminus \left\{ \bigcup_{t=1}^{G-1} \mathscr{I}_t^b \right\} \tag{7.59}$$

and call those individuals $b_1^{(G)}, \ldots, b_{n_G}^{(G)}$ and their collection $\mathscr{I}_G^b$.

5. For every resampled individual, $b_i^{(g)}$, we take all the acquisition decisions from that individual and call those decisions, $A_{b_i^{(g)}\kappa}^{(g)}$ where $\kappa = 1, \ldots, K_{b_i^{(g)}}$.
6. Calculate

$$F_\gamma = \frac{[\sum_{g=1}^{G} N_\gamma^{(g)b} (\hat{\gamma}_g^b - \bar{\gamma}^b)^2]/(G-1)}{[\sum_{g=1}^{G} N_\gamma^{(g)b} \hat{\gamma}_g^b (1-\hat{\gamma}_g^b)(1+(K_0^{(g)b}-1)\hat{\psi}_g^b)]/(N-G)} \tag{7.60}$$

where

$$\hat{\gamma}_g^b = \frac{\sum_{i=1}^{n_g} \sum_{\kappa=1}^{K_{b_i^{(g)}}} A_{b_i^{(g)}\kappa}^{(g)}}{\sum_{i=1}^{n_g} K^{b_i^{(g)}}} \tag{7.61}$$

and $N_\gamma^{(g)b} = \sum_{i=1}^{n_g} K_{b_i^{(g)}}$.

7. Repeat steps 2 through 6 some large number of times say M. Then our *p-value* for this test is

$$p = \frac{1 + \sum_{\varsigma=1}^{M} I_{\{F_\gamma \geq F\}}}{M+1}. \tag{7.62}$$

8. We then will reject the null hypothesis of equality of the G FTA's if the *p-value* is small. Having specified a significance level beforehand, we can reject if $p < \alpha$.

Example 7.25 For this example, we will compare the FTA of three different age groups. These three age groups are those between 20 and 29, those between 30 and 39 and those 40 and over. We will compare the FTA's of these three groups for left index finger attempts for Device 7 from the PURDUE-FTA database. There are $n_1 = 155$, $n_2 = 19$ and $n_3 = 12$ individuals in each of these three groups. As with all individuals in this database, these individuals contributed $K_0 = K_i = 6$ acquisition decisions each. The estimated FTA's from these observations are $\hat{\gamma}_1 = 0.0247$, $\hat{\gamma}_2 = 0.1228$ and $\hat{\gamma}_3 = 0.1528$. These quantities yield a value of $F = 4.6035$ for our test statistic. We randomize this data following the algorithm above to obtain an approximation to the distribution of F assuming that the null hypothesis, H_0, is true. This distribution is summarized for $M = 999$ randomizations in Fig. 7.14. We can calculate the *p-value* from these randomizations to be $p = 0.023$. Graphically, we can observe the proportion of values of F_γ that are at least $F = 4.6035$. The value of F is noted by the dashed vertical line in Fig. 7.14. Since in this case our *p-value* is small, we will reject the null hypothesis that these FTA's are equal.

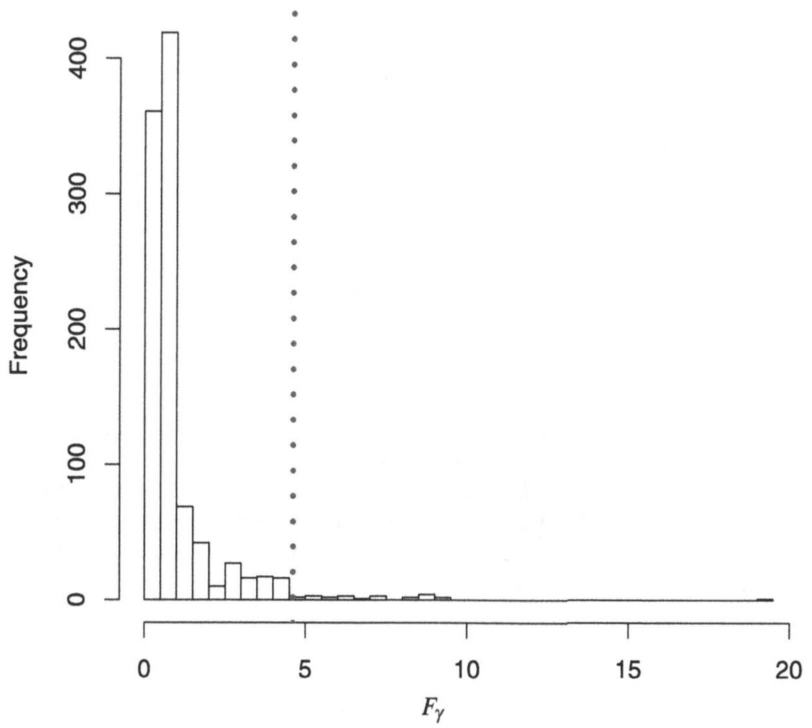

Fig. 7.14 Distribution of randomization F_γ's from testing the equality of three different age groups failure to acquire rate for left index finger acquisitions from Device 7 of the PURDUE-FTA database

7.2.3.2 Hypothesis Test for Multiple Paired FTA's

This section presents methods for the comparison of multiple FTA's that are matched or paired in some way. For this methodology, the unit for matching will be an acquisition attempt. Thus, there must be an equal number of acquisitions attempts per individual and an equal number of individuals per group for applying the methods provided here. A test where every individual attempts to have a fingerprint image acquired by each of ten devices on every visit would be an example where the methodology below is appropriate. In that case, each acquisition attempt from each individual is associated with nine other acquisition attempts from that same individual. We present both a bootstrap and randomization approach here.

Bootstrap Approach Here, we present a bootstrap method for testing whether there is a significant difference between multiple FNMR's. The hypotheses for this test are:

$$H_0 : \gamma_1 = \gamma_2 = \gamma_3 = \cdots = \gamma_G$$
$$H_1 : \text{not } H_0.$$

1. Calculate

$$F = \frac{[\sum_{g=1}^{G} N_\gamma^{(g)} (\hat{\gamma}_g - \bar{\gamma})^2]/(G-1)}{[\sum_{g=1}^{G} \hat{\gamma}_g (1-\hat{\gamma}_g)(1+(K_0-1)\hat{\psi}_g)]/(N-G)} \tag{7.63}$$

   for the observed data where

$$\bar{\gamma} = \frac{\sum_{g=1}^{G} \hat{\gamma}_g}{G}, \tag{7.64}$$

$$\hat{\gamma}_g = \frac{\sum_{i=1}^{n} \sum_{\kappa=1}^{K_i} A_{i\kappa}^{(g)}}{\sum_{i=1}^{n} K_i} \tag{7.65}$$

   and $N = GN_\gamma$.
2. Sample *with replacement* n individuals from the n individuals from whom data has been collected. Call these new individuals $b_1, \ldots, b_n$.
3. For each bootstrapped individual, b_i compile all m_{b_i} decisions from the first group, $g = 1$ on that individual. For each of the selected decisions in the first group, we will take the g decisions, one from each group that is paired with a selected decisions. Denote these new decisions by $A_{b_i\kappa}^{(g)}$ where $j = 1, \ldots, m_{b_i}$.
4. Calculate and store

$$F_\gamma = \frac{[\sum_{g=1}^{G} N_\gamma (\hat{\gamma}_g^b - \hat{\gamma})^2]/(G-1)}{[\sum_{g=1}^{G} \hat{\gamma}_g^b (1-\hat{\gamma}_g^b)(1+(K_0-1)\hat{\psi}_g^b)]/(N-G)} \tag{7.66}$$

   for the bootstrapped data where

$$\hat{\gamma}_g^b = \frac{\sum_{i=1}^{n} \sum_{j=1}^{K_{b_i}} A_{b_i\kappa}^{(g)}}{\sum_{i=1}^{n} K_{b_i}} \tag{7.67}$$

   and $N = GN_\gamma$.
5. Repeat steps 2 to 4 some large number of times, say M. Then the *p-value* can be calculated as

$$p = \frac{1 + \sum_{\varsigma=1}^{M} I_{\{F_\gamma \geq F\}}}{M+1}. \tag{7.68}$$

Example 7.26 For this example we are going to compare the FTA's of four fingers based upon their acquisition decisions from Device 7 of the PURDUE-FTA database. Since all $n = 186$ individuals in this database gave $K_0 = K_i = 6$ acquisition attempts for each of four fingers, we will use the paired approach. Then $N_\gamma = 1116$. The four fingers we will consider are the right middle, the right index, the left middle and the left index. The sample FTA for those four fingers are $\hat{\gamma}_1 = 0.0278$, $\hat{\gamma}_2 = 0.0081$, $\hat{\gamma}_3 = 0.0493$, and $\hat{\gamma}_4 = 0.0296$. Formally we would like to test

$$H_0 : \gamma_1 = \gamma_2 = \gamma_3 = \gamma_4$$

$$H_1 : \text{ not } H_0$$

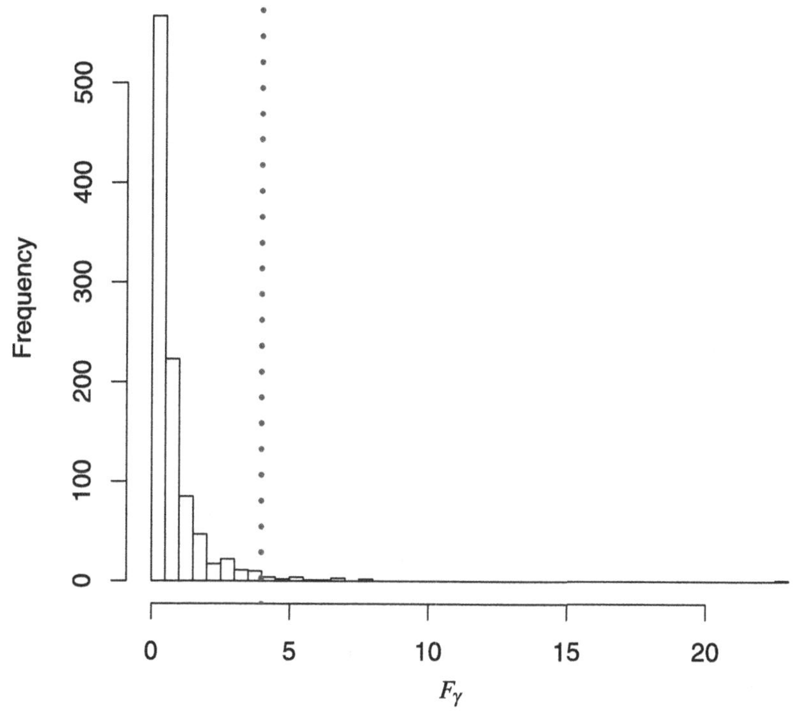

Fig. 7.15 Distribution of F_γ's for testing the equality of FTA's for four different fingers used on Device 7 from the PURDUE-FTA database

using a significance level of $\alpha = 0.05$ where γ_g, $g = 1, 2, 3, 4$ is the process FTA of each particular finger. Using the algorithm above, we randomized the decisions in this database $M = 999$ times. The resulting distribution of F_γ's is summarized in Fig. 7.15. The dashed vertical line is located at $F = 3.9756$ which is the value for F from the original data. Our *p-value* from this test is $p = 0.0200$ which is less than our significance level of 0.05. Therefore, we reject the null hypothesis that these four FTA's are equal and conclude that at least one of these FTA is significantly different from the others.

Example 7.27 In the PURDUE-FTA database, there are $G = 7$ different devices. This example will use the subsets bootstrap methodology given above to compare the FTA's for all of these devices based upon acquisition attempts from the right index finger of all $n = 186$ individuals. Each individuals i, contributed $K_0 = K_i = 6$ decisions which gives a total of $N_\gamma = 1116$ decisions for each of the seven devices. The decisions are paired by the order of an individual's acquisition attempts at a given device. The seven FTA's are 0.0090, 0.0224, 0.0251, 0.0197, 0.0341, 0.0179, and 0.0081, respectively. We bootstrapped the individuals in the database 999 times following the algorithm given above. A summary of the distribution of the F_γ's is given in Fig. 7.16. The *p-value* for this test is 0.259 based upon those values. The

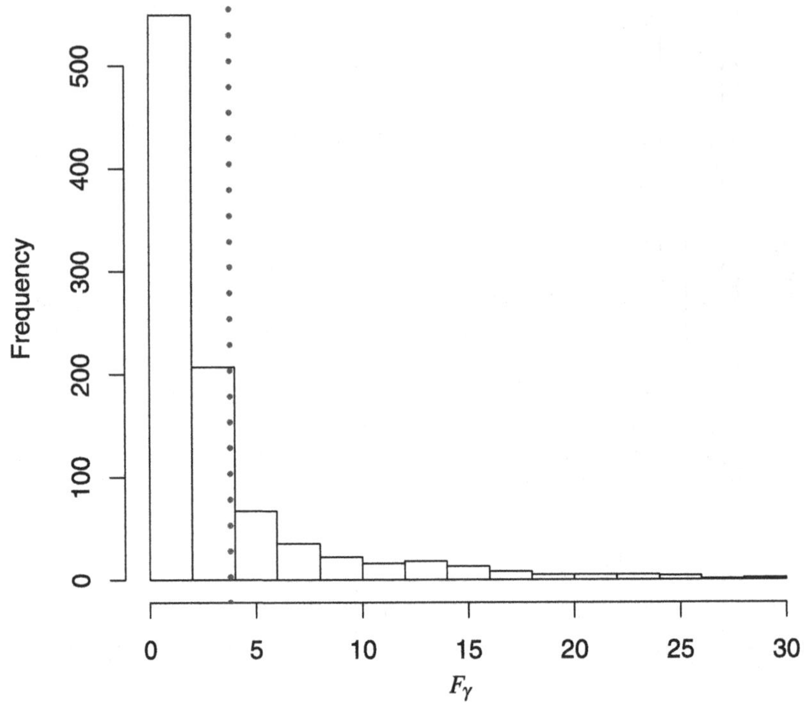

Fig. 7.16 Distribution of F_γ for comparing FTA's for right index finger attempts for all seven devices from PURDUE-FTA database

original test statistic is $F = 3.7708$ and it is represented by the dashed vertical line given in the aforementioned figure. Since our *p-value* here is large, we cannot reject the null hypothesis, H_0, that these process FTA's are all the same.

Randomization Approach Here we present a randomization test for multiple paired FTA's. Since we are here trying to assess whether we have equality of the FTA's between G groups on acquisition decisions that are paired, we must deal with the correlation induced by the pairing in our permutation scheme. Here we permute the individual acquisition decisions that are paired. That is, for the G matching decisions associated with the κth acquisition attempt by the ith individual, we randomly permute the order of the decisions, then assign the first decision from the permuted list to group 1, the second decision from the permuted list to group 2, etc. The logic here is that if FTA's are all equal then permuting the decisions should not produce test statistics that are drastically different from the value given by the test statistic calculated on the original data. Formally, we are testing:

$$H_0 : \gamma_1 = \gamma_2 = \gamma_3 = \cdots = \gamma_G$$

$$H_1 : \text{not } H_0.$$

Our algorithm for this test is the following:

1. Calculate

$$F = \frac{[\sum_{g=1}^{G} N_{\gamma}^{(g)}(\hat{\gamma}_g - \bar{\gamma})^2]/(G-1)}{[\sum_{g=1}^{G} \hat{\gamma}_g(1-\hat{\gamma}_g)(1+(K_0-1)\hat{\psi}_g)]/(N-G)} \tag{7.69}$$

   for the observed data where

$$\bar{\gamma} = \frac{\sum_{g=1}^{G} \hat{\gamma}_g}{G}, \tag{7.70}$$

$$\hat{\gamma}_g = \frac{\sum_{i=1}^{n} \sum_{\kappa=1}^{K_i} A_{i\kappa}^{(g)}}{\sum_{i=1}^{n} K_i} \tag{7.71}$$

   and $N = GN_{\gamma}$.
2. For the ith individual and the κth acquisition decision, take the G decisions—$A_{i\kappa}^{(1)}, A_{i\kappa}^{(2)}, \ldots, A_{i\kappa}^{(G)}$—that are associated with that particular individual i and that decision κ and randomly permute the order of those decisions.
3. Take the first acquisition decision from the permuted list and assign them to the first group. Denote that decision by $A_{i\kappa}^{(1)*}$. Take the second acquisition decision from the permuted list and assign it to the second group. Denote this decision by $A_{i\kappa}^{(2)*}$. Continue this for the remaining groups assigning a permuted decision to each group. Repeat this process for all decisions and all individuals.
4. Calculate and store

$$F_{\gamma} = \frac{[\sum_{g=1}^{G} N_{\gamma}^{(g)}(\hat{\gamma}_g^r - \bar{\gamma})^2]/(G-1)}{[\sum_{g=1}^{G} \hat{\gamma}_g^r(1-\hat{\gamma}_g^r)(1+(K_0-1)\hat{\psi}_g^r)]/(N-G)} \tag{7.72}$$

   where

$$\hat{\gamma}_g^r = \frac{\sum_{i=1}^{n} \sum_{\kappa=1}^{K_i} A_{i\kappa}^{(g)*}}{\sum_{i=1}^{n} K_i} \tag{7.73}$$

   and $N = GN_{\gamma}$.
5. Repeat steps 2 and 3 some large number of times, say M.
6. Then the *p-value* can be calculated as

$$p = \frac{1 + \sum_{\varsigma=1}^{M} I_{\{F_{\gamma} \geq F\}}}{M+1}. \tag{7.74}$$

7. Reject the null hypothesis if the *p-value* calculated in the previous step is small. If a significance level α has been determined, then reject the null hypothesis if $p < \alpha$.

Example 7.28 This example is a comparison of the right index finger FTA's for all seven devices from the PURDUE-FTA database. Since there are $G = 7$ devices, we

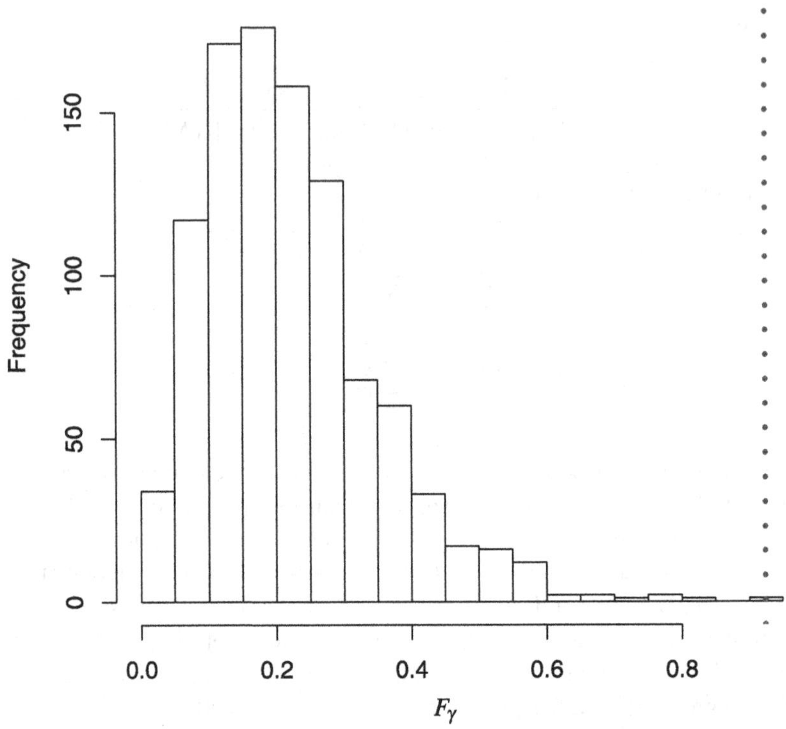

Fig. 7.17 Distribution of the F_γ for comparing seven right index finger FTA's for seven different devices from the PURDUE-FTA database

are formally testing the hypotheses

$$H_0 : \gamma_1 = \gamma_2 = \cdots = \gamma_7$$
$$H_1 : \text{ not } H_0$$

with a significance level of $\alpha = 0.01$. By choosing a significance level this small, we are asserting that there must be a great deal of evidence that the null hypothesis of equality of the FTA's is not true before we will reject that null hypothesis. The sample FTA's for these seven devices are 0.0089, 0.0224, 0.0251, 0.0197, 0.0341, 0.0179 and 0.0081, respectively. There were $n = 186$ individuals each of which had $K_i = K_0 = 6$ acquisition attempts per device. The pairing of these attempts is by their order on each device. We calculated the test statistic to be $F = 0.9239$. We randomized the original data following the schema above $M = 999$ times. A summary of the resulting distribution for F_γ is given in Fig. 7.17. The dashed vertical line on the right hand side of that graph is at the value $F = 0.9239$. Our *p-value* for this test is then $p = 0.001$. Since that value is less than our significance level $\alpha = 0.01$, then we have enough evidence to reject the null hypothesis, H_0, and con-

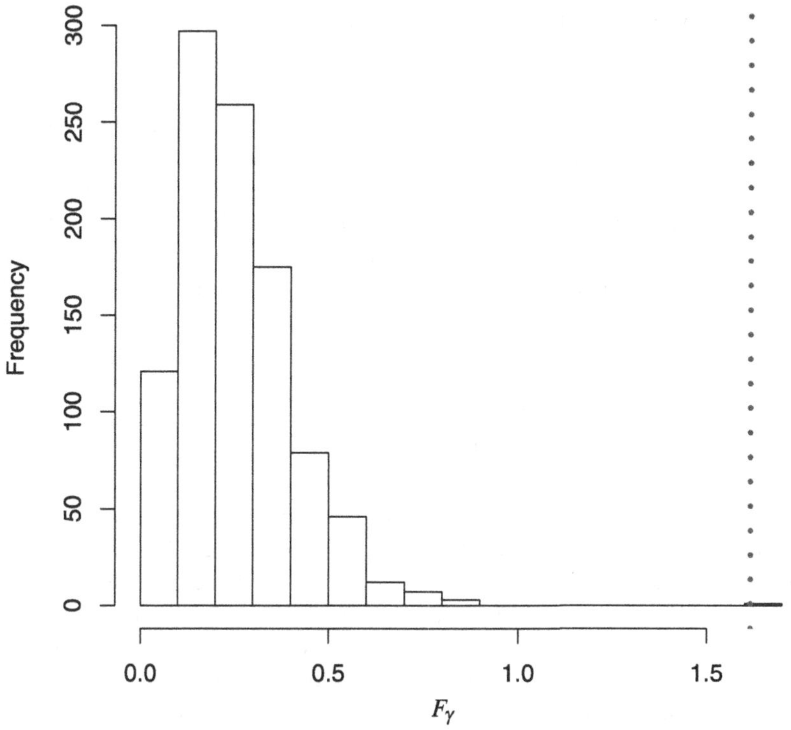

Fig. 7.18 Distribution of F_γ for comparing the left middle finger FTA's of seven fingerprint devices on males from the PURDUE-FTA database

clude that at least one of the device's FTA is significantly different from the others FTA's.

Example 7.29 The focus of this example will again be a comparison of the FTA's for all of the devices in the PURDUE-FTA database. This example, however, will focus on the men in that database. Thus, we are testing the equality of seven different left middle fingers FTA's each from a different fingerprint device. The $n = 128$ individuals/males are the same for each device. The pairing of decisions from each individual across devices is done on the order of the $K_0 = K_i = 6$ acquisition decisions that each individual contributed to the database. The estimated FTA's for each device are 0.0117, 0.0195, 0.0456, 0.0234, 0.0677, 0.0299, and 0.0417, respectively. The test statistic for the original data is $F = 1.6148$ which is represented on Fig. 7.18 by the dashed vertical line. That figure summarizes the distribution of the F_γ's. Using that distribution, we get a *p-value* of $p = 0.001$ based upon $M = 999$ randomized permutations of the observed data. With such a small *p-value*, we conclude that there is enough evidence to reject the null hypothesis, H_0, of equality of these FTA's.

7.3 Sample Size and Power Calculations

One important component of any test of a biometrics system is the number of individuals that are to be tested. In this section, we will discuss calculations and derivations that allows us to determine how many individuals need to be tested to meet a specific criterion. As mentioned in Chap. 2, any sample size calculation requires *a priori* specification of some parameters of the system. Below we will cover both sample size and power calculations for determining how many individuals to test. Sample size calculations are for confidence intervals of a given width, B, while power calculations are for achieving a particular power, $1-\beta$. We will start with sample size calculations.

7.3.1 Sample Size Calculations

Suppose that we want to make a confidence interval for a single process FTA. For example, we might want to estimate an FTA for a iris recognition system attached to an external entryway. Our goal for this estimation might be the creation of a 95% confidence interval that has a margin of error of 2% or 0.02. One consideration for this data collection will be the number of individuals that we will need. We will derive in this section calculations for determining how many individuals from whom we will need acquisition attempts to create a $100(1-\alpha)\%$ confidence interval of width $2B$ for a single process FTA.

Recall that our confidence interval for a single process FTA is

$$\hat{\gamma} \pm z_{1-\frac{\alpha}{2}}\sqrt{\frac{\hat{\gamma}(1-\hat{\gamma})[1+(K_0-1)\hat{\psi}]}{n_A\bar{K}}}, \tag{7.75}$$

where

$$\hat{\gamma} = \frac{\sum_{i=1}^{n_A}\sum_{\kappa=1}^{K_i} A_{i\kappa}}{\sum_{i=1}^{n_A} K_i}. \tag{7.76}$$

Then, we can set

$$B \overset{\text{SET}}{=} z_{1-\frac{\alpha}{2}}\sqrt{\frac{\hat{\gamma}(1-\hat{\gamma})[1+(K_0-1)\hat{\psi}]}{n_A\bar{K}}}. \tag{7.77}$$

We will want to take (7.77) and solve for n_A given a particular value of B. Solving for n_A, we get

$$n_A = \left\lceil \frac{z_{\alpha/2}^2\hat{\gamma}(1-\hat{\gamma})(1+(K_0-1)\psi)}{\bar{K}B^2} \right\rceil. \tag{7.78}$$

Now it is clear that the right hand size of (7.78) depends upon a variety of quantities of which we will not have complete *a priori* knowledge. These are K_0, $\bar{K}$, γ and ψ. (Note that by choosing the confidence level we determine $z_{\alpha/2}$.) As was noted in Chap. 2, we must estimate or specify some process parameter values in order to calculate the appropriate sample size. One simplification that can be made to (7.78) is to assume that all individuals will contribute the same number of acquisition decisions. Then, $K_0 = \bar{K}$ and (7.78) becomes

$$n_A = \left\lceil \frac{z_{\alpha/2}^2 \hat{\gamma}(1-\hat{\gamma})(1+(\bar{K}-1)\psi)}{\bar{K}B^2} \right\rceil. \tag{7.79}$$

As mentioned in Comment 2.9, we can conduct a small pilot study or use previous studies on similar processes to obtain estimates for γ and ψ. If neither or those are possible, then we must use our best judgement for specifying these parameters. Note that in a statistical sense, specifying larger values for γ and ψ would be conservative, i.e. would yield larger values for n_A. Determination of the average number of attempts per user, $\bar{K}$, is something that is somewhat under the control of the experimenter/tester and, therefore, we should have some idea of appropriate values for $\bar{K}$. To be statistically conservative, we could choose a slightly smaller value of $\bar{K}$, then we are likely to obtain, especially if there will be variability in the K_i's, the number of attempts per individual. Finally, it is worth mentioning that in some cases that the values of α and B are determined by external factors such as an oversight agency or an international standard. In those cases, we need to be especially cognizant of the requirements for designing and reporting the results of such a test.

7.3.2 Power Calculations

The power of a hypothesis test is the probability of rejecting the null hypothesis, H_0, when the alternative hypothesis, H_1, is true. As we did in Chap. 2, we will denote the power of a test by $1-\beta$. For example, we might be interested in determining how many individuals we would have to sample in order to have a power of 90% for rejecting a null hypothesis of $H_0: \gamma = 0.025$ against an alternative value for the process FTA of $\gamma = 0.01$ if we are using a significance of $\alpha = 0.05$. In that case, we would be required to find the number of individuals needed to reject a null hypothesis of $H_0: \gamma = 0.025$ (against an alternative of $H_1: \gamma < 0.025$) with probability 0.90 when the true process mean is $\gamma = 0.01$ assuming that the significance level of that test is $\alpha = 0.05$. As was the case for the sample size calculation in the preceding section, the power calculation requires us to specify certain parameters about the process *a priori*. In addition to those parameters specified in the sample size calculation, we need to specify the power, $1-\beta$ for the hypothesis test under consideration and to specify an alternative value for the process parameter that we are testing.

Recall that for a single FTA we have the following hypothesis test:

$$H_0 : \gamma = \gamma_0$$
$$H_1 : \gamma < \gamma_0.$$

To determine the number of individuals to test, it is necessary to specify a particular value in the alternative hypothesis. It is not possible to calculate power exactly unless an exact value is designated. We will call this value γ_a. In the example above $\gamma_a = 0.01$. The difference between γ_0 and γ_a is often known as the 'effect size' in some literature. This quantity $\gamma_0 - \gamma_a$ takes the place of the margin of error, B, in the sample size calculations. If the sample size is large, then we can assume that the sampling distribution of $\hat{\gamma}$ is approximately Gaussian. In order to determine the number of individuals n that is required, we must first specify that

$$P\left(\frac{\hat{\gamma} - \gamma_0}{s_{\hat{\gamma}0}} < -z_\alpha \middle| \gamma = \gamma_a\right) = 1 - \beta \tag{7.80}$$

where

$$s_{\hat{\gamma}0} = \sqrt{\frac{\gamma_0(1-\gamma_0)(1+(\bar{K}-1)\hat{\psi})}{n_A\bar{K}}}. \tag{7.81}$$

This is the formal statement of what we would like to occur. We would like to reject the null hypothesis, H_0, with probability $1 - \beta$ assuming that the alternative hypothesis, $\gamma < \gamma_0$, is true. In order to calculate an exact power, we need to specify some $\gamma_a < \gamma_0$. It then follows that

$$P\left(\hat{\gamma} < \gamma_0 - z_\alpha s_{\hat{\gamma}0} \mid \gamma = \gamma_a\right) = 1 - \beta. \tag{7.82}$$

If we can assume Gaussianity (due to large sample sizes), then we can write that

$$P\left(\frac{\hat{\gamma} - \gamma_a}{s_{\hat{\gamma}a}} < \frac{\gamma_0 - z_\alpha s_{\hat{\gamma}0} - \gamma_a}{s_{\hat{\gamma}a}}\right) = 1 - \beta \tag{7.83}$$

where

$$s_{\hat{\gamma}a} = \sqrt{\frac{\gamma_a(1-\gamma_a)(1+(\bar{K}-1)\hat{\psi})}{n_A\bar{K}}}. \tag{7.84}$$

Then the quantity on the right hand side of the inequality in this probability calculation is a standard Gaussian random variable. We can equate the quantity to the appropriate percentile, z_β, of the standard Gaussian distribution. We then have

$$z_\beta = \frac{\gamma_0 - z_\alpha s_{\hat{\gamma}0} - \gamma_a}{s_{\hat{\gamma}a}} \tag{7.85}$$

which can be expanded to be written as

$$z_\beta = \frac{\gamma_0 - z_\alpha\sqrt{\frac{\gamma_0(1-\gamma_0)(1+(\bar{K}-1)\hat{\psi})}{n_A\bar{K}}} - \gamma_a}{\sqrt{\frac{\gamma_a(1-\gamma_a)(1+(\bar{K}-1)\hat{\psi})}{n_A\bar{K}}}}. \tag{7.86}$$

Combining terms that contain n, we get

$$\frac{1}{\sqrt{n_A\bar{K}}}\Big(z_\beta\sqrt{\gamma_a(1-\gamma_a)(1+(K_0-1)\hat{\psi})} - z_\alpha\sqrt{\gamma_0(1-\gamma_0)(1+(K_0-1)\hat{\psi})}\Big) = \gamma_0 - \gamma_a. \tag{7.87}$$

We can then square both sides and solve for n which gives

$$n_A = \left\lceil \frac{(z_\alpha\sqrt{\gamma_0(1-\gamma_0)(1+(\bar{K}-1)\hat{\psi})} + z_\beta\sqrt{\gamma_a(1-\gamma_a)(1+(\bar{K}-1)\hat{\psi})})^2}{\bar{K}(\gamma_0-\gamma_a)^2} \right\rceil. \tag{7.88}$$

We note that the number of individuals, n_A, from whom acquisition decisions need to be collected as given in (3.83) is for testing a single FNMR. More sophisticated methods for determining the number of individuals needed to carry out a particular test given a particular power, $1-\beta$, are beyond the scope of this text.

7.4 Prediction Intervals

The focus of the previous methods in this chapter has been the FTA(s) for a process or processes. The goal of a prediction interval is to make inference about a different parameter. Instead, the goal of inference from a prediction interval is a future value of an FTA based upon a finite number of observations from the same process. One example of this might be the use of a prediction interval to predict what the FTA will be for the next month of acquisition attempts for a hand geometry system at an amusement park. In this section, we will focus on the creation of a prediction interval for a single future FTA. Let the unobserved FTA be denoted by $\hat{\gamma}^\diamond$. The estimation of this quantity will be based upon observations for acquisition decisions that we have already obtained and will require that the process under consideration is stationary in the sense of Definition 2.23.

Then,

$$\hat{\gamma}^\diamond = \frac{\sum_{i=1}^{n_A^\diamond}\sum_{\kappa=1}^{K_i^\diamond} A_{i\kappa}^\diamond}{\sum_{i=1}^{n_A^\diamond} K_i^\diamond} \tag{7.89}$$

is the future FTA where $n_A^\diamond$ represents the number of individuals that will be observed, K_i is the number of future acquisition decisions that the ith will record, and $A_{i\kappa}^\diamond$ is the κth future acquisition decision from the ith individual. All of these quantities are unobserved and so we will specify them. This may seem daunting but we can then use these tools to consider the impact of these quantities on the prediction interval which we derive below.

In order to determine the appropriate width of a prediction interval for $\hat{\gamma}^\diamond$, we need to specify the variance of that quantity. This can be written as

$$V[\hat{\gamma}^\diamond] = \hat{\gamma}(1-\hat{\gamma})\left[\frac{(1+(K_0-1)\hat{\psi})}{n_A\bar{K}} + \frac{(1+(K_0^\diamond-1)\hat{\psi})}{n_A^\diamond\bar{K}^\diamond}\right] \tag{7.90}$$

where

$$\bar{K}^\diamond = \frac{\sum_{i=1}^{n^\diamond} K_i^\diamond}{n_A^\diamond}. \tag{7.91}$$

Now the variance given in (7.90) is also dependent upon $K_0^\diamond$ which can be difficult to specify *a priori*. One approach that is reasonable is to assume that the $K_i^\diamond$'s will be the same for unobserved individuals, i. Then $K_0^\diamond = \bar{K}^\diamond$. Another alternative is for $K_0^\diamond$ to have the same relationship to $\bar{K}^\diamond$ that K_0 has to $\bar{K}$. We accomplish this by letting

$$K_0^\diamond = \bar{K}^\diamond\left[\frac{K_0}{\bar{K}}\right]. \tag{7.92}$$

We can now derive our prediction interval. Note that a $100(1-\alpha)\%$ prediction interval for an unobserved FTA, $\hat{\gamma}^\diamond$ is a $100(1-\alpha)\%$ confidence interval for $\hat{\gamma}^\diamond$. Assuming that we have sample sizes that are large enough for using the large sample confidence interval approach given in Sect. 7.2.1, then the sampling distribution for $\hat{\gamma}^\diamond$ will be approximately Gaussian. Then we can use the following to make a $100(1-\alpha)\%$ prediction interval for $\hat{\gamma}^\diamond$ is

$$\hat{\gamma} \pm z_{\alpha/2}\sqrt{\hat{\gamma}(1-\hat{\gamma})\left[\frac{(1+(K_0-1)\hat{\psi})}{n_A\bar{K}} + \frac{(1+(K_0^\diamond-1)\hat{\psi})}{n_A^\diamond\bar{K}^\diamond}\right]}. \tag{7.93}$$

We next present an example of applying this interval to sample data from an FTA process.

Example 7.30 To illustrate our methodology here, we will create a 90% prediction interval for the FTA of Device 5 from the PURDUE-FTA database. Our focus will be the FTA for acquisition attempts from the left middle finger on that device. Suppose that for this process we will have 6 attempts by each of 100 new individuals and we would like to determine a 90% prediction interval for those 600 acquisition attempts. Previously, we made a confidence interval for this process FTA in Example 7.3. In that example, we calculated a 90% confidence interval for the process FTA, γ, to be $(0.0299, 0.0812)$ which was centered on a sampled FTA of $\hat{\gamma} = 0.0556$. We will now create a 90% prediction interval for $\hat{\gamma}^{\diamond}$. To arrive at our prediction interval, we will have to specify values for $n^{\diamond}$, $\bar{K}^{\diamond}$, and $K_0^{\diamond}$. The original sample had $n = 186$ individuals, each of which contributed $K_0 = K_i = \bar{K} = 6$ acquisition decisions. From above, we will have $K_0^{\diamond} = \bar{K}^{\diamond} = 6$ which is the same as was obtained in the data collection. Additionally, we expect to have $n^{\diamond} = 100$ new individuals contribute acquisition decisions. Using $\hat{\psi} = 0.8361$ which we obtained in Example 7.3, we get a 90% prediction interval of $(0.0121, 0.0990)$ following (7.93). Then we can be 90% confident that the next 600 observations will yield an FTA of between 1.21% and 9.90%. Note that this interval is considerably wider than the 90% confidence interval for the process FTA of $(0.0299, 0.0812)$. This is to be expected since the prediction interval is dependent upon the variability in both the original sample and, also, the observed decisions. This additional uncertainty results in a wider interval.

7.5 Discussion

The focus of this chapter has been the development of inferential statistical methods for failure to acquire (FTA) rates. Estimation of FTA's is dependent upon the number of acquisition decisions made by a given bioauthentication system. This process of making acquisition decisions may be based upon multiple interactions with the system; however, at some juncture a single decision is rendered by the system. Having done that, we use those decisions for our inference. This chapter began with a discussion of the notation and correlation structure for FTA estimation. We then presented methods for estimation of a single FTA. Moving beyond a single FTA, we next proposed approaches that allowed for the comparison of two or more FTA's. We used large sample approaches as well as bootstrap and randomization approaches to accomplish these statistical techniques. Finally, we presented methodology for sample size and power calculations as well as for a prediction interval for a single FTA.

The fundamentals of FTA estimation are based upon a repeated measures correlation structure. Inherently, this is because we potentially have multiple acquisition decisions for a given individual. This correlation structure is shared with the false non-match rate since a single individual is involved in both cases.

Not all tests and situations are covered by the methodology here. That is not our goal. The aim here is to provide basic methodology that covers many of the data collections that are common in biometric authentication. As needed, our formulations may be extended and modified to match a particular circumstance. Additionally, we recommend choosing a confidence interval over a hypothesis test since the former provides more information about the quantity of interest. We recognize that hypothesis tests may be externally dictated in a particular scenario. For example, hypothesis testing methodology may be mandated as part of a test for a qualified products list. We also recommend when implementing a bootstrap or randomization approach that at least $M = 999$ replications are used. Large values of M are preferable since they lend themselves to stability of the tails of a distribution.

Part IV
Additional Topics and Appendices

Chapter 8
Additional Topics and Discussion

In this chapter we provide a summative discussion of the material covered in this book as well as a introduction to some other topics that were not covered in this book. We start with a discussion of the methods covered in this book.

8.1 Discussion

In this book, we have presented a variety of statistical methods for estimation and comparison of bioauthentication performance metrics. Many of these approaches are innovations. These tools will provide testers, practitioners and decision makers with additional useful information. This book has covered a range of tools for testing and evaluating the performance of biometric authentication systems. For the primary performance metrics of failure to enrol rate (FTE), failure to acquire rate (FTA), false non-match rate (FNMR), false match rate (FMR), equal error rate (EER) and receiver operating characteristic curve (ROC), we have provided a range of methods. Among this range are methods for estimation, as well as methods for comparison. For estimation of a single performance parameter, we have offered confidence intervals and hypothesis tests. Similarly, this book provides methodology for comparing each of the above metrics for two groups or for more than two groups. We have also proffered both large sample methods and non-parametric methods. Beyond confidence intervals and hypothesis tests, we have provided sample size and power calculations for FTE, FTA, FNMR and FTA. Similarly, we have introduced the concept of prediction intervals for those same metrics.

The focus of this book has been on elementary methods for standard circumstances that arise in bioauthentication testing. We have not tried to cover the gamut of all possible methods for every conceivable test. Rather, the aim has been to provide a foundational framework that can be expanded and modified to meet the needs of a given study. In some cases, there are other statistical methods that might also be appropriate.

In the case of the bootstrap methodology that we have used, recent research suggests that there is potential to improve these methods. See Efron and Tibshirani [29]

M.E. Schuckers, *Computational Methods in Biometric Authentication*, Information Science and Statistics,
DOI 10.1007/978-1-84996-202-5_8, © Springer-Verlag London Limited 2010

and Hesterberg [46] for a discussion of bootstrap research. Similarly, there has been a good deal of interest in the statistical research community on methods for estimation of binary proportions. This work suggests that there are improvements that can be gained by modifying the traditional methods that we have presented here. The interested reader should consult papers by Agresti and Coull [4] or Newcombe [72], for example.

The data analyses that are the focus of this book are based upon bioauthentication systems. The general approaches that we use for these methods can be applied more broadly to other type of classification or matching systems. In particular, the methodology that we have presented for ROC's is applicable to other classification methods that use the ROC as a measure of performance. Different data gathering approaches would yield different correlation structures and these would have to be accounted for in the methodology, particularly the bootstrap methodologies; however, the basic outlines of these procedures should remain.

Throughout this book, we have utilized several different approaches to inference. In particular, we have used large sample, bootstrap and randomization methods for a variety of metrics and in a variety of contexts. As long as the conditions are met for large sample methods, they are appropriate and straightforward. Bootstrap and randomization approaches require a greater computational burden which can be extensive for large analyses. This is especially true of the *two-instance bootstrap* for inference about FMR's and ROC's. In general, we favor the bootstrap methods though we recognize that the large sample methods are attractive.

Finally, we end this section with a series of recommendations for reporting of resulting from methods such as those given here:

- A confidence interval is more informative than a hypothesis test particularly for comparing two parameters.
- When using a hypothesis test, report the *p-value* explicitly rather than saying that $p < \alpha$.
- For each group in the study report the overall summary, usually a rate, the standard error of that summary, the number of individuals, and the total number of signals.
- When a bootstrap or randomization procedure is used, choose $M \geq 999$.
- Give explicit details about the procedures that led to the data collection.
- Be sure to state clearly the exact statistical methods that were used in the analysis.

8.2 Additional Topics

Below we discuss some related topics that were not previously covered in this book.

8.2.1 Non-bioauthentication Applications

The methods given in this book have a variety of non-bioauthentication applications. While we have focused on biometrics and biometric authentication in these pages, there are other classification areas that could benefit from this work. ROC curves are used in many different environments for assessing the performance of a classification system. The ROC methods that were discussed in Chap. 5 are versatile and can be implemented on any ROC or any collection of ROC's. In particular, the methodologies for determining if two or more ROC's are significantly different are extremely valuable and widely applicable in many classification settings. To use the ROC methods from this book, we must ensure that our bootstrapping (and, likewise, the underlying correlation structure) is appropriate for a given application.

Many classification applications involve a single measurement from a single source. The use of the *iid* bootstrap may be appropriate in those settings. Just as the *iid* bootstrap is appropriate for FTE data collections. For applications that involve multiple measurements from a single source, the 'subsets bootstrap' is appropriate. Generally, the sort of data collections that yield multiple measurements per source are known as *repeated measures* in the statistical literature. We used repeated measures methodology for estimation of false non-match rates and failure to acquire rates in Chaps. 3 and 7. The methodology that is unique to bioauthentication is the *two-instance bootstrap* which we use when we are matching signals from two different sources. This circumstance requires a distinct correlation structure that is unique to matching rather than classification.

8.2.2 Multiple Comparison Methods

The multiple comparison problem or the *multiplicity* problem is the result of running more than one hypothesis test. For every hypothesis test that is carried out, there is the chance of making a Type I error. By completing multiple tests, the chance of making *at least* one Type I error increases rapidly. This is why we test the hypothesis that $H_0 : \pi_1 = \pi_2 = \cdots = \pi_G$ rather than testing all of the possible pairs of two FNMR's. The drawback to this is that when we reject the null hypothesis we do not know which of the G rates are different from the others. This arises often in analysis of variance type problems like this. Consequently, there has been a good deal of research on statistical methods for these type of problems. See Hsu [49] for a thorough discussion of the topic. For example, in Example 7.26, we rejected the null hypothesis of equality of four FTA's. However, that conclusion only informs us that at least one of the FTA's is different from the rest. It does not tell us which FTA's are different from the others. Multiple comparison methods allow us to maintain the overall Type I error rate, α, while further investigating which of the multiple rates being tested are different from the others.

There have been some examples of the use of these methods in the biometric authentication literature. For example, Kukula [53] uses Tukey's method. In the testing of ROC's at multiple angles in Chap. 5, we used methodology based upon Holm [48]. Other commonly used methods include Bonferroni's, Scheffe's, Duncan's, Student Neumann-Keul's. These methods often are referred to as *post hoc* methods since they are applied after a rejection of the null hypothesis. The differences among these methods is in how each adjusts to the problem of multiple tests.

These methods should be applied to the hypothesis tests of three or more parameters that we have proposed in this text. Although we have not done so (other than the use of Holm's method to derive our test for multiple angles of the ROC curve tests), it is possible to utilize any of the above methods to determine *a posteriori* which of the group of parameters are significantly different from the others. They key to the application of these methods is appropriate variance calculations.

8.2.3 Statistical Significance and Effect Size

In this book, we have presented methodology, specifically hypothesis tests, for determining if a given set of differences—between an estimate and a specific value or between two or more estimates—are significantly different. We make a special note here that for a result to be useful it is important that the *practical significance* is also evaluated. By this, we mean that it is important to consider the size of the differences between the parameter estimates. It is possible for a very small difference to be statistically significant but not to be practically significant. For example, we might have a very large number of imposter decisions for estimation of a false match rate. If the estimated FMR is $\hat{\upsilon} = 0.009992$, it is possible that the number of decisions is so large as to render the estimated FMR significantly less than 0.01 based upon a hypothesis test. This will happen if the standard error of $\hat{\upsilon}$, $\sqrt{\hat{V}[\hat{\upsilon}]}$ is small. However, in most applications the difference between 0.009992 and 0.0100000 is not large enough to be practically significance. Thus, it is important to consider the so-called 'effect size', the magnitude of the differences between the estimates, when evaluating the outcome of the performance evaluation of a biometrics system.

8.2.4 Covariates and Generalized Linear Models

Covariates are additional variables that can help to explain the variability that is present in a response variable. Other disciplines use different terms for these variables. Covariates are sometimes known as factors, explanatory variables, predictor variables or independent variables. Response variables are known in some disciplines as dependent variables or outcome variables. In bioauthentication applications, covariates could include temperature, relative humidity, gender, or age. The goal of a generalized linear model is to describe the relationship between one or more covariates and the response variable. See McCullagh and Nelder[66] for more on GLM's. For the analyses of biometric performance metrics described in this book, we have primarily binary response variables including false match decisions, false non-match decisions, failure to reject and failure to enrol. Research in using generalized linear models (GLM's) and extensions of these models has been pioneered by the Computer Vision Group at Colorado State University in conjunction with Jonathon Phillips at NIST. See, for example, Givens et al. [36, 37] and Lui et al. [59]. The primary focus of their work has been factors that influence the failure non-match rate. Covariates such as age, race and gender among others have been studied through this work. The key to using generalized linear models is getting the variance component correct which depends upon the correlation structure. (Thus, we have paid a great deal of attention in this book to the correlation structure.) We can use the correlation structures here as a starting point for using generalized linear models to assess the impact of covariates on performance metrics besides the FNMR.

8.2.5 Data Collection Design

Another statistical topic that has not received a great deal of attention in the biometric authentication literature is *design of experiments*. This is particularly true of testing of biometric systems that is done in a technology testing or scenario testing mode. In most biometric testing, every individual goes through all of the possible levels for each factor that is being tested. There is often good reasons for this, as it is more expensive to bring in another individual than it is to collect another observation from an individual who is already part of the data collection. Collecting at all levels of all factors from every subject is akin to a full factorial design which can be inefficient. Fractional factorials that allow estimation of the impact of individual factors as well as some interactions effects are efficient uses of time and potentially allow for testing a larger number of factor combinations efficiently. In a fractional factorial design, each individual would not have data collected at all levels of all factors. Additionally, it is worth mentioning a series of designs suggested by Genichi Taguchi, so-called Taguchi experiments. These methods are particularly relevant for assessing design settings for devices that will work in a variety of environmental settings. Dean and Voss [20] have a short introduction to Taguchi methods. Since biometric systems are deployed across a range of temperatures, humidities, lighting conditions, etc. the use of Taguchi methods could benefit the designers of such systems.

Sequential testing is another area where savings in time and money are possible from using such methods. The basic premise behind this methodology is that testing of a hypothesis such as whether or not a metric or parameter is significantly below some boundary can occur at various points before all of the data is collected. Thus, if there is enough evidence to reject or fail to reject our hypothesis of interest after testing half of our subjects, then it is possible to halt the test. Work in this area has been quite extensive starting with the seminal research by Wald [100]. Schuckers [89] described a methodology for biometric authentication performance sequential testing using a parametric procedure based upon work by Lan and DeMets [55]. The basic idea for the Lan-DeMets paradigm is to create a 'spending function' of the significance level. That is, if we have a significance level of $\alpha = 0.05$, then we can 'spend' part of that significance level each time we carry out our hypothesis test on the observations that have already been collected. We continue to 'spend' until we can reject the null hypothesis or the test ends. Clinical trials have been using the Lan-DeMets methodology and extension of it for many years with considerable benefits. For examples, see Packer et al. [73] or Ridker et al. [80]. The approach proposed by Schuckers [89] is appropriate for FTA and FNMR testing. A binomial approach could be used for FTE testing. For applying sequential procedures to the specific responses and correlation structures of FMR, the work of Spiessens et al. [93] holds a good deal of promise.

8.2.6 Mean Transaction Time

One metric of a biometric system performance that has not been covered in this book is the mean transaction time (MTT). Statistical methodology for the MTT has not been addressed in the literature. Several authors report the MTT, for example Veeramachaneni et al. [97] or Mansfield et al. [64]; however, statistical inference for the MTT is still a matter for ongoing research. The United States Transportation Security Administration has made achieving a specific transaction time a precondition for a system to be on their qualified products list, See [2]. This document states—Table A-3 of that document—that the mean verification transaction time should be less than six seconds. One important issue in this area is the difference between mean identification transaction time and mean verification transaction time. The time it takes to evaluate a single comparison (of two biometrics signals) is likely to be *significantly* less than the time it takes to evaluate multiple comparisons. Briefly we will introduce the correlation structure for verification MTT.

The notation and the correlation structure that we present here are based upon verification transaction time. Assume that n_μ individuals have had their transaction time recorded and that each individual i had m_i transactions measured. We begin by letting T_{ij} be the transaction time for the jth verification access decision for the ith individual. This is the difference in time from a biometric presentation to the decision allowing or denying access. We assume here that the transaction time for a given individual (conditional on the overall mean) is not impacted by other individuals but that there is possibly a correlation between two transaction times on the same individual.

Formally, we let $E[T_{ij}] = \mu$ and $V[T_{ij}] = \sigma^2$. Thus, the mean transaction time for a process will be denoted by μ. As we have done elsewhere, we will assume that the transaction times arestationary. Further, we specify the correlation structure to be

$$Corr(T_{ij}, T_{i'j'}) = \begin{cases} 1 & \text{if } i = i', j = j' \\ \zeta & \text{if } i = i', j \neq j' \\ 0 & \text{otherwise.} \end{cases} \tag{8.1}$$

We estimate μ the mean transaction time using

$$\hat{\mu} = \frac{\sum_{i=1}^{n_\mu} \sum_{j=1}^{m_i} T_{ij}}{\sum_{i=1}^{n_\mu} m_i}. \tag{8.2}$$

In order to make inference about the mean transaction time for a biometric process, μ, we will need to estimate the variance in our estimation. To that end, we can calculate

$$
\begin{aligned}
V[\hat{\mu}] &= V\left[\frac{\sum_{i=1}^{n}\sum_{j=1}^{m_i} T_{ij}}{\sum_{i=1}^{n} m_i}\right] \\
&= V[N^{-1}\mathbf{1}^T\mathbf{T}] = N_\mu^{-2}V[\mathbf{1}^T\mathbf{T}] = N_\mu^{-2}\mathbf{1}^T\boldsymbol{\Sigma}_\mu\mathbf{1} = N_\mu^{-2}\sigma^2\mathbf{1}^T\boldsymbol{\Phi}_\mu\mathbf{1} \\
&= N_\mu^{-2}\sigma^2\left[N_\mu + \zeta\sum_{i=1}^{n} m_i(m_i-1)\right] \qquad (8.3)
\end{aligned}
$$

where $\boldsymbol{\Sigma}_\mu = Var[\mathbf{T}]$, $N_\mu = \sum_{i=1}^{n} m_i$, $\boldsymbol{\Phi}_\mu = Corr(\mathbf{T})$, $\mathbf{1} = (1, 1, \ldots, 1)^T$, $\hat{\mu}$ can also be written as $(\sum_{i=1}^{n} m_i)^{-1}\mathbf{1}^T\mathbf{T}$ and $\mathbf{T} = (T_{11}, \ldots, T_{1m_1}, T_{21}, \ldots, T_{2m_2}, \ldots, T_{n1}, \ldots, T_{nm_n})^T$.

The correlation model and the subsequent variance given above can serve as an initial attempt at methodology for verification MTT. Statistical methods for identification mean transaction time would require a more complicated structure since that they would have to explicitly account for the size of the database.

8.2.7 Bayesian Methods

Statistical methods are often denoted as either frequentist or Bayesian. The statistical methods that we have presented in this text could readily be described as frequentist. While there have been quite a few applications of Bayesian methods for biometric classification and matching, for example, Gonzalez-Rodriguez et al. [39], there have not been applications of Bayesian techniques to performance estimation and evaluation. The only implicit use of Bayesian methods is the 'Rule of 3' which is applied as an upper bound when there are no errors for a given error rate. More details on the 'Rule of 3' can be found in Jovanovic and Levy [52]. A discussion of this rule as seen from a Bayesian perspective is given in Louis [57]. There is a great deal of opportunity for the use of these methods for the evaluation of biometrics systems. The place where Bayesian methods might provide the most utility is in extrapolating performance from one type of testing to another. For example, Bayesian methods could assist in taking results from a technology test and extrapolate them to prediction operational performance. Because these methods have the flexibility to incorporate *prior* information, they can provide tools for predicting how a system's performance will change when going from one environment to another or when going from one type of testing to another.

8.2.8 The Biometrics Menagerie

Since Doddington's seminal paper on this topic, Doddington et al. [24], there has been a good deal of interest in what has been called the 'Biometric Menagerie' or

'Doddington's Zoo'. This includes Hicklin et al. [47], Wittman et al. [104], Rattani et al. [79], Poh and Kittler [75] as well as a series of papers by Dunstone and Yager including [27, 106, 107]. The basic thrust of this work is that it is possible to categorize certain users with regard to the ease or difficulty with which they match themselves or other individuals who are distinct from themselves. Thus, there are 'goats' who are individuals who are hard to match against themselves and 'sheep' who are easily matched against themselves. There is little doubt that some users will be harder to match and some users easier to match for any given system. This is the nature of any measurement process with noise.

It will certainly surprise some that we have not discussed this topic earlier with regard to the statistical methods for inference from our performance metrics. In terms of the evaluation of biometric system performance, there is not a need to consider the 'Zoo'. The collected data is what has been observed. The methods presented in the previous chapters of this text account for the variability among and between observations. That is not to say that this topic is unimportant. The variability in matching scores of users is of critical importance to developers of matching algorithms who would be wise to choose algorithms that yield distributions with short tails. The open research question which has not been empirically resolved is the existence of the 'Zoo' in the first place. In particular, is an individual a 'goat' for all systems of a particular modality? Do we have universality of animal type? Put another way, does the characterization of an individual as a particular type of animal depend solely on that individuals or does it depend upon other elements of the matching process such as environmental factors or whether the fingerprint reader is touchless. Further, this area of research is in need of some explicit definitions about what makes someone a 'goat' or a 'sheep' or a 'wolf'. Having mathematically rigorous definitions would allow for the creation of testable hypotheses regarding these characterizations.

8.2.9 Software

When considering the implementation of any computational statistical methods, the question of the appropriate choice of software arises. The primary software that was used in the creation of the output and the examples for this text was **R**. R is freeware that is available to download at www.r-project.org. R is a powerful tool for doing statistical analysis; however, it takes an extensive investment of time to learn. In 2010, with funding from the Center for Identification Technology (CITeR), Schuckers and Hou will introduce a new version of a stand-along software capable of performing all of the methodology given in this text. This system will be known as PRESSv2.0, where PRESS stands for **P**rogram for **R**ate **E**stimation and **S**tatistical **S**ummaries. The advantage of PRESSv2.0 will be that it will have a graphical user interface (GUI). As a consequence, users who do not have sufficient programming experience or who do not have a broad statistical background can utilize more readily the methods described in this text. PRESS2.0 will be available on the following url:

http://myslu.stlawu.edu/~msch/biometrics/press. (8.4)

Chapter 9
Tables

In this chapter, we provide four tables for the reader. We begin with a table of cumulative probabilities for the standard Gaussian distribution is given in Table 9.1. To use this table, we must recognize that the tenths place for the argument of our cumulative probability is found in the first column and the hundredth place is found among the columns. Thus, $P(Z < -2.75) = 0.0030$ and $P(Z < 1.26) = 0.8962$. The second table, Table 9.2, provides some useful percentiles for the Gaussian distribution along with their notation. The next table gives certain percentiles of the *t-distribution* for a variety of degrees of freedom. This is Table 9.3. Finally, we provide a table of percentiles for the χ^2 distribution in Table 9.4.

M.E. Schuckers, *Computational Methods in Biometric Authentication*,
Information Science and Statistics,
DOI 10.1007/978-1-84996-202-5_9,

Table 9.1 Standard Gaussian cumulative probabilities

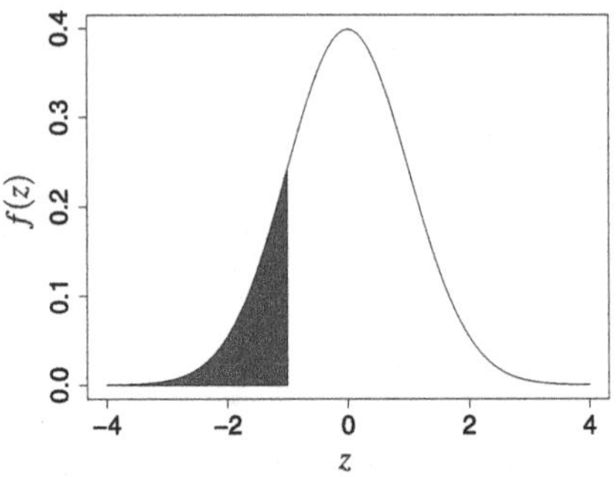

	0.00	0.01	0.02	0.03	0.04	0.05	0.06	0.07	0.08	0.09
−3.4	0.0003	0.0003	0.0003	0.0003	0.0003	0.0003	0.0003	0.0003	0.0003	0.0002
−3.3	0.0005	0.0005	0.0005	0.0004	0.0004	0.0004	0.0004	0.0004	0.0004	0.0003
−3.2	0.0007	0.0007	0.0006	0.0006	0.0006	0.0006	0.0006	0.0005	0.0005	0.0005
−3.1	0.0010	0.0009	0.0009	0.0009	0.0008	0.0008	0.0008	0.0008	0.0007	0.0007
−3.0	0.0013	0.0013	0.0013	0.0012	0.0012	0.0011	0.0011	0.0011	0.0010	0.0010
−2.9	0.0019	0.0018	0.0018	0.0017	0.0016	0.0016	0.0015	0.0015	0.0014	0.0014
−2.8	0.0026	0.0025	0.0024	0.0023	0.0023	0.0022	0.0021	0.0021	0.0020	0.0019
−2.7	0.0035	0.0034	0.0033	0.0032	0.0031	0.0030	0.0029	0.0028	0.0027	0.0026
−2.6	0.0047	0.0045	0.0044	0.0043	0.0041	0.0040	0.0039	0.0038	0.0037	0.0036
−2.5	0.0062	0.0060	0.0059	0.0057	0.0055	0.0054	0.0052	0.0051	0.0049	0.0048
−2.4	0.0082	0.0080	0.0078	0.0075	0.0073	0.0071	0.0069	0.0068	0.0066	0.0064
−2.3	0.0107	0.0104	0.0102	0.0099	0.0096	0.0094	0.0091	0.0089	0.0087	0.0084
−2.2	0.0139	0.0136	0.0132	0.0129	0.0125	0.0122	0.0119	0.0116	0.0113	0.0110
−2.1	0.0179	0.0174	0.0170	0.0166	0.0162	0.0158	0.0154	0.0150	0.0146	0.0143
−2.0	0.0228	0.0222	0.0217	0.0212	0.0207	0.0202	0.0197	0.0192	0.0188	0.0183
−1.9	0.0287	0.0281	0.0274	0.0268	0.0262	0.0256	0.0250	0.0244	0.0239	0.0233
−1.8	0.0359	0.0351	0.0344	0.0336	0.0329	0.0322	0.0314	0.0307	0.0301	0.0294
−1.7	0.0446	0.0436	0.0427	0.0418	0.0409	0.0401	0.0392	0.0384	0.0375	0.0367
−1.6	0.0548	0.0537	0.0526	0.0516	0.0505	0.0495	0.0485	0.0475	0.0465	0.0455
−1.5	0.0668	0.0655	0.0643	0.0630	0.0618	0.0606	0.0594	0.0582	0.0571	0.0559
−1.4	0.0808	0.0793	0.0778	0.0764	0.0749	0.0735	0.0721	0.0708	0.0694	0.0681
−1.3	0.0968	0.0951	0.0934	0.0918	0.0901	0.0885	0.0869	0.0853	0.0838	0.0823
−1.2	0.1151	0.1131	0.1112	0.1093	0.1075	0.1056	0.1038	0.1020	0.1003	0.0985
−1.1	0.1357	0.1335	0.1314	0.1292	0.1271	0.1251	0.1230	0.1210	0.1190	0.1170
−1.0	0.1587	0.1562	0.1539	0.1515	0.1492	0.1469	0.1446	0.1423	0.1401	0.1379
−0.9	0.1841	0.1814	0.1788	0.1762	0.1736	0.1711	0.1685	0.1660	0.1635	0.1611
−0.8	0.2119	0.2090	0.2061	0.2033	0.2005	0.1977	0.1949	0.1922	0.1894	0.1867
−0.7	0.2420	0.2389	0.2358	0.2327	0.2296	0.2266	0.2236	0.2206	0.2177	0.2148
−0.6	0.2743	0.2709	0.2676	0.2643	0.2611	0.2578	0.2546	0.2514	0.2483	0.2451
−0.5	0.3085	0.3050	0.3015	0.2981	0.2946	0.2912	0.2877	0.2843	0.2810	0.2776
−0.4	0.3446	0.3409	0.3372	0.3336	0.3300	0.3264	0.3228	0.3192	0.3156	0.3121
−0.3	0.3821	0.3783	0.3745	0.3707	0.3669	0.3632	0.3594	0.3557	0.3520	0.3483
−0.2	0.4207	0.4168	0.4129	0.4090	0.4052	0.4013	0.3974	0.3936	0.3897	0.3859
−0.1	0.4602	0.4562	0.4522	0.4483	0.4443	0.4404	0.4364	0.4325	0.4286	0.4247
−0.0	0.5000	0.4960	0.4920	0.4880	0.4840	0.4801	0.4761	0.4721	0.4681	0.4641

Table 9.1 (continued)

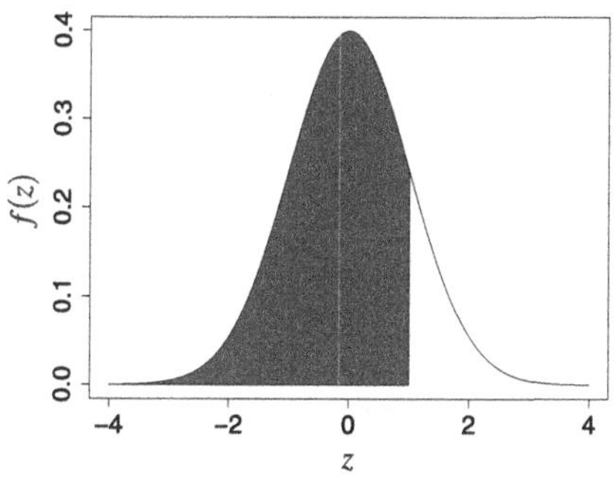

	0.00	0.01	0.02	0.03	0.04	0.05	0.06	0.07	0.08	0.09
0.0	0.5000	0.5040	0.5080	0.5120	0.5160	0.5199	0.5239	0.5279	0.5319	0.5359
0.1	0.5398	0.5438	0.5478	0.5517	0.5557	0.5596	0.5636	0.5675	0.5714	0.5753
0.2	0.5793	0.5832	0.5871	0.5910	0.5948	0.5987	0.6026	0.6064	0.6103	0.6141
0.3	0.6179	0.6217	0.6255	0.6293	0.6331	0.6368	0.6406	0.6443	0.6480	0.6517
0.4	0.6554	0.6591	0.6628	0.6664	0.6700	0.6736	0.6772	0.6808	0.6844	0.6879
0.5	0.6915	0.6950	0.6985	0.7019	0.7054	0.7088	0.7123	0.7157	0.7190	0.7224
0.6	0.7257	0.7291	0.7324	0.7357	0.7389	0.7422	0.7454	0.7486	0.7517	0.7549
0.7	0.7580	0.7611	0.7642	0.7673	0.7704	0.7734	0.7764	0.7794	0.7823	0.7852
0.8	0.7881	0.7910	0.7939	0.7967	0.7995	0.8023	0.8051	0.8078	0.8106	0.8133
0.9	0.8159	0.8186	0.8212	0.8238	0.8264	0.8289	0.8315	0.8340	0.8365	0.8389
1.0	0.8413	0.8438	0.8461	0.8485	0.8508	0.8531	0.8554	0.8577	0.8599	0.8621
1.1	0.8643	0.8665	0.8686	0.8708	0.8729	0.8749	0.8770	0.8790	0.8810	0.8830
1.2	0.8849	0.8869	0.8888	0.8907	0.8925	0.8944	0.8962	0.8980	0.8997	0.9015
1.3	0.9032	0.9049	0.9066	0.9082	0.9099	0.9115	0.9131	0.9147	0.9162	0.9177
1.4	0.9192	0.9207	0.9222	0.9236	0.9251	0.9265	0.9279	0.9292	0.9306	0.9319
1.5	0.9332	0.9345	0.9357	0.9370	0.9382	0.9394	0.9406	0.9418	0.9429	0.9441
1.6	0.9452	0.9463	0.9474	0.9484	0.9495	0.9505	0.9515	0.9525	0.9535	0.9545
1.7	0.9554	0.9564	0.9573	0.9582	0.9591	0.9599	0.9608	0.9616	0.9625	0.9633
1.8	0.9641	0.9649	0.9656	0.9664	0.9671	0.9678	0.9686	0.9693	0.9699	0.9706
1.9	0.9713	0.9719	0.9726	0.9732	0.9738	0.9744	0.9750	0.9756	0.9761	0.9767
2.0	0.9772	0.9778	0.9783	0.9788	0.9793	0.9798	0.9803	0.9808	0.9812	0.9817
2.1	0.9821	0.9826	0.9830	0.9834	0.9838	0.9842	0.9846	0.9850	0.9854	0.9857
2.2	0.9861	0.9864	0.9868	0.9871	0.9875	0.9878	0.9881	0.9884	0.9887	0.9890
2.3	0.9893	0.9896	0.9898	0.9901	0.9904	0.9906	0.9909	0.9911	0.9913	0.9916
2.4	0.9918	0.9920	0.9922	0.9925	0.9927	0.9929	0.9931	0.9932	0.9934	0.9936
2.5	0.9938	0.9940	0.9941	0.9943	0.9945	0.9946	0.9948	0.9949	0.9951	0.9952
2.6	0.9953	0.9955	0.9956	0.9957	0.9959	0.9960	0.9961	0.9962	0.9963	0.9964
2.7	0.9965	0.9966	0.9967	0.9968	0.9969	0.9970	0.9971	0.9972	0.9973	0.9974
2.8	0.9974	0.9975	0.9976	0.9977	0.9977	0.9978	0.9979	0.9979	0.9980	0.9981
2.9	0.9981	0.9982	0.9982	0.9983	0.9984	0.9984	0.9985	0.9985	0.9986	0.9986
3.0	0.9987	0.9987	0.9987	0.9988	0.9988	0.9989	0.9989	0.9989	0.9990	0.9990
3.1	0.9990	0.9991	0.9991	0.9991	0.9992	0.9992	0.9992	0.9992	0.9993	0.9993
3.2	0.9993	0.9993	0.9994	0.9994	0.9994	0.9994	0.9994	0.9995	0.9995	0.9995
3.3	0.9995	0.9995	0.9995	0.9996	0.9996	0.9996	0.9996	0.9996	0.9996	0.9997
3.4	0.9997	0.9997	0.9997	0.9997	0.9997	0.9997	0.9997	0.9997	0.9997	0.9998

Table 9.2 Selected percentiles of the standard Gaussian distribution

z	$P(Z \leq z)$	Notation
−3.7190	0.0001	$z_{0.9999}$
−3.2905	0.0005	$z_{0.9995}$
−2.5758	0.0050	$z_{0.995}$
−2.3263	0.0100	$z_{0.99}$
−2.0537	0.0200	$z_{0.98}$
−1.9600	0.0250	$z_{0.975}$
−1.6449	0.0500	$z_{0.95}$
−1.2816	0.1000	$z_{0.90}$
−0.8416	0.2000	$z_{0.80}$
−0.6745	0.2500	$z_{0.75}$
0.0000	0.5000	$z_{0.50}$
0.6745	0.7500	$z_{0.25}$
0.8416	0.8000	$z_{0.20}$
1.2816	0.9000	$z_{0.10}$
1.6449	0.9500	$z_{0.05}$
1.9600	0.9750	$z_{0.025}$
2.0537	0.9800	$z_{0.02}$
2.3263	0.9900	$z_{0.01}$
2.5758	0.9950	$z_{0.005}$
3.2905	0.9995	$z_{0.0005}$
3.7190	0.9999	$z_{0.0001}$

Table 9.3 t-distribution percentiles

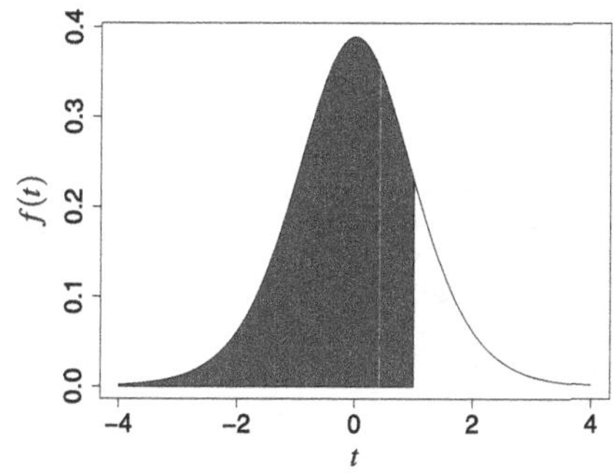

Degrees of freedom	0.5000	0.7500	0.8000	0.9000	0.9500	0.9750	0.9800	0.9900	0.9950	0.9995
1	0.000	1.000	1.376	3.078	6.314	12.706	15.895	31.821	63.657	636.619
2	0.000	0.816	1.061	1.886	2.920	4.303	4.849	6.965	9.925	31.599
3	0.000	0.765	0.978	1.638	2.353	3.182	3.482	4.541	5.841	12.924
4	0.000	0.741	0.941	1.533	2.132	2.776	2.999	3.747	4.604	8.610
5	0.000	0.727	0.920	1.476	2.015	2.571	2.757	3.365	4.032	6.869
6	0.000	0.718	0.906	1.440	1.943	2.447	2.612	3.143	3.707	5.959
7	0.000	0.711	0.896	1.415	1.895	2.365	2.517	2.998	3.499	5.408
8	0.000	0.706	0.889	1.397	1.860	2.306	2.449	2.896	3.355	5.041
9	0.000	0.703	0.883	1.383	1.833	2.262	2.398	2.821	3.250	4.781
10	0.000	0.700	0.879	1.372	1.812	2.228	2.359	2.764	3.169	4.587
11	0.000	0.697	0.876	1.363	1.796	2.201	2.328	2.718	3.106	4.437
12	0.000	0.695	0.873	1.356	1.782	2.179	2.303	2.681	3.055	4.318
13	0.000	0.694	0.870	1.350	1.771	2.160	2.282	2.650	3.012	4.221
14	0.000	0.692	0.868	1.345	1.761	2.145	2.264	2.624	2.977	4.140
15	0.000	0.691	0.866	1.341	1.753	2.131	2.249	2.602	2.947	4.073
16	0.000	0.690	0.865	1.337	1.746	2.120	2.235	2.583	2.921	4.015
17	0.000	0.689	0.863	1.333	1.740	2.110	2.224	2.567	2.898	3.965
18	0.000	0.688	0.862	1.330	1.734	2.101	2.214	2.552	2.878	3.922
19	0.000	0.688	0.861	1.328	1.729	2.093	2.205	2.539	2.861	3.883
20	0.000	0.687	0.860	1.325	1.725	2.086	2.197	2.528	2.845	3.850
25	0.000	0.684	0.856	1.316	1.708	2.060	2.167	2.485	2.787	3.725
30	0.000	0.683	0.854	1.310	1.697	2.042	2.147	2.457	2.750	3.646
40	0.000	0.681	0.851	1.303	1.684	2.021	2.123	2.423	2.704	3.551
50	0.000	0.679	0.849	1.299	1.676	2.009	2.109	2.403	2.678	3.496
100	0.000	0.677	0.845	1.290	1.660	1.984	2.081	2.364	2.626	3.390

Table 9.4 χ^2 distribution percentiles

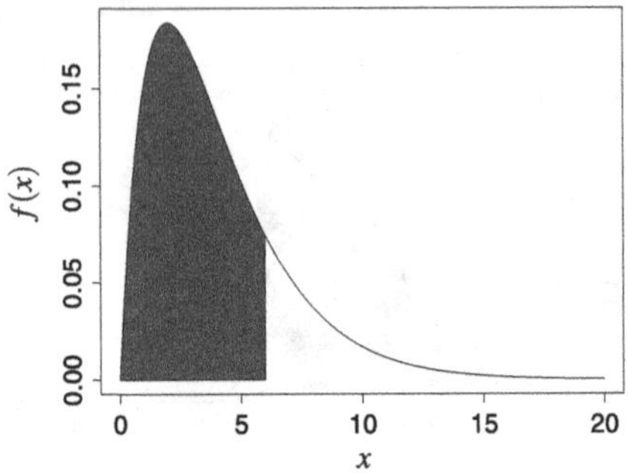

Degrees of freedom	0.5000	0.7500	0.8000	0.9000	0.9500	0.9750	0.9800	0.9900	0.9950	0.9995
1	0.455	1.323	1.642	2.706	3.841	5.024	5.412	6.635	7.879	12.116
2	1.386	2.773	3.219	4.605	5.991	7.378	7.824	9.210	10.597	15.202
3	2.366	4.108	4.642	6.251	7.815	9.348	9.837	11.345	12.838	17.730
4	3.357	5.385	5.989	7.779	9.488	11.143	11.668	13.277	14.860	19.997
5	4.351	6.626	7.289	9.236	11.070	12.833	13.388	15.086	16.750	22.105
6	5.348	7.841	8.558	10.645	12.592	14.449	15.033	16.812	18.548	24.103
7	6.346	9.037	9.803	12.017	14.067	16.013	16.622	18.475	20.278	26.018
8	7.344	10.219	11.030	13.362	15.507	17.535	18.168	20.090	21.955	27.868
9	8.343	11.389	12.242	14.684	16.919	19.023	19.679	21.666	23.589	29.666
10	9.342	12.549	13.442	15.987	18.307	20.483	21.161	23.209	25.188	31.420
11	10.341	13.701	14.631	17.275	19.675	21.920	22.618	24.725	26.757	33.137
12	11.340	14.845	15.812	18.549	21.026	23.337	24.054	26.217	28.300	34.821
13	12.340	15.984	16.985	19.812	22.362	24.736	25.472	27.688	29.819	36.478
14	13.339	17.117	18.151	21.064	23.685	26.119	26.873	29.141	31.319	38.109
15	14.339	18.245	19.311	22.307	24.996	27.488	28.259	30.578	32.801	39.719
16	15.338	19.369	20.465	23.542	26.296	28.845	29.633	32.000	34.267	41.308
17	16.338	20.489	21.615	24.769	27.587	30.191	30.995	33.409	35.718	42.879
18	17.338	21.605	22.760	25.989	28.869	31.526	32.346	34.805	37.156	44.434
19	18.338	22.718	23.900	27.204	30.144	32.852	33.687	36.191	38.582	45.973
20	19.337	23.828	25.038	28.412	31.410	34.170	35.020	37.566	39.997	47.498
25	24.337	29.339	30.675	34.382	37.652	40.646	41.566	44.314	46.928	54.947
30	29.336	34.800	36.250	40.256	43.773	46.979	47.962	50.892	53.672	62.162
40	39.335	45.616	47.269	51.805	55.758	59.342	60.436	63.691	66.766	76.095
50	49.335	56.334	58.164	63.167	67.505	71.420	72.613	76.154	79.490	89.561
100	99.334	109.141	111.667	118.498	124.342	129.561	131.142	135.807	140.169	153.167

References

1. Adler, A., Schuckers, M.E.: Calculation of a composite DET curve. In: Kanade, T., Jain, A., Ratha, N.K. (eds.) Proceedings of Audio- and Video-based Biometric Person Authentication 2005 (AVBPA2005). *Lecture Notes in Computer Science*, vol. 3546, pp. 279–288. Springer, Berlin (2005)
2. Administration, U.S.T.S.: Biometrics for airport access control. http://www.tsa.gov/assets/pdf/biometrics_guidance.pdf (2005)
3. Agresti, A., Caffo, B.: Simple and effective confidence intervals for proportions and differences of proportions result from adding two successes and two failures. The American Statistician **54**(4), 280–288 (2000)
4. Agresti, A., Coull, B.A.: Approximate is better than "exact" for interval estimation of binomial proportions. The American Statistician **52**(2), 119–126 (1998)
5. Bailly-Baillière, E., Bengio, S., Bimbot, F., Hamouz, M., Kittler, J., Marièthoz, J., Matas, J., Messer, K., Popovici, V., Porée, F., Ruiz, B., Thiran, J.P.: The BANCA database and evaluation protocol. In: 4th International Conference on Audio- and Video-Based Biometric Person Authentication, AVBPA. Springer, Berlin (2003)
6. Bengio, S., Mariéthoz, J.: The expected performance curve: a new assessment measure for person authentication. In: ODYS-2004, pp. 279–284 (2004)
7. Billingsley, P.: Probability and Measure, 2nd edn. Wiley, New York (1986)
8. Bistarelli, S., Santini, F., Vaccarelli, A.: An asymmetric fingerprint matching algorithm for java card^{TM}. In: Kanade, T., Jain, A., Ratha, N.K. (eds.) Proceedings of Audio- and Video-based Biometric Person Authentication 2005 (AVBPA2005). *Lecture Notes in Computer Science*, vol. 3546, pp. 279–288. Springer, Berlin (2005)
9. Bolle, R.M., Ratha, N.K., Pankanti, S.: Error analysis of pattern recognition systems—the subsets bootstrap. Computer Vision and Image Understanding **93**, 1–33 (2004)
10. Brockwell, P.J., Davis, R.A.: Time Series: Theory and Methods, 2nd edn. Springer, New York (1991)
11. Brown, L.D., Cai, T.T., DasGupta, A.: Interval estimation for a binomial proportion. Statistical Science **16**(2), 101–117 (2001)
12. Brown, L.D., Cai, T.T., DasGupta, A.: Confidence intervals for a binomial proportion and asymptotic expansions. The Annals of Statistics **30**(1), 160–201 (2002)
13. Campbell, G.: Advance in statistical methodology for the evaluation of diagnostic and laboratory tests. Statistics in Medicine **13**, 499–508 (1994)
14. Casella, G., Berger, R.L.: Statistical Inference. Duxbury, N. Scituate (2001)
15. Cochran, W.G.: The comparison of percentages in matched samples. Biometrika **37**(3/4), 256–266 (1950)
16. Coventry, L., Angeli, A.D., Johnson, G.: Biometric verification at a self service interface. In: Proceedings of the British Ergonomic Society Conference (2003)

M.E. Schuckers, *Computational Methods in Biometric Authentication*,
Information Science and Statistics,
DOI 10.1007/978-1-84996-202-5, © Springer-Verlag London Limited 2010

17. Crowder, M.J.: Beta-binomial anova for proportions. Applied Statistics **27**(1), 34–37 (1978). http://www.jstor.org/stable/2346223
18. Dass, S.C., Zhu, Y., Jain, A.K.: Validating a biometric authentication system: Sample size requirements. IEEE Transactions on Pattern Analysis and Machine Intelligence **28**(1), 19–30 (2007)
19. Daugman, J.: Biometric decision landscapes (2000)
20. Dean, A., Voss, D.: Design and Analysis of Experiments. Springer, Berlin (1999)
21. DeGroot, M.H., Schervish, M.J.: Probability and Statistics, 3rd edn. Addison–Wesley, Reading (2002)
22. Deming, W.E.: Some Theory of Sampling. Dover, New York (1950)
23. Dietz, Z., Schuckers, M.E.: Technical report: A proof of a central limit theorem for false match rates (2009). http://academics.hamilton.edu/mathematics/zdietz/Research/Schuckers/IEEE/CLM-repeated.pdf
24. Doddington, G.R., Liggett, W., Martin, A.F., Przybocki, M.A., Reynolds, D.A.: Sheep, goats, lambs and wolves: a statistical analysis of speaker performance in the NIST 1998 speaker recognition evaluation. In: ICSLP, pp. 608–611 (1998)
25. Doddington, G.R., Przybocki, M.A., Martin, A.F., Reynolds, D.A.: The NIST speaker recognition evaluation: overview methodology, systems, results, perspective. Speech Communication **31**(2–3), 225–254 (2000)
26. Drummond, C., Holte, R.C.: Cost curves: an improved method for visualizing classifier performance. Machine Learning **65**(1), 95–130 (2006)
27. Dunstone, T., Yager, N.: Biometric System and Data Analysis. Springer, Berlin (2009)
28. Edgington, E.S., Onghena, P.: Randomization Tests, 4th edn. Dekker, New York (2007)
29. Efron, B., Tibshirani, R.: An Introduction to the Bootstrap. Chapman & Hall/CRC Press, London/Boca Raton (1994)
30. Elliot, S., Kukula, E.: A definitional framework for the human-biometric sensor interaction model. Bspawp 060709, BSPA Laboratory Publications (2009)
31. Fairhurst, M., Deravi, F., Mavity, N., George, J., Silantzis, K.: Intelligent management of multimodal biometric transactions. In: Knowledge-Based Intelligent Information and Engineering Systems. *Lecture Notes in Computer Science*, vol. 2774, pp. 1254–1260. Springer, Berlin (2003)
32. Fawcett, T.: An introduction to ROC analysis. Pattern Recognition Letters **27**, 861–874 (2006)
33. Fleiss, J.L., Levin, B., Paik, M.C.: Statistical Methods for Rates and Proportions. Wiley, New York (2003)
34. Friedl, H., Stampfer, E.: Encyclopedia of environmetrics, chap. Jackknife Resampling, pp. 1089–1098. Wiley, New York (2002)
35. Garren, S.T., Smith, R.L., Piegorsch, W.W.: Bootstrap goodness-of-fit test for the beta-binomial model. Journal of Applied Statistics **28**(5), 561–571 (2001)
36. Givens, G., Beveridge, J.R., Draper, B.A., Bolme, D.: Statistical assessment of subject factors in the PCA recognition of human faces. In: IEEE CVPR 2003 Workshop on Statistical Analysis in Computer Vision Workshop (2003)
37. Givens, G.H., Beveridge, J.R., Draper, B.A., Bolme, D.: Using a generalized linear mixed model to study the configuration space of a PCA+LDA human face recognition algorithm. In: Articulated Motion and Deformable Objects. *Lecture Notes in Computer Science*, vol. 3179, pp. 1–11. Springer, Berlin (2004)
38. Golfarelli, M., Maio, D., Maltoni, D.: On the error-reject trade-off in biometric verification systems. IEEE Transactions on Pattern Analysis and Machine Intelligence **19**(7), 786–796 (1997)
39. Gonzalez-Rodriguez, J., Fierrez-Aguilar, J., Ramos-Castro, D., Ortega-Garcia, J.: Bayesian analysis of fingerprint, face and signature evidences with automatic biometric systems. Forensic Science International **155**(2), 126–140 (2005)
40. Grimmett, G., Stirzaker, D.: Probability and Random Processes, 3rd edn. Oxford University Press, London (2001)

41. Group, I.B.: Comparative biometric testing: Round 7 public report. www.biometricgroup.com/reports/public/reports/CBT7report.htm (2009)
42. Guyon, I., Makhoul, J., Schwartz, R., Vapnik, V.: What size test set gives good error rate estimates. IEEE Transactions on Pattern Analysis and Machine Intelligence **20**(1), 52–64 (1998)
43. Hahn, G.J., Meeker, W.Q.: Statistical Intervals: A Guide for Practitioners. Wiley, New York (1991)
44. Hall, P.: On the bootstrap and confidence intervals. The Annals of Statistics **14**, 1431–1452 (1986)
45. Hernández-Orallo, J., Ferri, C., Lachiche, N., Flach, P.A. (eds.) ROC Analysis in Artificial Intelligence, 1st Int. Workshop, ROCAI-2004. Valencia, Spain (2004)
46. Hesterberg, T.C.: Bootstrap tilting confidence intervals and hypothesis tests. Computing Science and Statistics **31**, 389–393 (1999)
47. Hicklin, A., Watson, C., Ulery, B.: The myth of goats: How many people have fingerprints that are hard to match. Tech. Rep. NISTIR 7271, National Institutes for Standards and Technology (2005)
48. Holm, S.: A simple sequentially rejective multiple test procedure. Scandinavian Journal of Statistics **6**(2), 65–70 (1979). http://www.jstor.org/stable/4615733
49. Hsu, J.: Multiple Comparisons: Theory and Methods. Chapman & Hall/CRC Press, London/Boca Raton (1996)
50. Jacod, J., Shiryaev, A.N.: Limit Theorems for Stochastic Processes. Springer, Berlin (2002)
51. Jain, A.K., Bolle, R., Pankanti, S. (eds.) Biometrics: Personal Identification in Networked Society. Kluwer Academic, Dordrecht (1999)
52. Jovanovic, B.D., Levy, P.S.: A look at the rule of three. The American Statistician **51**(2), 137–139 (1997)
53. Kukula, E.: Design and evaluation of the human-biometric sensor interaction method. Ph.D. thesis, Purdue University (2007)
54. Lahiri, S.N.: Resampling Methods for Dependent Data. Springer, Berlin (2003)
55. Lan, K.K.G., Demets, D.L.: Discrete sequential boundaries for clinical trials. Biometrika **79**(3), 659–663 (1983)
56. Li, S.Z. (ed.) Encyclopedia of Biometrics. Springer, Berlin (2009)
57. Louis, T.A.: Confidence intervals for a binomial parameter after observing no successes. The American Statistician **35**(3), 154 (1981)
58. Lüttin, J.: Evaluation protocol for the xm2fdb database (Lausanne Protocol). Communication 98-05, IDIAP, Martigny, Switzerland (1998)
59. Lui, Y.M., Bolme, D., Draper, B.A., Beveridge, J.R., Givens, G., Phillips, P.J.: A meta-analysis of face recognition covariates. In: IEEE Third International Conference on Biometrics: Theory, Applications and Systems (2009)
60. Ma, G., Hall, W.J.: Confidence bands for receiver operating characteristics curves. Medical Decision Making **13**, 191–197 (1993)
61. Macskassy, S.A., Provost, F.J., Rosset, S.: ROC confidence bands: an empirical evaluation. In: Raedt, L.D., Wrobel, S. (eds.) ICML, pp. 537–544. ACM, New York (2005)
62. Manly, B.F.J.: Randomization, Bootstrap and Monte Carlo Methods in Biology, 2nd edn. Chapman & Hall, London (1997)
63. Mansfield, T., Wayman, J.L.: Best practices in testing and reporting performance of biometric devices. www.cesg.gov.uk/site/ast/biometrics/media/BestPractice.pdf (2002)
64. Mansfield, T., Kelly, G., Chandler, D., Kane, J.: Biometric product testing final report. http://dematerialisedid.com/PDFs/PKTMP000.pdf (2001)
65. Martin, A., Doddington, G., Kamm, T., Ordowski, M., Przybocki, M.: The det curve in assessment of detection task performance. In: Proceedings of Eurospeech'97, pp. 1895–1898 (1997)
66. McCullagh, P., Nelder, J.A.: Generalized Linear Models. Chapman & Hall, New York (1983)
67. Modi, S.: Analysis of fingerprint sensor interoperability on system performance. Ph.D. thesis, Purdue University, West Lafayette, Indiana (2008)

68. Modi, S., Elliott, S.J., Kim, H.: Statistical analysis of fingerprint sensor interoperability performance. In: Proceedings of the 3rd International Conference on Biometrics: Theory, Applications, and Systems, BTAS 09 (2009)
69. Moore, D.F.: Modeling the extraneous variance in the presence of extra-binomial variation. Applied Statistics **36**(1), 8–14 (1987)
70. Newcombe, R.: Interval estimation for the difference between independent proportions: comparison of eleven methods. Statistics in Medicine **17**, 873–890 (1998)
71. Newcombe, R.G.: Improved confidence interval for the difference between binomial proportions based on paired data. Statistics in Medicine **17**, 2635–2650 (1998)
72. Newcombe, R.G.: Logit confidence intervals and the inverse sinh transformation. The American Statistician **55**(3), 200–202 (2001)
73. Packer, M., O'Connor, C.M., Ghali, J.K., Pressler, M.L., Carson, P.E., Belkin, R.N., Miller, A.B., Neuberg, G.W., Frid, D., Wertheimer, J.H., Cropp, A.B., DeMets, D.L.: The prospective randomized amlodipine survival evaluation study group: effect of amlodipine on morbidity and mortality in severe chronic heart failure. New England Journal of Medicine **335**(15), 1107–1114 (1996)
74. Poh, N., Bengio, S.: Database, protocol and tools for evaluating score-level fusion algorithms in biometric authentication. Pattern Recognition Journal (2005)
75. Poh, N., Kittler, J.: A biometric menagerie index for characterising template/model-specific variation. In: Tistarelli, M., Nixon, M.S. (eds.) ICB. *Lecture Notes in Computer Science*, vol. 5558, pp. 816–827. Springer, Berlin (2009)
76. Poh, N., Martin, A., Bengio, S.: Performance generalization in biometric authentication using joint user-specific and sample bootstraps. IEEE Transactions on Pattern Analysis and Machine Intelligence (2007)
77. Popper, K.: The Logic of Scientific Discovery. Routledge, London (2002)
78. Rao, C.R.: Linear Statistical Inference and Its Application, 2nd edn. Wiley–Interscience, New York (2002)
79. Rattani, A., Marcialis, G.L., Roli, F.: An experimental analysis of the relationship between biometric template update and the Doddington's zoo: a case study in face verification. In: Image Analysis and Processing—ICIAP 2009. *Lecture Notes in Computer Science*, vol. 5716/2009, pp. 434–442. Springer, Berlin (2009)
80. Ridker, P.M., Goldhaber, S.Z., Danielson, E., Rosenberg, Y., Eby, C.S., Deitcher, S.R., Cushman, M., Moll, S., Kessler, C.M., Elliott, C.G., Paulson, R., Wong, T., Bauer, K.A., Schwartz, B.A., Miletich, J.P., Bounameaux, H., Glynn, R.J.: The PREVENT investigators: long-term, low-intensity warfarin therapy for the prevention of recurrent venous thromboembolism. New England Journal of Medicine **348**(15), 1425–1434 (2003)
81. Ridout, M.S., Demétrio, C.G.B., Firth, D.: Estimating intraclass correlation for binary data. Biometrics **55**, 137–148 (1999)
82. Schervish, M.J.: Theory of Statistics. Springer, New York (1995)
83. Schuckers, M.E.: Estimation and sample size calculations for correlated binary error rates of biometric identification rates. In: Proceedings of the American Statistical Association: Biometrics Section [CD-ROM]. American Statistical Association, Alexandria, VA (2003)
84. Schuckers, M.E.: Using the beta-binomial distribution to assess performance of a biometric identification device. International Journal of Image and Graphics **3**(3), 523–529 (2003)
85. Schuckers, M.E.: Theoretical statistical correlation for biometric identification performance. In: Proceedings of the International Conference on Acoustics, Speech, and Signal Processing (ICASSP) (2008)
86. Schuckers, M.E.: Theoretical statistical correlation for biometric identification performance. In: Proceedings of the IEEE International Conference on Acoustics, Speech, and Signal Processing (ICASSP) (2008)
87. Schuckers, M.E.: A parametric correlation framework for the statistical evaluation and estimation of biometric-based classification performance in a single environment. IEEE Transactions on Information Forensics and Security **4**, 231–241 (2009)

88. Schuckers, M.E., Minev, Y.D., Adler, A.: Curvewise det confidence regions and pointwise EER confidence intervals using radial sweep methodology. In: Lee, S.W., Li, S.Z. (eds.) Advances in Biometrics, International Conference, ICB, 2007, pp. 376–395. Springer, Berlin (2007)
89. Schuckers, M.E., Sheldon, E., Hartson, H.: When enough is enough: early stopping of biometrics error rate testing. In: Proceedings of the IEEE Workshop on Automatic Identification Advanced Technologies (AutoID) (2007)
90. Shao, J., Wu, C.F.J.: A general theory for jackknife variance estimation. The Annals of Statistics **17**(3), 1176–1197 (1989)
91. Shen, W., Surette, M., Khanna, R.: Evaluation of automated biometrics-based identification and verification systems. Proceedings of the IEEE **85**(9), 1464–1478 (1997)
92. Snedecor, G.W., Cochran, W.G.: Statistical Methods, 8th edn. Iowa State University Press, Ames (1995)
93. Spiessens, B., Lesaffre, E., Verbeke, G.: A comparison of group, sequential methods for binary longitudinal data. Statistics in Medicine **22**(4), 501–515 (2003)
94. National Institute of Standards, Technology: Biometric Score Set Release 1. http://www.itl.nist.gov/iad/894.03/biometricscores/ (2004)
95. Vardeman, S.B.: What about the other intervals? The American Statistician **46**(3), 193–197 (1992). http://www.jstor.org/stable/2685212
96. Veaux, R.D.D., Velleman, P.F., Bock, D.E.: Intro Stats. Addison–Wesley, Reading (2009)
97. Veeramachaneni, K., Osadciw, L.A., Varshney, P.K.: An adaptive multimodal biometric management algorithm. IEEE Transactions on Systems, Man, and Cybernetics, Part C: Applications and Reviews **35**(3), 344–356 (2005)
98. Vitaliano, P.P.: The relative merits of two statistics for testing a hypothesis about the equality of three correlated proportions. Journal of the American Statistical Association **74**(365), 232–237 (1979)
99. Wackerly, D.D., Mendenhall, W. III, Scheaffer, R.L.: Mathematical Statistics with Applications, 6th edn. Duxbury, N. Scituate (2002)
100. Wald, A.: Sequential Analysis. Dover, New York (1947)
101. Wayman, J.L.: Confidence interval and test size estimation for biometric data. In: Proceedings of IEEE AutoID'99, pp. 177–184 (1999)
102. Wayman, J., Jain, A., Maltoni, D., Maio, D. (eds.) Biometric Systems: Technology, Design and Performance. Springer, Berlin (2005)
103. Williams, D.A.: The analysis of binary responses from toxicological experiments involving reproduction and teratogenicity. Biometrics **31**, 949–952 (1975)
104. Wittman, M., Davis, P., Flynn, P.: Empirical studies of the existence of the biometric menagerie in the FRGC 2.0 color image corpus. In: 2006 Conference on Computer Vision and Pattern Recognition Workshop (CVPRW'06) (2006)
105. Wu, J.C.: Studies of operational measurement of ROC curve on large fingerprint data sets using two-sample bootstrap. Tech. Rep. NISTIR 7449, National Institute of Standards and Technology (2007)
106. Yager, N., Dunstone, T.: Worms, chameleons, phantoms and doves: New additions to the biometric menagerie. In: 2007 IEEE Workshop on Automatic Identification Advanced Technologies (AutoID'07) (2007)
107. Yager, N., Dunstone, T.: The biometric menagerie. IEEE Transactions on Pattern Analysis and Machine Intelligence **99**(1) (2008). http://doi.ieeecomputersociety.org/10.1109/TPAMI.2008.291
108. Zhou, X.H., McClish, D.K., Obuchowski, N.A.: Statistical Methods in Diagnostic Medicine. Wiley, New York (2002)

Index

M.E. Schuckers, *Computational Methods in Biometric Authentication*, Information Science and Statistics,
DOI 10.1007/978-1-84996-202-5, © Springer-Verlag London Limited 2010